T0251014

Molecular Biology
of
Symbiotic
Nitrogen Fixation

Editor

Peter M. Gresshoff
Professor
Plant Molecular Genetics
Institute of Agriculture
University of Tennessee
Knoxville, Tennessee

CRC Press
Taylor & Francis Group
Boca Raton London New York

CRC Press is an imprint of the
Taylor & Francis Group, an **informa** business

First published 1990 by CRC Press
Taylor & Francis Group
6000 Broken Sound Parkway NW, Suite 300
Boca Raton, FL 33487-2742

Reissued 2018 by CRC Press

© 1990 by CRC Press, Inc.
CRC Press is an imprint of Taylor & Francis Group, an Informa business

No claim to original U.S. Government works

Library of Congress Cataloging-in-Publication Data

The Molecular biology of symbiotic nitrogen fixation / editor, Peter
    M. Gresshoff.
        p.   cm.
    Bibliography: p.
    Includes index.
    ISBN 0-8493-6188-5
    1. Nitrogen—Fixation.   2. Molecular biology.   3. Nitrogen-fixing
microorganisms.   I. Gresshoff, Peter M., 1948-   .
QR89.7.M65   1990
589.9'504133—dc19   88-38286

A Library of Congress record exists under LC control number: 88038286

Publisher's Note
The publisher has gone to great lengths to ensure the quality of this reprint but points out that some imperfections in the original copies may be apparent.

Disclaimer
The publisher has made every effort to trace copyright holders and welcomes correspondence from those they have been unable to contact.

ISBN 13: 978-1-315-89564-2 (hbk)
ISBN 13: 978-1-351-07474-2 (ebk)

Visit the Taylor & Francis Web site at http://www.taylorandfrancis.com and the
CRC Press Web site at http://www.crcpress.com

PREFACE

Symbiotic nitrogen fixation and nodulation have long represented a fascinating area of science for me. When I was a boy in postwar Berlin, nodulation became a relevant phenomenon when I was told of the importance of lupins, their nodules, and the bacteria in them, which increased the nitrogen status of the field where we flew paper kites.

Today the process of nodulation and nitrogen fixation has become more than just part of agricultural productivity. Many legumes are crop plants, but perhaps an objective view of the situation will highlight the fact that our more detailed understanding of the mechanism and inheritance of nodulation in general has done little to improve these crop plants beyond what simple trial and error could have achieved. Similar statements could also be made about research on photosynthesis or hormone action. However, there is more to science than direct technical application. The analysis of the nodulation and nitrogen fixation processes has produced a large data base leading to related advances in bacterial genetics, plant and bacterial gene regulation, developmental biology, metallo-protein chemistry, plant diseases, microbial ecology, reaction mechanisms, evolutionary research, and cellular energetics. This one biological process thus amalgamates prokaryotic and eukaryotic research at levels ranging from subatomic particles (electron flow) to the entire biosphere.

These considerations made symbiotic nitrogen fixation an inspiring subject area of biological research. On the one side, this system has the extreme relevance and potential to increase agricultural production, and, through this, input into all areas of our societies, whether they are supported by high technology agriculture or subsistence farming. Legumes and other nodulated and nitrogen-fixing plants play an important part in the success of these production systems.

It is fascinating to note the diversity of structure that this symbiosis has produced. Yet there are fundamental restraints which have led to preserved commonalities. For example, we know that in all symbioses carbon molecules are provided by the plant, while nitrogen is returned from the prokaryote to be assimilated by the host. In all symbioses one sees the "concern" about oxygen regulation — yet note how variably this problem is solved in different biological systems. In all systems, the plant and the bacterium communicate by the exchange of small molecules.

Furthermore, the variety of symbioses tells us about cell/cell interaction. The prokaryote must recognize, stimulate, and interact with the eukaryote. Likewise, the eukaryote must receive and respond to these interactive signals. Many symbioses require the development of specialized structures. so that new molecules dealing with cellular and tissue differentiation need to be expressed. This generates a relevance in the field of genetics and gene regulation, especially in regards to development. Both the eukaryote and the bacterium differentiate. Generally, the bacterium is not nitrogen fixing in the free-living state. The symbiotic state requires and causes several biochemical and thus genetic regulatory changes that are needed to elicit the nitrogen fixation mechanisms.

The complexities can even be greater, if one considers the interactions occurring in the soil. Here gene-flow and competition with other microbes, genetic plasticity, and environmental adaptation generate a highly complex situation, as yet only partially understood.

Symbiotic nitrogen fixation and related plant developmental changes thus represent a meeting ground for plant physiologists, microbiologists, biochemists, geneticists, agronomists, cell biologists, and chemists. The data base dealing with this subject matter is now so immense that it is impossible to integrate it all into a common perspective. We, as scientists and teachers, are therefore faced with the dilemma of passing on the *essence* of the subject to new generations of students and researchers. This book is meant to be such a vehicle for the transfer of the "essence" of nitrogen fixation.

This book was motivated by Professor Anton Quispel's important book, *The Biology of Nitrogen Fixation*. His text was different from others of its time, because here was a summary of the knowledge of the day, brought together by authors who showed an immense interest in the subject. The book became essential reading for many students, and even today, I go back to it from time to time as a valuable source.

However, since the early 1970s a lot has happened in the technologies that are available to researchers in nitrogen fixation. The expansion of molecular techniques, coupled with the recognition of the importance of nitrogen fixation to world agriculture brought about by the 1973 oil-crisis, has meant that perhaps there is a need for an "update" of Quispel's text. Hence, the title of this work was chosen, stressing the molecular biology of symbiotic nitrogen fixation. The structure of this book totally reflects my own preferences. What I felt was needed was an in-depth analysis and status report on the subject in 1987/88.

I find the short articles in the regular conference proceedings a perfect update, but often length restrictions do not allow integration beyond the authors' work. Likewise, I felt a book should provide some opinion, some personal thoughts, rather than reading like the "7 o'clock news", with a constant stream of unedited facts. This book, like others, will have oversights and duplications. It is hoped that the reader will tolerate these and appreciate the value of the personal evaluation of published fact.

I am particularly conscious of the absence of chapters dealing with nitrogenase chemistry per se, *Klebsiella* genetics, *Azotobacter* and *Azospirillum,* and even the *Sesbania/Azorhizobium* nodule symbiosis. Numerous reasons contributed to this — the major one being a preferred focus on symbiosis rather than nitrogen fixation as a whole.

Chapter 1, which originally was to be only a "foreword", gives a general introduction into the field of genetics and molecular biology. This was written by one of the men who actually lived through the breathtaking developments of the last 40 years and who himself was responsible for the genetic characterization of the first transmissible plasmid, i.e., the F-factor of *Escherichia coli*. To the specialist, this contribution may be too general, but I felt it was of great value to the nonspecialist in molecular genetics, such as the agronomist, patent lawyer, investment banker, or struggling undergraduate, who may not be as familiar with the technical jargon as the practicing research worker.

Chapter 2 deals with vector technology and is written by two scientists whose constructions have been so widely used (i.e., all the pSUP plasmids). In Chapter 3, Jacek Plazinski describes the present state of knowledge of the *Azolla-Anabaena* symbiosis, taking both a genetic view as well as providing a descriptive, biological perspective, the result of his association with Professor Brian Gunning, who researched extensively the *Azolla* cell biology and root morphology.

Chapter 4 starts the analysis of the plant root nodule symbiosis with a detailed look at the actinomycete *Frankia* and its nonlegume partners. One question which I always had over the years and for which I have yet to find an answer is, does *Frankia* also form nodules on a legume? If so, which one? If it doesn't, why not? Perhaps we do not recognize a *Frankia*-induced legume nodule as we would suspect a *Rhizobium* infection.

Chapter 5 provides a thorough overview of the legume nodule biochemistry. Emphasis is given to the cell biology and how it relates to biochemistry and genetics. Edward Appelbaum condensed the vast data base on *Rhizobium/Bradyrhizobium* genetics in Chapter 6. This was an immense task, especially as the literature is so rapidly expanding. The reader hopefully will consult the proceedings of the Acapulco and Cologne Meetings in 1988 to get the newest data. Chapter 7 was written by Bernard Carroll and Anne Mathews. Their recent work on both supernodulation and nonnodulation mutants of soybean allowed them to make a summary of our knowledge of the effects of nitrate on the nodule symbiosis. Such a statement is much needed, and I hope that the reader values the innovative synthesis.

The parallel approach of molecular biological investigation of nodule functioning utilizes the straight molecular approach. The fundamental work by Professor D. P. S. Verma has led to the expansion of nodulin research. The Wageningen laboratory recently contributed substantially to molecular studies of nodulation, especially in their investigation on early nodulins. Chapter 8 summarizes their perception.

The final chapter deals with the nonlegume *Parasponia*, which forms nodules with *Bradyrhizobium* and *Rhizobium* species. The bacteria also elicit functional nodules in legumes (like siratro); hence, an elegant experimental system exists, allowing the comparison of nodule function.

The core of the text is aimed at the research worker in the field of nitrogen fixation, but, despite its specialization, does not lose the emphasis on teaching, both as a direct reference book and as a backbone for a graduate course on the subject.

The closing part of the book includes a subject index and a glossary of terms. The latter was included *not* for the expert, for whom many of the definitions will be too general, but for the newcomer; I hope that the quick survey of key terms will help in the reading of this book.

Like all things, one must accept that the book will not retain its up-to-dateness forever. By the nature of the editing and publication process, time will already have been lost. In view of the rapid expansion of our data base and new technologies, such as plant transformation, *in vitro* mutagenesis, anti-sense strand mutagenesis, ultrasensitive analytical techniques, and advances in developmental genetics, progress over the next decade will probably surpass that of the last. The contributors to this book collectively hope that their efforts will be of value to all those who share an interest in the molecular biology of symbiotic nitrogen fixation.

Canberra/Knoxville 1988   **Peter M. Gresshoff**

THE EDITOR

Peter Michael Gresshoff, Ph.D., D.Sc., holds the endowed Racheff Chair of Excellence in Plant Molecular Genetics (College of Agriculture) at the University of Tennessee in Knoxville.

Dr. Gresshoff, a native of Berlin (Germany), graduated in genetics/biochemistry from the University of Alberta in Edmonton in 1970 and then undertook his postgraduate studies at the Australian National University in Canberra, Australia, where he obtained his Ph.D. in 1973 and his D.Sc. in 1989.

He completed his postdoctoral work as an Alexander von Humboldt Fellow (1973 to 1975) at the University of Hohenheim (F.R.G.) and Research Fellow (1975 to 1979) in the Genetics Department (R.S.B.S.), headed by Professor William Hayes. He was appointed Senior Lecturer of Genetics in the Botany Department at the Australian National University in 1979, where he built up an internationally known research group investigating the genetics of symbiotic nitrogen fixation. He assumed his present position in January 1988, continuing the research direction by focusing on the macro- and micromolecular changes involved in nodulation.

He was awarded the Alexander von Humboldt Fellowship twice (1973 to 1975 and 1985 to 1986) and is a member of the editorial board of the *Journal of Plant Physiology*. He has received major research grants from biotechnology firms and the Australian federal government. He has published over 100 refereed publications and has contributed to many international congresses and symposia. He has been awarded membership in Phi Kappa Phi and Sigma Xi and is a member of the Knoxville Chamber of Commerce. He is a dedicated teacher and researcher, who believes in technology transfer and innovative science. His current major research interest concerns the characterization of the soybean genes controlling supernodulation and nonnodulation as well as the molecules that control these genes.

CONTRIBUTORS

Antoon D. L. Akkermans, Ph.D.
Department of Microbiology
Wageningen Agricultural University
Wageningen, The Netherlands

Edward R. Appelbaum, Ph.D.
Senior Research Scientist
Agrigenetics Advanced Science Company
Madison, Wisconsin

Gregory L. Bender, Ph.D.
Plant-Microbe Interaction Group
Australian National University
Canberra, Australia

Ton Bisseling, Ph.D.
Department of Molecular Biology
Agricultural University
Wageningen, The Netherlands

Bernard J. Carroll, Ph.D.
Botany Department
Australian National University
Canberra, Australia

Peter M. Gresshoff, Ph.D., D.Sc.
Professor, Racheff Chair of Excellence
Department of Plant Molecular Genetics
University of Tennessee Institute of
  Agriculture
Knoxville, Tennessee

William Hayes, D.Sc.
Professor (retired)
Department of Botany
Australian National University
Canberra, Australia

Ann M. Hirsch, Ph.D.
Associate Professor
Department of Biology
University of California
Los Angeles, California

Anne Mathews, Ph.D.
Molecular and Population Genetics Group
Research School of Biological Sciences
Australian National University
Canberra, Australia

Robert B. Mellor, Dr.Phil.
Associate Professor
Department of Botany
University of Basel
Basel, Switzerland

Jan-Peter Nap, Ph.D.
Department of Molecular Plant Breeding
Research Institute ITAL
Wageningen, The Netherlands

Philippe Normand, Ph.D.
Research Associate
C. R. B. F.
University Laval
Laval, Quebec, Canada

Jacek Plazinski, Ph.D.
Research Fellow
Plant Cell Biology Group
Research School of Biological Sciences
Australian National University
Canberra, Australia

Ursula B. Priefer, Ph.D.
Department of Genetics
University of Bielefeld
Bielefeld, West Germany

Kieran F. Scott, Ph.D.
Visiting Fellow
Research School of Biological Sciences
Australian National University
Canberra, Australia

Reinhard Simon, Ph.D.
Department of Genetics
University of Bielefeld
Bielefeld, West Germany

Pascal Simonet, Ph.D.
Laboratoire de Biologie des Sols
Université Claude-Bernard Lyon I
Villeurbanne, France

Dietrich Werner, Prof. Dr.
Department of Botany
Philipps University
Marburg, West Germany

TABLE OF CONTENTS

Chapter 1

GENETICS AS A TOOL TO UNDERSTAND STRUCTURE AND FUNCTION

William Hayes

> The ambition of molecular biology is to interpret the essential properties of organisms in terms of molecular structures. — *Jacques Monod, Nobel Lecture, 1965.*

Although the hereditary basis of structure and function had been known since the rediscovery of Mendel's principles in 1900, and enzymes were first shown to be proteins following the crystallization of urease by Sumner in 1926, it was not until 1941 that Beadle and Tatum, following their genetic analysis of several hundred defined, auxotrophic mutants of the prototrophic fungus *Neurospora crassa*, proposed their one gene/one enzyme hypothesis that, although not widely accepted until several years later, marks the beginning of biochemical genetics.

The initial outcome of this hypothesis was analysis of intermediary metabolism by isolating mutants having a specific amino acid, B group vitamin, or nucleotide requirement, crossing those of the same group to see whether the same or different genes of the pathway were involved, and carrying out complementation tests in heterokaryons. Each allelic group was then tested for growth on presumptive precursors of the required end product. The first such pathway to be analyzed in this way was that of arginine synthesis in *Neurospora* by Srb and Horowitz in 1944.[1] Fifteen arginine-requiring mutants were isolated and were found to comprise seven different allelic groups of which one responded only to arginine, two to either arginine or citrulline, and four to ornithine, citrulline, or arginine. It was therefore clear that ornithine and citrulline are arginine precursors, and since two groups responded to citrulline but not to ornithine, the synthetic sequence must be ornithine-citrulline-arginine.

An important outcome of the one gene/one enzyme hypothesis was that Tatum proceeded to extend his *Neurospora* studies to bacteria and isolated large numbers of auxotrophic mutants in *Escherichia coli*. From mixtures of pairs of multiply deficient mutants, he and Lederberg then recovered rare prototrophic progeny by selecting for prototrophy on minimal medium. This discovery of bacterial sexuality and recombination[2] and the subsequent isolation in the early 1950s of strains which yielded heterozygotes and recombinants at high frequency enormously facilitated conjoint biochemical and genetic studies and marks the real beginning of molecular biology as an established science.

Then followed a dramatic decade during which molecular biology reached its zenith, ushered in by the elucidation of DNA structure by Watson and Crick and the first amino acid sequence of a protein (insulin) by Sanger and Thompson, in 1953. This was followed by a series of key discoveries concerning the genetic basis of protein structure and how it is translated and its synthesis regulated. Foremost among these were Crick's sequence and colinearity hypotheses, namely, that the folded structure of globular proteins, and, hence, their specificity, is determined by the sequence of amino acids in their polypeptide chains, and that the sequence of bases (or base pairs) in the DNA chain is colinear with, and specifies the sequence of amino acids in, the derivative protein. Both hypotheses were confirmed experimentally by the demonstration that the sequences of, and linear distances between, the amino acid substitutions in nine mutant polypeptides of the *E. coli* tryptophan synthetase A protein, established by sequence analysis, were equivalent to those of the DNA mutation sites revealed by genetic crosses.[3]

The solving of the genetic code as a sequence of nonoverlapping nucleotide triplets or codons, each coding for a specific amino acid, is a fascinating example of the imaginative integration of theoretical, genetic, and biochemical studies. Since the sequence of only four bases must code for 20 amino acids, it was clear that, from an alphabet of four letters, only $16(4^2)$ two-letter words but $64(4^3)$ three-letter words could be made, so that a triplet code seemed the most likely. Then followed a brilliant but complicated genetic experiment based on the fact that it has been shown, for a small region of the phage T4 chromosome, that acridine-induced mutations were mainly due to either deletion ($-$) or addition ($+$) of a single base pair and that these could be distinguished. A single $-$ or $+$ mutation disrupted protein synthesis; however, a second mutation of opposite sign, located close to the first, resulted in a "pseudowild" reversion. This suggested that the code was read in one direction from a fixed starting point, so that the deletion or addition of a single base would alter the reading frame and lead to misreading of all the subsequent codons. For example, if the code was ABC.ABC.ABC.ABC.ABC and the second B was deleted, it would now be read as ABC.ACA.BCA.BCA.BCA. If, however, another base, D, was inserted close to the first deletion, the code would now be read as ABC.ACA.BCA.DBC.ABC.ABC; the initial reading frame would be restored, and only those codons between the deleted and added bases would be mistranslated, leading to only a few altered amino acids in the protein and a pseudowild phenotype. The crucial experiment was to construct (by genetic recombination) phage strains, each containing either two or three closely linked mutations of the same type, either base deletions ($- -$ and $- - -$) or additions ($+ +$ and $+ + +$). The result was that all 14 double mutant strains of $- -$ or $+ +$ type were mutant; on the contrary, six triple mutants, five of $+ + +$ type and one of $- - -$ type, all displayed the pseudowild phenotype.[4].

The identification of the nucleotide triplets that code for each amino acid depended on the development, during the same decade, of an understanding of the basic mechanisms and structures involved in translating the genetic information into protein. Kinetic studies and the fact that RNase taken up by cells completely blocked protein synthesis revealed that the DNA was not directly involved but that the base sequences of its "sense" strand were transcribed into a single-stranded RNA copy. This "messenger" RNA (mRNA) then became attached to complex ribosomal particles in the cytoplasm which were richly associated not only with amino acids, but also with soluble(s) RNA oligonucleotides. Once again it was Crick who suggested, in 1958, that the amino acids did not recognize their mRNA codons directly but through the intermediary of the sRNA molecules (now called "transfer" or tRNA), each of which attached specifically to a particular amino acid and carried an anticodon triplet which recognized its mRNA codon. Thus, as the ribosomes move along the mRNA, the amino acids are assembled in their correct sequence. Contiguous amino acids are then joined by peptide bonds, and the tRNAs are released to participate in another similar reaction.

Let us now return to the biochemical nature of the code which, in essence, was solved by the use of cell-free systems containing ribosomes, tRNAs and enzymes, ATP, and an ATP-generating system, to which were added ^{14}C-labeled amino acids and synthetic polyRNA molecules of known sequence to act as mRNA. The first outcome of this method, reported by Nirenberg at a meeting in Moscow in 1961, was the exciting discovery that a pure polymer of uracil (U), which substitutes for thymine (T) in mRNA and therefore represents adenine in the DNA, yielded a polypeptide composed only of phenylalanine residues, implying that the mRNA codon for phenylalanine is UUU. Thereafter the codons for the remaining 19 amino acids were rapidly deciphered by variations of this general method. The most precise and effective of these was the addition to 20 tubes, containing the ribosomal cell-free system and a particular synthetic trinucleotide, of a mixture of all 20 amino acids of which a single one, different for each tube, was radioactive. The ribosomes were then isolated by filtration, and the particular amino acid bound was identified by its radioactivity.[5] It turned out that 61 of the 64 possible triplets code for amino acids so that the code is "degenerate", some amino acids such as serine, leucine, and arginine having as many as six different codons while only methionine has a single one. The remaining three triplets, UAA, UAG, and UGA, were later identified as "chain-terminating" triplets; they are found at the end of mRNA codes for polypeptides, often in pairs or groups of three, and disrupt translation so that the completed polypeptide is released from the ribosome.[6]

Since many codons share two bases with chain-terminating triplets, it follows that a mutational base change, instead of resulting in a normally inocuous substitution of one amino acid by another of similar charge, may yield only an abbreviated and inactive polypeptide fragment. Although the genetic code is a universal one, its degeneracy means that different species may use different triplets for the same amino acid, as shown, for example, by the fact that the DNA base compositions may range in cytosine + guanosine content from about 25 to 80%. This variation is reflected in the codon specificities of transfer RNA molecules, so that an anticodon used for a particular amino acid in one species may fail to recognize a different codon for the same amino acid in another species. This leads to a possible flaw in genetic engineering experiments, leading to faulty translation following interspecies gene transfer.

We cannot leave the dramatic decade that followed the discovery of DNA structure without mentioning the key that led to it, namely, X-ray crystallography, which was based on the pragmatic concept that if you know the three-dimensional structure of a molecule you will know how it works. The structure of DNA is, indeed, an outstanding example of this concept. One of the earliest revelations of the method was that polypeptide chains do not usually exist as fully extended molecules but are coiled into α-helices (secondary structure), not only in fibrous proteins, but also in straight stretches of globular proteins in which, at certain regions, the chain is folded back on itself to give a *tertiary* structure. In general, these foldings are due to bondings between the side chains of amino acids, the most important of which are disulfide bonds between two cysteine residues.

Although the first complex protein to be defined by crystallography was myoglobin (see Reference 7), egg-white lysozyme was the first *enzyme* whose amino acid sequence and structure were elucidated and the active site of which was identified with a surface cleft into which the substrate specifically fits.[8] It is worth noting here that while many enzymes and hormones will tolerate mutations along straight runs of an α-helix or, for example, the removal of many acids from the end of a terminal chain, without loss of activity, a single amino acid substitution close to the active site may abolish function. A good example is sickle-cell anemia, which follows a single amino acid alteration in hemoglobin, while mutations involving cysteine residues may lead to drastic distortions of structure.

So far we have visualized protein function in terms of the complicated three-dimensional folding of a polypeptide chain with a highly specific amino acid sequence which, at the genetic level, reflects a linear sequence of codons in the sense strand of the DNA.

X-ray crystallography then revealed that hemoglobin has a *quaternary* structure, consisting of four protein *subunits* of two chemical types (α and β), each closely resembling the myoglobin molecule, organized into a compact symmetry.[9] Most large biological globular proteins are quaternary aggregates of polypeptide chains, each several hundred amino acids long. In some proteins the subunits are identical and usually specified by a single gene; in others they are different. For example, the *E. coli* β-galactosidase enzyme contains four identical polypeptide chains, tryptophan synthetase is composed of two pairs of dissimilar chains coded by different genes, while the *Neurospora crassa* glutamic dehydrogenase enzyme contains six to eight subunits. If such quaternary proteins are purified and the individual subunits separated by lowering the pH, at first they display negligible or no functional activity at normal pH, even when they are all identical, but activity gradually returns as aggregates reform. Thus, the formation of active enzymes is not coded for *directly* by the DNA, but by the structure of the individual subunits. Similarly, if the disulfide bonds of globular proteins are reduced and broken, the polypeptide chains become linear, but refold automatically into their tertiary structure when normal conditions are restored.

The quaternary structure of proteins may lead to complications in the interpretation of genetic analysis, even when subunits are identical and are determined by a single gene. Thus, in complementation tests, the restoration of the wild phenotype when two mutant genomes affecting the same function are introduced into the same cell normally implies that the mutations are in different genes; the heterokaryon has one good gene of each type. However, if the protein is an aggregate of two or more identical subunits, mutations at different locations in the *same* gene may produce nonoverlapping defects in the subunits which can then aggregate randomly and stabilize one another; thus, a functional protein is produced, but it is usually less active than the wild type. This is called "intraallelic complementation".

On the other hand, if the two subunits differ, heterozygotes for mutant alleles at each locus may inherit four different types of molecule which do not reflect the genotype. A good example of this is human hemoglobin of which the α and β subunits are determined by unlinked genes, so that their assembly must occur *after* their synthesis. Cases have been reported of individuals who have inherited a defect in the α chain ($\alpha_2^- \beta_2^+$) from one parent and in the β chain ($\alpha_2^+ \beta_2^-$) from the other. Their red blood cells contain four different types of hemoglobin from random reassortment of the individual chains, often in approximately equal amounts; the two parental types ($\alpha_2^- \beta_2^+$ and $\alpha_2^+ \beta_2^-$), a normal "recombinant" type ($\alpha_2^+ \beta_2^+$), and molecules defective in both chains ($\alpha_2^- \beta_2^-$).[10]

Living cells are factories which have evolved to produce a wide range of complex end products from a limited input of raw materials. Since this evolution has proved successful in a highly competitive environment, they must have developed intricate genetic mechanisms for organizing and controlling end-product synthesis with a minimum of wastage and energy. Most of our present knowledge comes from studies of the genetic regulation of enzyme synthesis in *E. coli* in the 1960s, but the general principles that have emerged certainly apply, to some extent at least, to eukaryotes.

It had been known since the 1930s that enzymes appeared to fall into two categories. They were either "inducible', being synthesized only in the presence of their substrates, or "constitutive" and produced independently of substrates. Subsequent genetic analysis of the *E. coli*-inducible enzyme β-galactosidase, which breaks down lactose to glucose and galactose, disclosed that mutations in a small chromosomal region which we will call *R*, distinct from the gene *Z*, which species the enzyme, resulted in constitutive production and that these mutations were recessive in heterozygotes with wild type, i.e., the heterozygotes were inducible. It was therefore clear that gene *R* coded for a "repressor" that switched off transcription of the *Z* gene and was subsequently shown to be a protein. Other constitutive mutations were then found that mapped very close and upstream to the 5' end of the *Z* gene; these proved to be dominant and were postulated to be the site of attachment of the repressor, termed the "operator" (*O*), later confirmed by the binding of radioactive repressor to the *O* region. Further genetic studies of this small region of chromosome revealed mutations which mapped between the repressor and operator loci without involving either, but nevertheless reduced β-galactosidase synthesis. This turned out to be the site where RNA polymerase molecules first attach before starting to transcribe mRNA and was termed the "promoter" (*P*) (see Figure 1). Thus, the attachment of repressor effectively blocks transcription.

In this system the "natural" inducer is lactose, but other β-galactosides are also inducers without being substrates and vice versa. How, then, does the inducer act? Many enzymes are known which have two (or more) combining sites of which one is specific for its substrate and the other usually for small molecules such as amino acids, whose attachment alters the tertiary structure of the enzyme so that it loses its primary function. Such enzymes are called "allosteric". In the case of β-galactosidase, galactoside-inducer molecules specifically combine with such a secondary site on the repressor and inactivate it.

Most of these studies and the concepts derived from them were the work of Jacob and

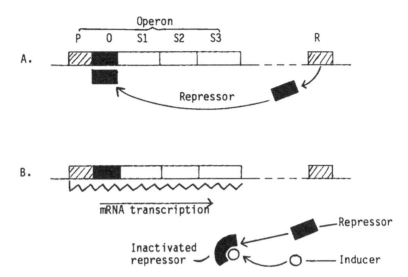

FIGURE 1. The induction and repression of an operon. P indicates the promotor, O the operator, and R the repressor locus. S1, S2, and S3 are three functional genes of the operon which are coordinately expressed. (A) Shows the repression of enzyme synthesis by attachment of repressor to the operator, blocking transcription by RNA polymerase; (B) shows the initiation of messenger RNA synthesis following attachment of the inducer to the repressor which it specifically inactivates.

Monod at the Pasteur Institute, Paris (review see Reference 11) for which they were awarded a Nobel Prize. Early on, they also found that two other genes concerned with lactose fermentation, of which one is a permease, were located adjacent to the β-galactosidase Z gene and that all three genes were transcribed together on the same mRNA molecule and were coordinately expressed and regulated. Such systems they called "operons", which are commonly found in bacteria but not in eukaryotic cells, although the general principles of regulation are widely applicable (Figure 1). Another type of automatic control mechanism, first revealed by the biochemical genetics of tryptophan synthesis in *E. coli*, is "feedback inhibition" which, like repression, is a vital factor governing cellular economy. In this case, when the end product of synthesis begins to accumulate, it interacts allosterically with the first enzyme of the pathway so that not only its own synthesis, but that of all the intermediates is switched off.

Finally, we come to a brief account of how genes are identified and of the elements of genetic engineering. Genes determining normal structure and function are identified by observing the altered phenotype of spontaneous or induced mutations and mapping their location by "classic" genetic crosses with strains of known genotype on which we will not elaborate here. This is relatively simple in the case of bacteria which have only one chromosome and are haploid, so that recessive mutations are immediately expressed while positive selection of rare mutants or recombinants is often possible

In plants, however, recessive mutations, which are much more frequent than dominant ones, induced in the seeds of self-fertilizing diploids by irradiation or chemical mutagens, are expressed only by the M2 generation of seeds planted for phenotype, an operation involving considerable time and labor. Again, selective techniques for plant mutants are rare but can be very effective when available as, for example, in the isolation of nitrate reductase-deficient mutants which, unlike wild type, grow in the presence of chlorate which they are unable to break down to the toxic chlorite. The introduction of cell culture systems whereby freshly isolated leaf protoplasts from normal or haploid plants are grown *in vitro* offers many advantages. Thus, very large numbers of cells can be mutagenized and selection applied, while complementation is readily tested by protoplast fusion in the presence of polyethylene glycol (PEG). However, plant cells in continuous culture tend to be unstable and to generate "somaclonal variants", while the capacity to regenerate whole plants from them may readily be lost.[12]

The essence of genetic engineering is the cloning of particular genes from one organism and their functional incorporation into the genetic material of another. This is accomplished by means of two basic tools: conjugal plasmids and restriction enzymes. Conjugal plasmids are small, self-replicating circles of DNA present in the cytoplasm of many bacteria, especially the normal intestinal flora of animals and man. They promote conjugation and their own transfer to other bacteria and are readily separable from chromosomal DNA by gel electrophoresis.

Restriction enzymes comprise a wide range of specific endonucleases produced by various bacteria, each of which recognizes short, randomly occurring DNA sequences, usually four to six pairs long, which are absent from the DNA of the bacteria producing them and their lysogenic phages. These enzymes, therefore, act as a potent defense against infection by foreign DNA, such as that of virulent phages, which is cut at the restriction sites.

If DNA extracted from cells is purified and treated with a combination of restriction enzymes, each enzyme will cut the molecules into small fragments at its specific site. These fragments will vary widely in size, but those carrying a particular gene will all be of equal length since they are bounded by the same target sites. The fragments of different length can then be separated into discrete bands by their electrophoretic mobility.

The identification of the particular fragment carrying the gene or genes determining the phenotype under investigation is a more complex problem. If a particular chromosome, or part of a chromosome, has been shown to be involved by recombinational mapping and can be isolated, radioactive cDNA copies of it will recognize and hybridize with the band on the gel that contains its DNA homologue. Alternatively, a radioactive cDNA copy of mRNA is equally effective, but it is only rarely that a particular mRNA is produced locally in a relatively pure state. If, however, the aim is to clone the gene determining a particular protein, the addition of a specific antiserum to an *in vitro* transcriptional system in a radioactive medium will bind to the protein as it is transcribed so that its associated mRNA can be isolated.

When the band containing the desired DNA fragments has been identified, it needs to be inserted into a new replicon, such as a bacterial plasmid. The DNA circles of a suitable plasmid are opened with a restriction enzyme which cuts in one place only, and then closed again by ligase after inserting the DNA fragments to be cloned. These hybrid plasmids (and hence the term "recombinant DNA") are then introduced into bacteria such as *E. coli* in which they multiply with their host. Thus, large bacterial populations, each carrying one or more hybrid plasmids, can be grown. If the foreign gene in the plasmid encodes a valuable product, such as a human hormone, for example, the bacteria will synthesize and excrete it, provided that the genetic regulatory sequences that promote its transcription were included in the clone and that the bacterial translational system (e.g., tRNAs) is suitable. This valuable aspect of genetic engineering has recently become widely commercialized and forms the basis of modern biotechnology.

Plasmids may be used directly in comparative genetic studies of bacteria since many plasmids will promote conjugation between, and then multiply in, different bacterial species so that complementation tests are readily performed. Again, for example, nitrogen-fixation (*nif*) genes from the bacterium *Klebsiella pneumoniae*, which can fix nitrogen in the free-living state, can be transferred by conjugation to *E. coli*. Hybrid plasmids containing such *nif* genes allowed the analysis of homologous genes in other bacteria such as *Rhizobium* which mediate nitrogen fixation in symbiosis with leguminous plants but are not easily switched on to fix nitrogen in free-living culture. In fact, the genes responsible for nodule formation (*nod*) and symbiotic nitrogen fixation (*fix*) in most fast-growing *Rhizobium* strains reside, not in the bacterial chromosome, but in large indigenous plasmids which are lost at high temperature.

The final step in genetic engineering is the incorporation of donor genes, isolated and amplified in bacterial plasmids, into the genomes of eukaryotic cells in which they will be expressed and inherited by progeny cells. It happens that a common soil bacterium, *Agrobacterium tumefaciens* which produces tumors in dicotyledonous plants, normally does this in nature and offers a tool for artificial gene transfer. Briefly, *A. tumefaciens* carries a temperature-sensitive tumor-inducing (Ti) conjugal plasmid. Following infection of the plant, a small fraction of the plasmid DNA (the T fraction) is transferred to the plant cells and integrated into their DNA which is thereafter replicated and inherited by daughter cells. Intensive genetic analysis of the Ti plasmid has identified a number of genes concerned with tumorogenesis on the T-DNA,

Recently, "disarmed" (i.e., pathogenicity-deleted) plasmids have been produced and have proved successful in the transfer of foreign genes. For example, in one of the early experiments, infection of a tobacco plant with a plasmid in which part of the T-DNA had been replaced by a yeast alcohol dehydrogenase gene yielded cells which grew into plants. Moreover, after self-pollination, the plants produced seeds that generated healthy tobacco plants whose cells yielded multiple copies of T-DNA containing the inserted yeast gene which, however, was not expressed.[13]

One of the difficulties in the effective transfer of genes between plants of different species or genera is the problem of their expression. This can usually be solved by introducing a considerable fraction of DNA on either side of the required gene, containing its own regulatory sequences. Thus, using a disarmed plasmid, a 7.5-kb fragment containing the gene determining a chlorophyll-binding protein from wheat, which is a monocotyledon, was transferred to protoplasts of the tobacco plant, *Nicotiana tabacum,* as well as of petunia, both dicotyledons. Whole plants were then regenerated from the protoplasts, and the wheat gene was found to be expressed and regulated normally.[14] (An excellent and detailed review of plant genetic engineering is by Chilton, Reference 13.)

Before leaving the topic of gene vectors it should be noted that, in addition to plasmids, temperate bacterial transducing phages, such as *lambda* in *E. coli*, are widely used. Moreover, "libraries' or "banks'', especially of phage *lambda*, carrying a wide selection of gene clones from various bacterial species such as *Rhizobium,* are now available.

I have previously mentioned, en passant, general methods of inducing mutations in bacteria and plant seeds by irradiations and chemical mutagens. However, the resulting mutants arise randomly so that the identification and isolation of specific phenotypes is laborious. In recent years, greatly simplified ways have been devised for induction and genetic localization of mutations. Foremost among these has been the discovery and development of *transposons*. These comprise a range of naturally occurring linear segments of DNA, varying from a few hundred to many thousands of nucleotides long and located in apparently silent regions of the chromosomes of prokaryotes and eukaryotes. They have the bizarre property of occasionally excising themselves spontaneously and "jumping" across the cytoplasm to insert not only into other regions of the same chromosome, but into other chromosomes, as well as into DNA viruses and plasmids in the cell. Insertion into a functional gene produces a fairly stable mutation, but the transposon can jump out again, usually leaving the gene intact as before.

The DNA of these transposable elements in bacteria as well as in plasmids and phages is terminated at both extremities by inverted nucleotide sequences and appears to have no homology with the chromosomal sites into which they insert themselves, while insertion occurs independently of the normal recombination enzymes. If a chromosomal fragment or

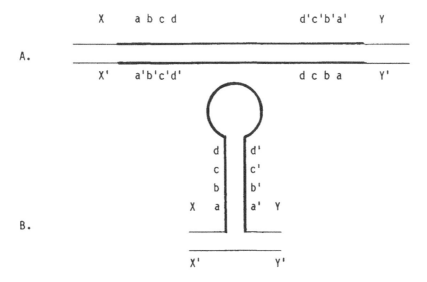

FIGURE 2.   The structure of transposons. The bold lines indicate the strands of transposon DNA, and the light lines (X, Y) those of the host DNA into which they are inserted. The letters aa', bb', and c represent the base pairs of the inverted repeat sequences of DNA at the transposon termini. (A) Shows transposon DNA inserted into the host chromosome; (B) shows a denatured single strand of the transposon, reannealed to one of the homologous strands of wild-type host DNA, revealing the "lollipop" structure as seen by electron microscopy.

plasmid containing a transposon is denatured to separate its DNA strands which are then allowed to reanneal with the strands of an homologous wild-type fragment and are visualized by electron microscopy, the transposon strand is seen as a single-stranded loop attached to the normal DNA by a double-stranded stalk formed by renaturation of the inverted sequences at the transposon ends — the aptly named "lollipop" structure! (See Figure 2). The mechanism of transposition is explained by the action of a specific endonuclease-ligase enzyme, coded by the transposon DNA, which recognizes the inverted sequences at its termini and cuts them, as well as the DNA between adjacent base pairs in the recipient chromosome, and then joins them together again.

A serious medical and veterinary problem for many years has been the spread, by transmissible plasmids, of bacterial resistance to virtually all the antibiotics in clinical use, thus nullifying the effect of antibiotic therapy. In fact, plasmids with a wide host range have been isolated in which the determinants of resistance to at least ten different antibiotics are encoded in the same genome. It has now been shown that the resistance gene is carried by transposons which jump from one plasmid to another and from plasmid to chromosome and back again. These transposons can, of course, also cause mutations. In fact, it has been estimated that in *E. coli* 25% of all spontaneous mutations are transposon induced.

Enough has been said about the general aspects of transposons. How can we use them for the induction, localization, and isolation of mutations? As an example let us consider the identification and cloning of nonnodulating (*nod⁻*) mutations in a fast-growing *Rhizobium* strain by a transposon, Tn5, which carries a kanamycin-resistant (*kan*ʳ) gene and inserts randomly.[15]. The transposon is transferred from *E. coli* to the *Rhizobium* strain by a conjugative "suicide" plasmid in which it is inserted, but which is unable to replicate in *Rhizobium* so that only those cells in which the transposon has jumped from the plasmid to the cellular genome will be resistant to kanamycin. *Kan*ʳ colonies are, therefore, selected and screened by a rapid plant assay system, for a defective phenotype which may, of course, result from mutations in more than one gene.[16] Single colonies are now purified, checked for kanamycin resistance, and their DNA isolated, treated with an appropriate restriction enzyme known not to cut the transposon DNA, and fractionated by gel electrophoresis. The band containing the mutant DNA is then identified by hybridization to ^{32}P-labeled Tn5 DNA. The particular region of mutant DNA can then be identified by hybridization with known *Rhizobium* sequences obtained from a clone bank. As a final test, the homologous wild-type region of DNA, carried on a conjugative plasmid, is transferred to the mutant bacteria whose defect it should correct by complementation (for details, see References 15 and 17).

The central theme of this book is the interaction between nitrogen-fixing microorganisms and the plants with which they associate symbiotically. Until very recently virtually nothing was known about the biochemical genetics of these interactions, such as the nature and sequence of the plant products that are presumed to switch on the genes known to be concerned with various aspects of nodulation, several of which have been mapped, for example, on a small region (14 kb) of the *Rhizobium trifolii* symbiotic plasmid. The main difficulty of such studies has been the tedious and time-consuming *in vivo* testing of potential inducers for their specific effects on nodulation. This has been enormously simplified by inserting, in turn, in each of the four *nod* genes on the 14-kb fragment of DNA, a kanamycin-resistant transposon which also carries part of the *E. coli* lactose operon containing the β-galactosidase Z gene (see above). Each fusion complex was then inserted into a conjugal plasmid and introduced into an *R. trifolii* strain which has lost its symbiotic plasmid, and exposed to the plant products suspected of being Nod inducers. If effective, the inducer would switch on the operator of the mutant *nod* gene and mRNA transcription would then proceed along the hybrid DNA to the lactose operon, leading to the production of β-galactosidase, identified by a simple colorimetric assay. Using this method it has recently been shown that only one *nod* gene (*nodD*) is actively expressed by free-living *Rhizobium* in artificial culture, but that the other three genes are also induced when clover plants are grown in the medium.[18] Subsequent work using *lac* gene fusions has identified and characterized various phenolic compounds from clover roots which either stimulate or inhibit the induction of *R. trifolii nod* genes.[19] Similar findings were obtained from root exudates of alfalfa, pea, and soybean (see Chapter 6).

It is important to realize that genetic experiments not infrequently reveal paradoxical aspects of the phenomena under investigation which had not previously been anticipated. Two examples from microbial genetics should make this clear. This first concerns the growth and maturation of a virulent bacteriophage (T4) in *E. coli*. Basically, the phage resembles an angular tadpole with an icosahedral protein head containing the DNA and a long protein tail to which are attached fibers that specifically bind the phage to the cell wall. A complex sheath surrounding the tail core then contracts and drives the DNA into the cell where it replicates and determines the synthesis of the structural proteins of the progeny phages which can be recognized in cell lysates by electron microscopy.

Recombination analysis of temperature-sensitive phage mutants which can grow at normal temperatures revealed more than 40 genes concerned with morphogenesis. However, the lysates of cells infected by different mutants at the high restrictive temperature were found to contain heads but no tails, tails but no heads, heads and tails but no tail fibers, and so on. It then turned out that when the lysates from different defective groups, one lacking heads and the other tails, for example, were mixed *in vitro*, active infective phage was often produced with high efficiency. Thus, it became apparent that at least some of the complex structural components of the phage were capable of self-assembly. Of course, in many cases more intricate patterns were found. For example, the phage head is constructed of subunits of a single protein determined by one gene, but its actual shape can be profoundly modified by mutations in other genes.[20]

The second example highlights the importance of doing the wrong experiment and recounts the discovery of two sexes, gene donors and recipients, in *E. coli* conjugation. The experiment was devised to initiate studies on the kinetics of recombination by finding the time, after mixing two different auxotrophic mutants, that recombinants were formed. A streptomycin-resistant (Sm^r) mutant of strain A was mixed with a sensitive (Sm^s) strain of B, and streptomycin, which rapidly killed strain B, was added at intervals to the mixture. Sm^r prototrophs were then selected. The experiment worked perfectly; no recombinants were isolated during the first 2 h, but thereafter began to appear and increased linearly to a plateau. As a nominal control, the experiment was then repeated by reversing the resistance pattern of the two parental auxotrophs (i.e., A.Sm^s × B.Sm^r). This time, however, the results were quite different: about the same number of recombinant colonies emerged from all the samples, even when streptomycin was added immediately after mixing. Further experiments then showed that while the sensitive A strain treated with streptomycin to a negligible survival and subsequently mixed with B produced recombinants, B strains similarly treated were completely sterile. Therefore, it was evident that A strains transfer their genes to B, the "female" in which the whole process of recombination and segregation occurred. It later transpired that gene transfer by the male A strain was mediated by a conjugal plasmid, little affected by streptomycin which specifically inhibits protein synthesis.[21].

Finally, a word about repeated DNA sequences and their possible role in genetic instability in plants. Whereas the genomes of prokaryotes contain amounts of DNA commensurate with their estimated number of functional genes, the DNA in the individual cells of eukaryotes is normally much in excess of that expected from their comparative size and complexity, and covers a 1000-fold span. For example, the genomes of some amphibians contain about 10^{11} DNA base pairs per cell as compared with about 10^9 bp for fish and mammals and 4×10^6 for *E. coli*.

When the DNA from eukaryotes is broken up and the various fractions denatured and their separated strands then allowed to reanneal, it is found that some of the fractions complete renaturation much more quickly than others. This rapid reassociation was found to be due to a preponderance of redundant or repeated polynucleotide sequences, the strands of which recognize one another much more quickly than those from a more complicated series of different sequences.[22] Such repeated and apparently superfluous DNA sequences are absent from the chromosomes of bacteria and viruses. However, it has been found that in most plants, more than 75% of all DNA sequences that are 50 bp or longer is repetitive DNA. Their origin is unknown, but it is thought that they may arise from aberrant replication during mitosis in the germ cell line.

The genomes of plants are known to be unusually unstable, and this has been ascribed, in part at least, to the high proportion of repeated sequences in their DNA. There are several possible reasons for this. One is that redundant copies of particular genes are free to accumulate mutations which would be lethal if only a single copy were present; thus, novel genotypes may arise which are of adaptive value and will spread through the interbreeding population by natural selection. Again, a multiplicity of repeated sequences will lead to frequent incorrect pairing and recombination during meiosis with resulting inheritable chromosomal transpositions and deletions. Moreover, many of the repeats are, in fact, transposons whose effects have already been described. In conclusion, there is now evidence that the growth of plants under various conditions of stress induces changes in their repetitive DNA. This has been construed as explaining an earlier "Lamarckian" finding that inbred strains of flax, grown from seeds in different and stressful nutrient environments, yielded plants of abnormal and distinguishable phenotypes; however, the seeds from these plants, now grown under *normal* conditions, reproduced their parental abnormalities (for review see Reference 23).

Enough has been said to reveal the growth and complexity of modern molecular biology, as well as the key role that genetics is playing in enhancing our understanding of the symbiotic interactions between bacteria and plants in nitrogen fixation.

REFERENCES

1. **Srb, A. M. and Horowitz, N. H.,** The ornithine cycle in *Neurospora* and its genetic control, *J. Biol. Chem.*, 154, 129, 1944.
2. **Lederberg, J. and Tatum, E. L.,** Genetic recombination in *E. coli, Nature (London)*, 158, 558, 1946.
3. **Yanofsky, C., Carlton, B. C., Guest, J. R., Helinski, D. R., and Henning, U.,** On the colinearity of gene structure and protein structure, *Proc. Natl. Acad. Sci. U.S.A.*, 51, 266, 1964.
4. **Crick, F. H. C., Barnett, L., Brenner, S., and Watts-Tobin, R. J.,** General nature of the genetic code for proteins, *Nature (London)*, 192, 1227, 1961.
5. **Nirenberg, M., Leder, P. Berbfield, M., Brimacombe, R., Trubin, J., Rotman, F., and O'Meal, S.,** RNA code words and protein synthesis. VII. On the general nature of the RNA code, *Proc. Natl. Acad. Sci. U.S.A.*, 53, 1161, 1965.
6. **Sarabhai, A. S., Stretton, A. O. W., Brenner, S., and Bolle, A.,** Colinearity of the gene with the polypeptide chain, *Nature (London)*, 201, 13, 1964.
7. **Kendrew, J. C.,** The structure of globular proteins, in *Biological Structure and Function*, Vol. 1, Goodwin, T. W. and Lindberg, O., Eds., Academic Press, London, 1960, 5.
8. **Johnson, L. N. and Phillips, D. C.,** Structure of some crystalline lysozymes — inhibition complexes determined by X-ray analysis at 6Å resolution, *Nature (London)*, 206, 751, 1965.
9. **Perutz, M. F., Rossman, M. G., Cullis, A. F., Muirgead, H., Will, G., and North, A. C. T.,** Structure of haemoglobin, *Nature (London)*, 185, 416, 1960.
10. **Baglioni, C. and Ingram, V. M.,** Four adult haemoglobin types in one person, *Nature (London)*, 189, 465, 1961.
11. **Jacob, F. and Monod, J.,** Genetic mapping of the elements of the lactose region in *Escherichia coli, Biochem. Biophys. Res. Commun.*, 18, 693, 1965.
12. **Blonstein, A. D.,** Auxotroph isolation *in vitro*, in *Plant Gene Research*, Blonstein, A. D. and King, P. J., Eds., Springer-Verlag, Vienna, 1986, 259.
13. **Chilton, M. D.,** A vector for introducing new genes into plants, *Sci. Am.*, 248(6), 36, 1983.
14. **Lamppa, G. Nagy, F. and Chua, N. M.,** Light-regulated and organ-specific expression of a wheat cab gene in transgenic tobacco, *Nature (London)*, 316, 750, 1985.
15. **Rolfe, B. G. and Djordjevic, M. A.,** Genetic engineering of *Rhizobium*, in *Reviews in Rural Science 6, Biotechnology and Recombinant DNA Technology in the Animal Production Industries*, Leng, R. A., Barker, J. S. F., Adam, D. B., and Hutchinson, K. J., Eds., 1985, 176.
16. **Rolfe, B. G., Gresshoff, P. M., and Shine, J.,** Rapid screening for symbiotic mutants of *Rhizobium* and white clover,, *Plant Sci. Lett.*, 19, 277, 1980.
17. **Scott, K. F., Hughes, J. E., Gresshoff, P. M., Beringer, J. E., Rolfe, B. G., and Shine, J.,** Molecular cloning of *Rhizobium trifolii* genes involved in symbiotic nitrogen fixation, *J. Mol. Appl. Genet.*, 1, 315, 1982.
18. **Innes, R. W., Kimpel, P. I., Plazinski, J., Canter-Cremers, M., Rolfe, B. J., and Djordjevic, M. A.,** Plant factors induce expression of nodulation and host-range genes in *Rhizobium trifolii, Mol. Gen. Genet.*, 201, 426, 1985.
19. **Djordjevic, M. A., Redmond, J. W., Batley, M., and Rolfe, B. G.,** Clovers secrete specific phenolic compounds which either stimulate or repress *nod* gene expression in *Rhizobium trifolii, EMBO J.*, 6, 1173, 1987.
20. **Kellenberger, E.,** The genetic control of the shape of a virus, *Sci. Am.*, 215, 32, 1966.
21. **Hayes, W.,** The mechanism of genetic recombination in *Escherichia coli, Cold Spring Harbor Symp. Quant. Biol.*, 18, 75, 1953.
22. **Britten, R. J., and Kohne, D. E.,** Repeated sequences in DNA, *Science*, 161, 529, 1968.
23. **Marx, J. L.,** Instability in plants and the ghost of Lamarck, *Science*, 224, 1415, 1984.

REFERENCES

Chapter 2

VECTOR TECHNOLOGY OF RELEVANCE TO NITROGEN FIXATION RESEARCH*

Reinhard Simon and Ursula B. Priefer

TABLE OF CONTENTS

---

* Abbreviations: Ap = ampicillin; Cb = carbenicillin; Cm = chloramphenicol; Gm = gentamicin; Hg = mercury; Km = kanamycin; Nm = neomycin; Pn = penicillin; Sm = streptomycin; Sp = spectinomycin; Su = sulfonamide; Tc = tetracycline; Tm = tobramycin; Tp = trimethoprim.

I. GENERAL INTRODUCTION

Considerable progress in the molecular analysis of symbiotic nitrogen fixation, achieved within the last few years, is directly related to advances in methods of recombinant DNA technology, generally suitable for Gram-negative bacteria. Indeed, many of these methodologies have been especially developed for research in nitrogen fixation.

Genetic manipulative procedures employed currently include general and site-directed mutagenesis for the analysis of genetic organization by gene and operon inactivation; *in vivo* or *in vitro* cloning systems which allow the stable introduction and maintenance of the gene under study into host organisms of interest; methodologies for inserting a test gene within the gene or operon of interest to monitor gene regulation; and techniques to modulate gene expression, e.g., by placing the gene under study under the control of a foreign promoter.

Essential for such analytical studies is the availability of versatile cloning and vector systems. Here we review the types of vector systems and genetic manipulations most commonly employed currently and give some examples of their application in nitrogen-fixation research.

II. TRANSPOSON MUTAGENESIS AND GENOMIC INTEGRATION OF CLONED DNA

A. INSERTIONAL MUTAGENESIS USING TRANSPOSONS

An indispensable requirement for the genetic analysis of any bacterium is the availability of mutations in the gene(s) of interest. Mutants arise spontaneously with low probability or may be obtained by treatment with chemical or physical mutagens. However, the alternative, transposon mutagenesis, offers major advantages compared to these classic procedures.

Kleckner et al.[1] have summarized numerous concepts and techniques on how transposons can be used in virtually every type of genetic manipulation of bacteria. Using transposons many types of experiments can be greatly simplified, and, in turn, a series of subsequent manipulations depend on them. These new methods in bacterial genetics were pioneered as usual in *Escherichia coli* and closely related organisms, but have been adapted rapidly to other Gram-negative bacteria. In recent years, the use of transposons as mutagenic agents has become one of the cornerstones of the genetic analysis of nitrogen-fixing organisms.

1. General Features and Advantages of Transposon Mutagenesis

Transposable elements are defined as genetic entities that are capable of promoting their own translocation from one site to another within or between different replicons, usually without leaving their original location and without the requirement of extensive DNA homology at the sites of insertion. Using physical and genetic criteria, transposable elements in bacteria are subdivided into three groups:

1. Insertion sequences (IS elements), which are small DNA segments (normally less than 2 kb) encoding no apparent phenotypic determinants unrelated to their transposition
2. Transposons (Tn elements), which are larger in size, since they carry additional genes determining specific phenotypes, e.g., antibiotic resistances
3. Certain bacteriophages (mutator phages), like the classic example of Mu, which uses transposition reactions to replicate its DNA, and upon lysogenization insert randomly into the host genome

For detailed description of the biology of prokaryotic transposable elements (nomenclature, physical and genetic properties, ecological and evolutionary considerations, and models proposed to explain mechanisms of translocation), the reader is referred to a series of recent reviews that have covered the subject thoroughly.[2-7]

Not long after their detection, the translocatable drug-resistance elements (transposons) especially changed from mere laboratory curiosities to useful tools for *in vivo* genetic engineering. The first and most obvious use of transposons results from their mutagenic activity, as schematically shown in Figure 1.

The mutation caused by the insertion of a drug-resistance transposon is accompanied by the acquisition of a selectable marker. Thus, after transposon mutagenesis, a cell carrying an insertional mutation is positively selectable; nonmutated cells will not survive the selection by drug resistance for the presence of the transposon. This is in striking contrast to classic mutagenesis procedures. For example, in practice it is not possible to apply chemical or physical mutagens at doses which would result in one mutation per surviving cell; low-level mutagenesis results in a high proportion of unmutated cells. On the other hand, heavy mutagenic treatment results in the high frequency of multiple mutational events, an undesirable point in genetic analysis.

Such considerations are of special importance if there is no straightforward test system available to detect the presence or absence of a specific gene function. A mutation in a gene, e.g., for nodule development or symbiotic nitrogen fixation in *Rhizobium*, usually has no phenotypic consequence for cell growth of the mutant clone. Therefore, laborious and time-consuming tests with appropriate host plants are necessary to screen for such mutants. In contrast to conventional mutagenesis procedures, an optimal transposon mutagenesis experiment results in 100% mutants, all of which carry a single-site transposon insertion at different loci. This drastically reduces the number of plant infection assays to be performed in order to identify bacterial mutants with altered plant-interactive phenotypes. For such reasons alone a transposon would be the mutagenic agent of choice if there is no easily scorable phenotypic change in a potential mutant.

Transposon-induced mutations offer additional advantages as compared to point mutations. Due to physical disruption of the DNA sequence, insertion of a transposon within a

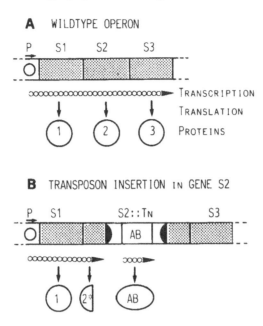

FIGURE 1. Effect of transposon insertion on gene expression. (A) In the wild-type operon, the structural genes (S1-S3) are transcribed from promoter (P) resulting in proteins (1—3); (B) transposon insertion interrupts the continuity of the operon resulting in a truncated protein S* and absence of protein 3. The insertion is selected for by the transposon-mediated antibiotic resistance (AB).

structural gene usually leads to complete loss of its function; there is no leakiness, i.e., low-level expression or partial function of the encoded protein. Moreover, termination signals carried by the transposon are also responsible for the strong polar effect on genes located downstream from the insertion site within the same operon. The mutation is also usually extremely stable; reversion by precise excision of the inserted transposon occurs at very low frequencies and can be easily recognized since it is accompanied by the loss of the transposon-mediated drug resistance.

Many uses result from the direct physical association between the mutational and drug-resistance phenotype in a transposon-induced mutant. For example, the mutation can be positively selected upon transfer into a new host following transduction or conjugation, or the DNA fragment containing the mutated gene can be cloned and identified simply by selection for the transposon-mediated resistance. A transposon insertion also provides the target gene with a defined pattern of new restriction enzyme sites allowing precise physical mapping of the mutation. One limitation of transposon mutagenesis is, of course, that insertions into essential genes cannot be isolated, but this is usually not a problem in the genetics of nitrogen fixation.

2. Transposons Used for Mutagenesis

Numerous drug-resistance transposons have been isolated from natural sources, but by far the most popular transposon used in mutagenesis of non-*E. coli* bacterial species is Tn5. This may be due partly to historical reasons: Tn5 was one of the first transposons shown to be useful for *in vivo* genetic engineering,[1,7] and the first transposon donor vector that has gained widespread use outside of *E. coli* and especially in *Rhizobium* was loaded with Tn5.[7]

The search for alternatives to Tn5 has been superfluous because of the optimal or, at least, satisfactory properties of Tn5. The advantageous features of Tn5 have been specifically addressed in many research papers and general reviews (e.g., References 7 to 9) and can be summarized as follows. The high frequency of transposition of Tn5 is of special importance, particularly when the transfer frequency of the transposon carrier vector into the recipient is much below 100%, as is the case for many of the nonenteric bacteria. As a result of the generally low insertional specificity of Tn5, insertions are distributed almost randomly. Usually all possible mutants should be present in a few thousand clones selected after mutagenesis. It should be noted, however, that cases of moderate[6,10] or even high specificity[11,12] of Tn5 insertions have also been reported. Tn5 encodes an aminoglycoside 3′ phosphotransferase II[13,14] which confers resistance to the commonly used antibiotics Nm and Km. In addition, it encodes resistance to metalloglycopeptide antibiotics[15-17] like bleomycin, followed by a streptomycin phosphotransferase gene that is expressed to various degrees in several non-*E. coli* species.[18-22] Thus, if selection for resistance to Nm or Km is not sufficient (e.g., because of high frequency of spontaneous resistance in the recipient), the presence of Tn5 can be verified unambiguously by using one of the other resistances instead of or in addition to Km/Nm.

Other noteworthy properties of Tn5 (as reviewed in Reference 7) are the low probability of genome rearrangements and deletions upon transposition, the insertion stability of Tn5 once established in a genome, and the strong polarity of Tn5-induced mutations, although a few exceptions have been reported (e.g., References 7,9,23,41).

Finally, Tn5 is best characterized at the molecular level; its complete DNA sequence (5818 bp) is known.[16,24,25] There are convenient sites for several conventionally used restriction enzymes to map the position of Tn5 insertion; some sites are absent so that the cloning of DNA fragments containing Tn5 is possible with commonly used enzymes (e.g., *Eco*RI); and there are restriction sites which facilitate the construction of new derivatives by cloning DNA fragments into Tn5 without affecting transposition. Thus, a number of different Tn5 derivatives have been constructed which combine its advantageous transposition properties with other useful features.

In Tn5-132 the Nm resistance is replaced by the Tc resistance gene from Tn10.[14] Tn5-Tc contains, in addition to Km/Nm, the Tc resistance gene from plasmid RP4.[26] The central part of Tn5 has been exchanged for the drug-resistance determinants from Tn7 (Tp, Sm, Sp) to construct Tn5.7,[27] or from Tn1697 to yield Tn5-GmSpSm.[28] Similarly, Tn5-233 carries the Gm/Km and Sp/Sm resistance markers from plasmid pSa.[29] In Tn5-751 the Tp resistance gene from plasmid R751 is inserted as an additional marker.[30] Other useful derivatives of Tn5 will be described in the next section.

As mentioned above, transposons other than Tn5 have not gained such widespread use. In many cases when other transposons have been used, clear-cut disadvantages are apparent from the literature as far as generalized mutagenesis is concerned. A few examples will illustrate this point. Various degrees of undesired insertional specificity, ranging from regional preference to single insertion sites, have been reported for transposons Tn1[31,32] and Tn7.[33-38] Tn10[1] and its new derivatives (carrying, e.g., other resistance markers or the indicator gene *lacZ*[39]) are extremely useful tools in *E. coli* and *Salmonella*. Unfortunately, transposition frequencies are either very low or are even not detectable in several other organisms.[36,40,107] Tn501(Hg) has been successfully used for insertion mutagenesis, e.g., in *Pseudomonas aeruginosa*[42,43] whereas in other bacterial strains its transposition frequency was far too low.[36] From the literature it is also clear that Tn5 has gained the widest acceptance in work with non-*E. coli* species, but to date there is no systematic overview available of the properties of a representative series of transposons in many different Gram-negative species. Thus, other transposons may well be useful for specific purposes or, at least, worth testing in a particular recipient of interest.

3. Special Purpose Transposons

Upon insertion the above transposons provide the target replicon, the mutated operon, or gene with a selectable drug-resistance marker. Additionally, there are a series of *in-vitro* constructed transposons which offer further applications.

a. Mobilization Transposons

The three derivatives Tn5-*oriT*[44] Tn5-A1,[40] and Tn5-Mob[45,46] are basically identical. They were constructed by cloning the IncP-specific recognition site for mobilization (i.e., passive conjugative transfer) into the central part of Tn5.

Insertion of these transposons provides the target replicon with the ability of being efficiently mobilized between different Gram-negative bacteria by the broad host-range transfer functions of RP4 or other IncP-type transfer plasmids. In principle two applications are possible: mobilization of bacterial chromosomes or plasmid transfer. Random Tn5-Mob insertions into the host chromosome create a series of Hfr-like donors which transfer its genome in the presence of RP4 at high frequencies into a recipient cell, thus facilitating the establishment of chromosomal linkage maps.[46]

In nitrogen-fixing organisms, indigenous plasmids very often carry genetic information of interest. Thus, as in other instances, the genetic analysis of "cryptic" plasmids may be greatly facilitated by their transfer individually into appropriate recipients. Hence, problems arising from the lack of selectable markers and absence of natural transfer systems on plasmids can be overcome by Tn5-Mob insertion followed by RP4-mediated conjugation. The mobilization of plasmids facilitates the identification and analysis of plasmid-encoded symbiotic functions in different strains, construction of new plasmid combinations, definition of host range of plasmid replication, and studies of incompatibility reactions between plasmids. Finally, transfer of plasmid DNA into a plasmid-free strain, e.g., a totally cured *Agrobacterium tumefaciens* derivative,[47] allows subsequent plasmid purification without contamination by other extrachromosomal DNA molecules.

Some of these experimental manipulations have been successfully used in *Rhizobium*. For example, by Tn5-Mob-mediated transfer, the two megaplasmids of several *R. meliloti* strains of different geographic origin have been analyzed,[48] or the symbiotic plasmids of a *R. phaseoli* strain[49] and of the broad host range *Rhizobium* MPIK3030[50] have been identified.

b. Promoter Probe Transposons

After many years of intensive research with nitrogen-fixing bacteria, a large collection of invaluable mutants, mainly induced by transposons, have been isolated, and many genes or gene clusters of interest have been cloned and studied in detail. Yet, a study of these mutants and the available sequence data cannot explain the complex circuits of regulation of all these genes. Thus, the analysis of gene expression under defined conditions or in particular genetic backgrounds becomes more and more important for a better understanding of the processes involved in nitrogen fixation or development of a symbiotic relationship.

The most serious problem in studies of gene expression is very often the lack of simple assays for any of the individual gene products of interest. These difficulties have been overcome by techniques in which easily scorable indicator genes are fused to the promoter to be analyzed.

One strategy for this purpose is to construct *in-vitro* gene fusions by cloning the gene(s) under study into special vectors as will be discussed in Section III. A way of creating *in-vivo* gene fusions has become possible in recent years by the construction of various promoter probe transposons.

Figure 2 shows schematically the principle of construction of a transposon-mediated gene fusion. If the expression of a gene X, for which there is no test system available, needs to be analyzed to answer questions about the activity of its promoter, a transposon with the

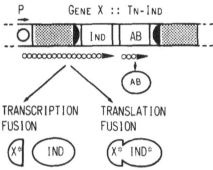

FIGURE 2. Transposon-mediated gene fusion. (A)
Gene X is expressed from its promoter (P) resulting
in protein X. The activity of the promoter should be
analyzed, but there is no test system for the gene
product X available. (B) A promoter probe transposon
is inserted into gene X. The transposon encodes an
antibiotic resistance (AB) which allows selection for
its presence. In the inserted orientation shown, the
promoterless indicator gene (Ind) is under the control
of the promoter of gene X. The resulting gene fusions
are discussed in the text.

following characteristics is inserted directly into or closely downstream of it. The transposon
as usual carries an antibiotic resistance marker to select for its presence. Close to one end
there is a gene whose product can easily be monitored or quantitatively measured. Transcription could proceed into the transposon to express the indicator gene. As shown in Figure
2, there are two ways of generating fusions, depending on the transposon used.

1. Translational fusion: the indicator gene not only lacks its promoter, but also its translational initiation signals; the resulting gene product may be a fusion of a truncated
   gene X polypeptide with a functional indicator gene product, provided the reading
   frame has not been disrupted by the insertion.
2. Transcriptional fusion: the indicator gene still carries its own translational start signals
   and is individually translated.

The variety of gene fusion techniques using this principle can be briefly summarized as
follows:

i. The Indicator Genes

The most widely used indicator gene is the *lacZ* gene of *E. coli* which specifies the
enzyme β-galactosidase. There are numerous indicator media available for detecting β-

galactosidase expression, and several functional analogues of the disaccharide lactose, in which the glucose moiety is substituted by a chromogenic compound, can be used as substrate for the enzyme. For example, *o*-nitrophenyl-β-D-galactopyranoside (ONPG) yields, upon hydrolysis, a yellow color; this is the basis of one of the most sensitive enzyme assays available and allows very simple and accurate quantitative measurements of gene expression.

Because of this advantageous property, most of the promoter probe transposons that have been constructed and used to date carry *lacZ* as an indicator gene. For detailed information about the numerous theoretical and methodological aspects of *lac*-gene fusions the reader is referred to recent reviews.[51-54]

In principle, any other promoterless gene or operon whose gene product(s) is(are) readily assayable may be used for gene fusion experiments, and there are several systems other than that of *lacZ* that have also been used successfully in various Gram-negative bacteria. These include the following genes: *cat* (chloramphenicol transacetylase),[55,56] *galK* (galactokinase),[57] *npt* (neomycin phosphotransferase of transposon Tn5),[58] *phoA* (alkaline phosphatase),[59] and the promoterless lux-operon of *Vibrio fischeri*.[60,61]

ii. Transposons Used to Insert Indicator Genes

Genetic analyses using movable indicator genes have been pioneered in *E. coli* with derivatives of bacteriophage Mu carrying the promoterless Lac operon, as has been reviewed by Silhavy and Beckwith.[51] These techniques have also been rapidly adapted to the free-living nitrogen-fixing strain *Klebsiella pneumoniae*.[62,63]

There is also a variety of different Mu-based *lac*-fusion elements[64-69] available that might be worth considering for specific purposes. For example, in order to identify and analyze the genes that are induced by plant substances in several *Rhizobium* species, transposable Mu-*lac* elements have been used successfully.[70-73]

Conventional transposons have also been modified to carry indicator genes. Although not as yet used extensively in research in nitrogen fixation, these transposons may well have some advantages compared to the phage Mu derivatives since they are less likely to induce undesirable genetic alterations like deletions or rearrangements which are frequently encountered upon Mu transposition.[74]

There are *lac*-fusion transposons which are based on Tn10,[39] Tn5,[75] and Tn3.[76] The latter, designated Tn3-HoHo1, has been used in *R. leguminosarum* to study gene expression involved in exopolysaccharide and melanin synthesis.[77,78]

Other indicator genes have also been inserted into transposons:

1.   Tn5-VB32 contains a promoterless Nm-resistance gene; its insertion downstream of an active promoter can be positively selected by Nm-resistance.[79]
2.   Tn5*phoA* contains an alkaline phophatase gene without its signal peptide, allowing identification by insertional mutagenesis of genes whose products are secreted.[80]

Finally, Tn4431 contains a promoterless lux-gene cassette which enables the study of gene expression upon insertion using the bioluminescence phenotype as assay system.[61a]

B. RANDOM TRANSPOSON MUTAGENESIS

1. General Aspects

There are several basic requirements for transposon mutagenesis experiment. The transposon of choice must be carried on a suitable vector so that it can be propagated in *E. coli*. In addition, the transposon to be used must have a reasonably high transposition frequency together with a low insertional specificity. Naturally, the recipient strain to be mutagenized should not have intrinsic or high frequency spontaneously occurring resistance to the antibiotic(s) for which the transposon carries the resistance gene(s).

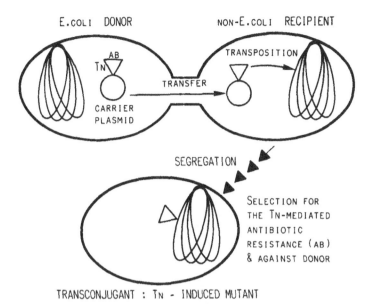

FIGURE 3. Transposon mutagenesis using a suicide plasmid. From the *E. coli* donor, the suicide plasmid, carrying the antibiotic resistance (AB) transposon (Tn), is introduced into the recipient to be mutagenized, in which its replication is not possible. The transposon is rescued by transposition into the recipient genome.

There must also be an efficient method for the introduction of the vector carrying the transposon into the recipient bacterium under study. Especially with bacteria other than *E. coli* or *Salmonella*, the conjugative transfer of suitable plasmids from cell to cell is the most feasible way of introducing a transposon into the recipient.

The desired translocation event from the carrier replicon onto the recipient host genome occurs spontaneously and usually at low frequencies. The event can be identified by selection for the transposon-mediated drug resistance only if the original transposon carrier vector is lost from the cell.

Thus, in summary, to carry out transposon mutagenesis, a carrier vehicle is needed which introduces the transposon at high frequency into the target cells, but which fails to become established in the recipient. The vehicles displaying such properties are often referred to as "suicide plasmids". Figure 3 shows a general scheme for transposon mutagenesis using a suicide plasmid. On the basis of the mechanism of their suicide, the currently used donor plasmids of nonenteric bacteria may be divided into three classes.

1. The classic suicide plasmid is an IncP-type resistance plasmid harboring a Mu prophage that prevents vector establishment in many non-*E. coli* recipients.
2. There are also temperature-sensitive promiscuous replicons, the use of which is of course limited to bacteria able to grow at elevated temperatures.
3. More universally applicable donor plasmids combine broad host-range transfer or mobilization functions with a narrow host-range of replication.

This diverse group of vehicles consists mainly of *in-vitro* constructions, but also naturally occurring plasmids have been found which exhibit these properties.[81-83]

The following section will provide a critical review of the currently available transposon donor vehicles, but restricted mainly to those that have been successfully used in nitrogen-fixing bacteria.

2. Vector Systems for Transposon Mutagenesis
a. Mu-Mediated Suicide of the Vector Plasmid

The first suicide plasmids were found more or less by chance. The initial idea was to insert bacteriophage Mu into the promiscuous IncP plasmid RP4, to introduce the phage conjugally into naturally Mu-resistant soil bacteria like *Rhizobium, Agrobacterium,* or *Pseudomonas,* and to take advantage of its properties in the new hosts. As a result of this, it was found that the Mu prophage carried by the plasmid severely reduced its ability to become established in some bacteria.[84-86] This phenomenon was explained by Mu-specific gene function(s) as well as by restriction mechanisms that primarily affected Mu but not plasmid DNA.[86] Consequently, this Mu-dependent suicide system has been used to construct a transposon delivery system of wide applicability; the IncP-type resistance plasmid pPH1JI (Gm,Sp) was loaded with phage Mu and transposon Tn5 to yield plasmid pJB4JI.[8,87] Using pJB4JI as a Tn5 donor vehicle, many experimental approaches have been initiated primarily in *Rhizobium* genetics by the isolation of symbiotic mutants. Additionally, a long and diverse list of other organisms has also been successfully mutagenized by the transfer of pJB4JI (see in References 7 and 10). However, a substantial amount of data has accumulated showing disadvantages or failure of the Mu-based suicide plasmid. For example, pJB4JI shows a high frequency of survival in species of *Pseudomonas,*[30,88,89] *Xanthomonas,*[38] *Rhodopseudomonas capsulata,*[90] and *Azospirillum brasilense.*[91] Also, stably replicating deletion derivatives of the suicide plasmid have been identified in several *Rhizobium* strains,[92-95] which makes the selection of Tn5 mutants very difficult. A further complication has also been observed after pJB4JI transfer to *Rhizobium meliloti.* In a high proportion of clones, Mu sequences have transposed together with Tn5 onto the recipient genome complicating further genetic and physical analyses of the mutants.[94,96]

b. Combination of Broad Host-Range Transfer with Narrow Host-Range Replication Systems

The basic idea of combining in one replicon both narrow host-range maintenance functions and interspecific transfer proficiency has been realized by cloning transfer genes from different sources into *E. coli* specific vector plasmids.

i. IncP-Type Transfer Genes

The plasmid pRK2013 arose from work undertaken to study replication functions of the plasmid RK2.[97] It contains the entire segment of the broad host-range transfer genes of RK2 linked with a Km resistance determinant and cloned into the narrow host-range replicon ColE1. The selective marker is located on Tn903, and this transposon has been introduced into a number of plant-associated pseudomonads using pRK2013 as a suicide vector.[98] Derivatives of pRK2013 have also been used as suicidal carrier replicons for other transposons. After pRK2013∷Tn7 transfer, site-specific transposition of Tn7 into a megaplasmid of *R. meliloti* has been demonstrated.[33] The Tn5 derivatives Tn5-233 and Tn5-235 have also been inserted into the *R. meliloti* genome via pRK2013 transfer.[29]

Other pRK2013-based delivery vectors[38] are pBEE7 (Tn7), pBEE10 (Tn10), pBEE104 (Tn10-HH104, overproducing transposase[99]), pBEE132 (Tn5-132), pUW964 (Tn5),[100] and pMD100 (Tn501).[101]

In summary, like the parent plasmid RK2, pRK2013 can be transferred into a wide range of Gram-negative bacteria with high frequency, and any recipient that does not allow ColE1 replication may be mutagenized with transposons introduced by pRK2013.

ii. IncW- and IncN-Type Transfer Genes

A DNA fragment from the IncW plasmid R388 containing the genes for conjugal transfer has been cloned into the narrow host-range *E. coli* vector pBR322,[102] and the resulting hybrid was loaded with Tn5 to construct pWI281.[89] Nonenteric bacteria able to act as recipients for the IncW-type transfer system can be mutagenized using this suicide plasmid.[89]

Similarly, by cloning the promiscuous transfer genes from the IncN plasmid pCU1 into the *E. coli* narrow host-range p15A-derived plasmids pACYC184 and pACYC177,[103] a further series of suicidal vectors has been constructed: pGS9 (Tn5), pGS18 (Tn1), pGS27 (Tn9), and pGS16 (Tn10).[40] Due to the advantageous properties of Tn5, pGS9 especially has been successfully used to isolate transposon insertion mutants of various strains of *R. meliloti* and *R. leguminosarum*,[40] slow- and fast-growing strains of *R. japonicum*,[104] and *Azotobacter vinelandii*.[105]

c. Mobilization of Transposon Carrier Vehicles

In the methods discussed so far, the transposon donor plasmids harbored a complete set of transfer genes, i.e., they are self-transmissible. However, the gene products of a broad host-range transfer system can also act in *trans* and promote the mobilization of plasmids that carry the origin of transfer (*oriT*) as has been shown already during the analysis of RK2 conjugation.[106]

By cloning the IncP-type specific recognition site for mobilization (Mob-site = *oriT*) into usual *E. coli* vectors like pACYC184 and pACYC177,[103] or pBR325,[102] a series of vehicles was constructed designated pSUP vectors.[26,45,46,107] These small multicopy plasmids can be mobilized at very high frequencies by the transfer functions of IncP plasmids.

For optimal use of the pSUP vectors, special *E. coli* mobilizing strains have been constructed that carry derivatives of RP4 integrated into their chromosomes. From these donors, the mobilizable vectors are transferred to a recipient cell without concomitant self-transmission of the helper plasmid that provides the transfer functions. The pSUP vectors can be introduced into all bacteria that can act as recipients for RP4, but their replication host range is restricted to *E. coli* and some close relatives. Consequently, they have been used as suicidal carrier vehicles for transposon Tn5 mutagenesis in many different Gram-negative bacterial species.[10,26,107]

With pSUP102(= pACYC184-Mob) and pSUP202(= pBR325-Mob) as carriers, it has been shown that the insertion site of Tn5 in the vector molecules significantly influences the transposition frequency in a given recipient: transposition frequencies differing up to 100-fold have been found in a series of vectors with different Tn5 insertion sites.[107] One reason for the marked difference in the transposition frequencies could be due to the promoter activity in the border sequences of the Tn5 insertion site[108] since the rate of transcription is, of course, not equally distributed within a replicon and is probably also strain dependent.

d. Concluding Remarks

Transposon mutagenesis has now become a standard method since it offers tremendous advantages for genetic analyses of Gram-negative bacteria. Although the various techniques using transposons in nonenteric strains have been developed within the last several years, it is now almost impossible to compile a complete list of references that would cover all important results of transposon mutagenesis. Nevertheless, the knowledge of the behavior of the different suicide vectors and transposons within the large variety of Gram-negative bacteria is still very incomplete. Thus, in most cases the optimal transposon delivery system for a new recipient bacterium of interest cannot be theoretically predicted but still has to be experimentally developed with appropriate modifications. Most obvious are strain-dependent variations in transfer frequency and the efficiency of the suicide system. However, the frequency of transposition of a given transposon may also vary significantly from strain to strain. Furthermore, as shown for the Tn5 carrying pSUP vectors, the insertion site of the transposon within the plasmid has an influence on the frequency of its transposition in a target cell.[107] Therefore, well-established suicide vector systems may still be optimized

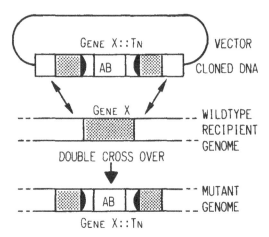

FIGURE 4.   Localized transposon mutagenesis. The vector contains a cloned DNA fragment including gene X which has been mutagenized by the insertion of a transposon (gene X::Tn). Following transfer to the recipient, the wild-type gene X is replaced by recombination between the homologous DNA sequences. The double crossover event is identified by selection for the transposon-mediated antibiotic resistance (AB) and loss of the vector molecule.

simply by testing derivatives of a given vector having the transposon inserted at different sites.

However, even if a chosen vector for Tn5 mutagenesis seems to work satisfactorily in a bacterial strain under study, experience has shown that it is still advisable to check the mutant clones for possible anomalies, e.g., one of the 1.5-kb terminal repeats of Tn5, called IS50, may have transposed independently from a genuine Tn5 insertion, or the whole vector may have been inserted together with Tn5 into the host genome.[7,109-112] Such events have been reported to take place in mutagenesis experiments with pSUP::Tn5 vectors, but depending on the transposon, such anomalies are also likely to occur with other suicide vectors. Rare cases of independent IS50 transpositions have been found by DNA hybridization tests, e.g., in *R. meliloti*.[107] Tn5-mediated vector insertion has been found to occur with widely varying frequencies. For example, the frequency of vector insertion in species of *Pseudomonas* is not significant,[113] whereas about 5 to 30% of Tn5-induced mutants of *R. meliloti*[107] or *R. japonicum*[114] have also the vector molecule inserted. In an extreme situation, found in one *Rhizobium* species,[115] all mutant clones resulted from concomitant insertion of Tn5 and vector DNA. This again emphasizes strain-dependent variations inherent in any Tn5 mutagenesis system and cautions about the necessity of careful tests with any new recipient to be mutagenized.

C. GENOMIC INTEGRATION OF CLONED DNA

1. Localized Transposon Mutagenesis

An extremely useful method for fine-structure genetic analysis results from combining *in-vitro* cloning with *in-vivo* transposon mutagenesis. This technique, termed site-specific or localized transposon mutagenesis, homogenotization, marker exchange, or gene replacement, was first applied in non-*E. coli* bacteria for the study of symbiotic genes of *Rhizobium*[116,117] and is now being widely used in many different Gram-negative bacteria other than *E. coli* (for review see References 9 and 118).

The principle of site-directed transposon mutagenesis can be briefly summarized as follows (Figure 4): a cloned DNA fragment is mutagenized by transposon insertion in *E.*

coli, using one of the well-established standard procedures, e.g., with a λ::Tn5 vector or with a chromosomally located transposon.[9,118] The insertion site can easily be mapped by restriction fragment analysis of the isolated plasmid DNA. The mutagenized DNA fragment is transferred back to its original host, where sequence homology allows recombination on both sides of the transposon insertion. This double crossover finally leads to the exchange of the wild-type gene for the transposon-mutated one. This technique not only enables the insertion of a transposon at a predetermined site in a target cell, but also allows very accurate analyses of genes and operons by comparing the physical map with the phenotype resulting from the gene replacement.

The transfer of the transposon-mutated DNA segment back into its original host may be achieved by broad host-range or narrow host-range vectors. In the technique first described by Ruvkun et al.[116,117] the IncP-type cloning vehicle pRK290 was used as a carrier for the mutated gene to be introduced. Since pRK290 and similar vectors (see Section III) replicate stably in the non-*E. coli* recipient, they have to be displaced in a second mating by an incoming plasmid of the same incompatibility group. Concomitant selection for the transposon rescue and the second plasmid allows the isolation of transconjugants in which the desired homogenotization event has taken place.

Alternatively, mobilizable narrow host-range cloning vectors may be used for this type of experiment, e.g., pSUP102 (= pACYC184-Mob), pSUP202 (= pBR325-Mob),[26,107] or pRK2013.[29,97] As discussed above, these vehicles are unable to maintain themselves autonomously in the recipient and therefore require no further mating to be displaced from the transconjugants. After selection for the presence of the transposon, the clones can be screened for loss of the vector molecule, i.e., for sensitivity to a vector-specific resistance gene, to distinguish between a single crossover (resulting in genomic integration of the vector) and the desired gene replacement by the double crossover.

To verify the fidelity of the marker exchange, i.e., to rule out possible transpositions or DNA rearrangements, transconjugant clones should be checked by appropriate hybridization experiments as has been described, e.g., by Ditta.[118]

It should be noted here that the use of promoter probe transposons instead of normal transposons offers a tremendous number of further possible applications of this type of experiment, as has been discussed for the generalized transposon mutagenesis.

2. Localized Insertion of Cloned Marker Genes and Unselectable Modifications

The gene replacement technique described above is not restricted to the use of a transposon to modify a cloned DNA fragment. Provided the vector used can be reintroduced into the recipient under study, any of the sophisticated manipulations usually performed in *E. coli* to dissect the function of cloned genes can be applied. Then the manipulated DNA can be inserted into the original host of the cloned DNA by homologous recombination. A discussion of all these possibilities would certainly not be within the scope of this article, but a few theoretical examples will be given to illustrate this point.

Instead of *in-vivo* transposon mutagenesis, a cloned DNA sequence can be altered *in-vitro* by inserting a restriction fragment that carries a selectable marker. Similarly, deletions can be generated by replacing internal restriction fragments of the cloned DNA with such a marker fragment.

Numerous so-called cartridges or cassettes, i.e., small DNA fragments carrying a selectable gene flanked by appropriate restriction sites, have been constructed for this purpose (see, for example, the Km-cartridges of pUC4K[119] and derivatives thereof[120]), or may be taken simply from natural sources, such as Tn5 or resistance plasmids.

Whereas after homogenetization the above-described insertion of a transposon usually results in a polar mutation, the genomic insertion of a cloned resistance marker normally allows further transcription of the operon downstream of the insertion site. If a polar mutation needs to be introduced into a hybrid plasmid and finally into the host genome, the so-called Ω interposon can be used for *in-vitro* insertion. This 2-kb fragment not only carries selectable markers flanked by useful restriction sites, but also strong termination signals[121,122]

The experimental protocol outlined here can be readily adopted to the generation of transcriptional or translational fusions, facilitating the analysis of gene expression at an exactly predetermined site. Appropriate indicator genes that could be used for such experiments have been already discussed above. New gene fusion cartridges for specific purposes can easily be constructed by standard cloning procedures. For example, Mulligan and Long[123] combined a *lacZ* indicator gene[124] with a Sp resistance marker and utilized this cassette to study the expression of nodulation genes in *R. meliloti*.

The final homogenotization event of an *in-vitro*-manipulated, cloned-DNA fragment into the recipient genome is identified by the acquisition of the marker gene used and loss of the vector molecule, as has been described for the localized transposon mutagenesis.

Jagadish et al.[125] have discussed an extremely useful alternative, namely, how to use the mobilizable narrow host-range vectors of the pSUP series[26,107] for introducing nonselectable modifications of cloned fragments, i.e., without adding an easily recognizable phenotypic trait, into a recipient genome. Such modifications include deletions, additions, or inversions of DNA fragments, and even alterations of single base pairs within the cloned DNA. The latter possibility is of special importance for site-specific mutagenesis of DNA signal sequences (e.g., in order to increase a particular promoter activity) or for the so-called protein design experiments which involve point mutations within a gene resulting in a defined amino acid exchange in the encoded polypeptide.

Alterations of these types are introduced into the host DNA in two steps. After mobilization of the modified hybrid plasmid into the recipient, transconjugants are isolated by primary selection for a vector-specific antibiotic resistance. In these transconjugants the whole construct has been recombined into the genome by a single crossover event, generating a duplication of the target DNA sequence. This DNA segment duplication allows secondary homologous recombinations to occur which can finally lead to excision of the vector molecule together with the wild-type sequence, leaving the modified sequence in the genome. Since the vector used cannot autonomously maintain itself, its segregation can be monitored simply by screening for the loss of its resistance marker.

III. CLONING AND STABLE MAINTENANCE OF DNA IN A WIDE RANGE OF BACTERIA

Reintroduction of the cloned DNA into the natural host is often a precondition for a more detailed study of gene function, especially when the activity under investigation normally occurs in a particular environment and under specific conditions as is the case for symbiotic nitrogen-fixing organisms. Hence, cloning and expression of genes in *E. coli* vector systems are often of only limited use, since their narrow host range precludes replication in nonenteric bacteria.

One solution to this problem is the construction of cloning vectors able to replicate in bacteria other than *E. coli*. Basic replicons for this purpose may be plasmids that are indigenous in the host under study. However, these vector systems are then restricted to only a limited number of species.

Usually, one would like to study the genes of interest not only in the original background, but also in other organisms. It is therefore much more preferable to have a cloning system capable of establishment in a wide range of bacteria. Moreover, since conjugation is currently the most effective gene transfer method, an ideal gene cloning system should not only exhibit a broad replication host range, but also possess the ability of being transmitted via bacterial matings.

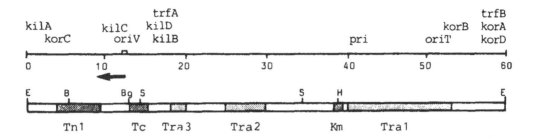

FIGURE 5. Physical and genetic map of IncP plasmid RK2 (according to Thomas et al.[131]). The plasmid genome is linearized at its unique EcoRI site and calibrated in kb coordinates. Transfer (Tra) and resistance genes are referred to in the text as well as *pri*, *oriT*, and *oriV*. The direction of vegetative replication is indicated by the arrow. Gene *trfA* encodes an essential *trans*-acting replication product; additional maintenance determinants consist of genes lethal to the host (*kil* genes) and their regulatory elements (*kor* genes). Restriction sites are B = *Bam*HI, Bg = *Bgl*II, E = *Eco*RI, H = *Hind*III, S = *Sal*I.

Fortunately, natural plasmids with a broad host range are available which can maintain themselves in a wide variety of prokaryotic species as well as being either self-transmissible or mobilizable. Their versatile characteristics have made them especially attractive for gene cloning and expression in species other than *E. coli*, and they have proved extraordinarily useful for the transfer of cloned DNA across species barriers.

A. BROAD HOST-RANGE PLASMIDS

Natural plasmids which display replication and maintenance proficiency in a diversity of prokaryotic species have been isolated from various Gram-negative bacteria and belong to different incompatibility groups. The most important representatives of the IncP, IncQ, and IncW groups will be described in the following in more detail.

1. IncP Plasmids

The conjugative IncP group of plasmids (*Pseudomonas aeruginosa* IncP-1) has been extensively studied. The best-known representatives RP1,[126] RP4[127] R68,[128] and RK2[129] are identical[130] and will therefore not be discussed separately. The map of RK2[131] is shown in Figure 5. It has a size of 60 kb, exists in five to eight copies per *E. coli* chromosome,[132] and carries genes conferring resistance to Ap/Cb/Pn (on Tn1), Tc, and Km/Nm.

Replication of RK2 depends essentially on the gene product of gene *trfA* and on a single origin (*oriV*) for unidirectional replication.[133,134] Detailed studies have shown that the minimal RK2 replicon (*trfA* function, *oriV*) is able to replicate not only in *E. coli* and *Pseudomonas*, but also in other bacteria such as *Agrobacterium tumefaciens*, *Rhizobium meliloti*, and *Azotobacter vinelandii*. However, the requirements for stable maintenance of RK2 involve the interaction of several other regions and vary among different bacterial species.[135]

IncP plasmids are not only capable of stable maintenance in most Gram-negative bacterial species, but their transfer system is also particularly well adapted to overcome species barriers. The host range of the conjugative system appears to be even broader than that of the replicative system, since RK2 has been reported to be transferred into the Gram-negative anaerobe *Bacteroides fragilis*, but appears not to be maintained in this host.[136] This clearly demonstrates that the promiscuity of IncP plasmids is not merely a reflection of a wide conjugal transfer host range. In addition to their self-transmissibility, IncP plasmids are also able to mobilize a wide variety of small, nonconjugative plasmids, including ColE1, pSC101, and RSF1010. A large portion of the plasmid is occupied by regions involved in conjugal DNA transfer: Tra1, Tra2, and Tra3 are clusters of genes encoding transactive transfer functions; *oriT* represents the origin of transfer replication; *pri* codes for a DNA primase which is involved in priming the single DNA strand transferred during conjugation.

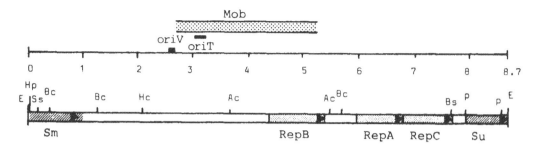

FIGURE 6.   Genetic and physical map of IncQ plasmid RSF1010. The plasmid genome (8.7 kb in size) is opened at its single *Eco*RI site: The location and transcriptional direction of resistance and replication genes (indicated by arrows) are taken from Haring et al.[146] as well as the position of the *ori*V region. The maximum extension of the region required for mobilization including the *ori*T sequence is indicated according to Brasch and Meyer.[147] Restriction sites are Ac = *Acc*I, Bc = *Bcl*I, Bs = *Bst*EII, E = *Eco*RI, Hp = *Hpa*I, P = *Pst*I, Ss = *Sst*I.

2. IncQ Plasmids

Bacterial plasmids belonging to the incompatibility group IncQ (corresponding to *P. aeruginosa* IncP-4) are represented by a number of similar or identical replicons isolated in a wide range of bacterial hosts. Some examples are R300B,[137] R1162[138,139] and RSF1010.[140] They are all relatively small (8.7 kb) replicons, encoding resistance to Sm and Su.[137] Their copy number has been estimated to be 10 to 50 copies in *E. coli*.[137,140] They are not self-transmissible, but may be mobilized by certain conjugative plasmids with various efficiencies. Mobilization studies have shown that IncI plasmids (e.g., CoIIb) can promote transfer of RSF1010 and related replicons at sufficient frequencies, whereas mobilization by IncF representatives occurred less efficiently.[140] Mobilization with IncP plasmids occurs at a very high efficiency (100%) and, at the same time, also allows their introduction into strains of *Pseudomonas* and other nonenteric bacteria.[138,141,142]

A physical and genetic map of RSF1010 is shown in Figure 6. A unique origin of replication (*ori*V) has been mapped at position 2.6 kb, at which replication in *E. coli* and *P. aeruginosa* is initiated.[143,144] Replication may proceed bidirectionally or unidirectionally.[143] Linked to this *ori*V sequence is the expression of plasmid incompatibility and the copy number control system.[138] In addition to this *cis*-active site, three essential replication regions have been identified, *repA*, *repB*, *repC*,[144-146] which are separated from *ori*V by a region not essential for replication but including functions necessary for plasmid mobilization (Mob, *ori*T).[147]

These data mean that in RSF1010 about 50% of the genome is occupied by genes or regions which are essential for plasmid replication and maintenance and that only half of the replicon is dispensable.

3. IncW Plasmids

The incompatibility W group comprises conjugative plasmids with broad host range. Representatives are pSa,[148] R388,[149] and R7K,[150] all of which are between 30 to 35 kb in size. Heteroduplex analyses have shown that they have a common homologous region of about 20 kb that contains genes involved in replication and transfer function.[151] The best-studied member of this group is pSa,[152,153] which was isolated from *Shigella*, although the data available are not as extensive as those for IncP and IncQ plasmids.

pSa confers resistance to Cm, Km, Gm, Tm, Sm, Sp, and Su, has a length of 29.6 kb, and exists at three to five copies per cell. Resistance to Tm, Km, and Gm is encoded by a region located around position 18 kb, and it is believed that they are specified by the same enzyme. The map of pSa, according to Tait et al.,[153] is shown in Figure 7. The replication functions of pSa are confined to a region between positions 18.5 and 22.2 kb. A 1.9-kb

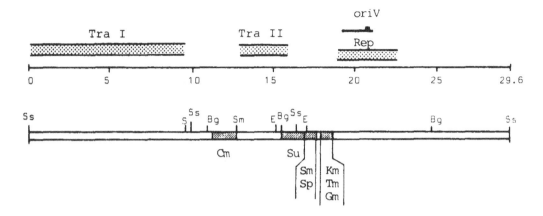

FIGURE 7. Genetic and physical map of the IncW plasmid pSa (according to Tait et al.[153]). The molecule is linearized at one of its *Sst*II sites. The regions involved in plasmid transfer (TraI, TraII) as well as the area encoding replication functions are indicated in the upper part. The 1.9-kb region sufficient for autonomous replication and carrying the bidirectional *ori*V is indicated by a solid bar. Restriction sites are Bg = *Bgl*II, E = *Eco*RI, S = *Sal*I, Ss = *Sst*II, Sm = *Sma*I.

fragment within this region, carrying the bidirectional origin of replication and *trans*-acting replication functions, is capable of autonomous replication in *E. coli* and other bacterial hosts.[154]

The regions responsible for conjugative transfer are separated by the Cm resistance gene. Deletions in region Tra II were found to reduce the efficiency of plasmid transfer by 40 to 60%, whereas deletions in region Tra I resulted in almost complete loss of transferability.

It has been demonstrated that the presence of pSa in oncogenic strains of *A. tumefaciens* interferes with the formation of crown gall tumors.[155] According to the investigations of Tait et al.,[153] this inhibition phenomenon is correlated to transfer region I.

B. BROAD HOST-RANGE CLONING VECTORS

The plasmids described above have useful and interesting characteristics, but they are not especially suitable for *in-vitro* cloning and manipulation. Therefore, they have been modified to meet the requirements of versatile cloning vectors.

1. General Characteristics and Methodology.

In addition to their broad host-range replication and maintenance properties, broad host-range vectors should fulfill the same requirements as for *E. coli*-specific cloning vectors. They should be (1) small in size in order to facilitate isolation and cloning procedures, (2) contain unique cleavage sites for commonly used restriction enzymes, (3) code for at least one (preferably more) trait which can be selected for in a variety of different organisms, and (4) permit the detection of recombinant molecules by insertional inactivation.

Of course, they should also allow a sufficiently high efficiency of introduction into the host cells. Therefore, versatile broad host-range cloning vectors should be transmissible by conjugation into a wide range of bacterial species. A high frequency of transfer is not only advantageous for the introduction of single recombinant molecules, but also a prerequisite for mobilization of a large number of clones, e.g., screening a gene library.

Furthermore, in order to reduce the potential hazard of promiscuous and uncontrolled spread, and to keep the cloning vectors as small as possible, it is desirable to have the broad host-range conjugative system separated from the replication system, i.e., an optimal vector should be mobilizable but not self-transmissible.

Most broad host-range systems in use to date are therefore binary vector systems: the actual cloning vector, characterized by its ability to be efficiently mobilized into and stably replicated in a wide range of bacteria, and the helper plasmid, which provides the functions necessary for conjugal transfer. The unique ability of IncP plasmids to mediate DNA transfer between almost all Gram-negative bacteria and, moreover, the ability to mobilize nonself-transmissible plasmids such as IncQ derivatives proved extremely helpful in this respect. Ideally, the establishment of the helper plasmid in the target organism is prevented by using derivatives with restricted replication host range, e.g., pRK2013,[97] or pRK2073,[156] containing transposon Tn7 in the Km gene of pRK2013. An elegant alternative is the use of mobilizing donor strains, which carry the plasmid RP4 integrated in the chromosome. These transfer techniques have already been outlined in Section II.

Thus, cloning experiments with broad host-range vectors are usually a two-step procedure. Cloning and manipulation of the DNA of interest is carried out in *E. coli*, applying the wide array of genetic procedures available; subsequently, the recombinant molecule is transferred by means of helper plasmids to any host organism of choice, where its function can be analyzed in detail. If elimination of the recombinant plasmid is desired (e.g., to prove that an observed phenotype was actually caused by the introduced DNA fragment), this can be easily achieved by introducing a plasmid of the same incompatibility group. This two-host cloning procedure is of special importance for the construction and characterization of gene libraries in broad host-range cosmid vectors, as no *in-vitro* phage-packaging system has been developed for soil bacteria.

Within the last few years a large number of broad host-range cloning vectors have been constructed, too many to be discussed here. Only those which have mainly been successfully applied in the genetic analysis of nitrogen-fixing bacteria are described below in more detail.

2. General-Purpose Cloning Vectors

These types of vectors include normal plasmid vectors suitable for cloning of small DNA fragments as well as cosmid vectors which are more appropriate for the construction of gene banks because their larger insert capacity minimizes the number of hybrid molecules required. Representatives discussed below are listed in Table 1.

a. IncP Derivatives

Although IncP plasmids have been employed for cloning *Rhizobium* DNA *in vitro*,[157-159] their large size is a serious drawback to routine use as a cloning vehicle. Therefore, smaller derivatives have been constructed which meet the requirements outlined above.

Probably the best-known and widely used IncP-derived vector is pRK290 (20 kb), which still retains the RK2-encoded Tc resistance, the *oriT* (Mob site), and those regions essential for broad host-range replication.[160] It has single *Eco*RI and *Bgl*II cloning sites and is mobilizable at high frequencies, but it does not allow insertional inactivation. A gene bank of *R. meliloti* strain F34 was constructed in pRK290 with an average size of cloned fragments of 19 kb, representing 98% of the *R. meliloti* genome.[160] Clones homologous to the nitrogenase structural genes of *K. pneumoniae* have been identified and were the basis for an extensive study and characterization of the *R. meliloti nifHDK* gene cluster.[161,162] pRK290 and slightly modified derivatives of it (e.g., pRK292[162] which carries the Cm gene and a *Cla*I site of pACYC184) have been used extensively for cloning and characterizing symbiotic genes.

The use of pRK290 is, of course, not restricted to *R. meliloti*. It also has been successfully used in many other *Rhizobium* strains, including *Bradyrhizobium* sp.(*Parasponia*)[164] and the wide host-range strain MPIK3030.[165] It has also been shown to function in *Azospirillum* by the construction of *nif* partial diploids.[166]

Even more progress has been achieved in the genetics of symbiotic nitrogen-fixing organisms with the construction of pLAFR1 (21.6 kb), derived from pRK290 by inserting

TABLE 1
List of General Broad Host-Range Cloning Vectors

Vector	Basic replicon	Size (kb)	Single cloning sites	Select. marker	Insert. inact.	Ref.
pRK290	RK2	20	Bg,E	Tc	—	160
pLAFR1	RK2	21.6	E	Tc	—	167
pVK101	RK2	21.3	S	Km	Tc	177
			H,X,Bg	Tc	Km	
			E	Km,Tc	—	
pVK100	RK2	23	S	Km	Tc	177
			X,H	Tc	Km	
			E	Tc,Km	—	
pVK102	RK2	23	S	Km	Tc	177
			X,H	Tc	Km	
pRK404	RK2	10.6	H,P,B	Tc	LacZ	176
pKT230	RSF1010	11.9	E,Ss	Km	Sm	192
			H,X,Xm	Sm	Km	
			B,Bs	Km,Sm	—	
pKT247	RSF1010	11.5	E,Ss	Ap	Sm	192
pMMB33	RSF1010	13.75	E,B,Ss	Km	—	198
pSUP106	RSF1010	10	E	Tc	Cm	201
			C,H,B,S	Cm	Tc	

Note: The table summarizes the characteristics of some of the cloning vectors described in the text. The restriction sites are abbreviated as follows: B = *Bam*HI, Bg = *Bgl*II, Bs = *Bst*EII, C = *Cla*I, E = *Eco*RI, H = *Hind*III, Kp = *Kpn*I, P = *Pst*I, Pv = *Pvu*II, S = *Sal*I, Sa = *Sac*I, Ss = *Sst*I, X = *Xho*I, and Xm = *Xma*I.

the λ cos site,[167] thus allowing much more efficient cloning of *Rhizobium* DNA as compared to plasmid vectors. Its usefulness has clearly been demonstrated by the work of Long et al.,[168] who used a *R. meliloti* gene bank constructed in the "costramid" pLAFR1 to directly complement Nod⁻ mutants. The cosmid isolated by this method, pRmSL26, carrying the common nodulation genes of *R. meliloti,* has formed the basis for numerous hybridization and interspecific complementation tests.

Consequently, pLAFR1 or pLAFR3 (with an additional multiple cloning site[168a]) has been the cosmid vector of choice in the construction of gene banks of various *Rhizobium* strains, including *R. leguminosarum,*[169,170] *R. meliloti* 41,[171] *Bradyrhizobium japonicum,*[172,173] *R. phaseoli,*[174] and *Azorhizobium sesbania* ORS571.[175]

One disadvantage of pRK290 and pLAFR1 is their lack of insertional inactivation and their limited number of cloning sites. Therefore, derivatives have been constructed with additional selectable markers and single restriction sites without affecting their advantageous characteristics. Examples include pRK293, pRK310, and pRK311[176] and the series of pVK vectors,[177] which carry an additional Km resistance gene, which allows not only insertional inactivation, but also offers additional cloning sites. pVK101 (21.3 kb) and the cosmids pVK100 and pVK102 (both 23 kb) have proved to be useful in a variety of cloning strategies. For example, complementation studies of Nif mutants of the broad host-range cowpea *Rhizobium* have been successfully carried out[178] using pVK100 as a cloning vehicle and mobilization from the *E. coli* mobilizing strain SM10. pVK102 has been used to construct a cosmid bank of plasmid pRjaUSDA193, the symbiotic plasmid of *R. fredii* USDA193,[179] and mutants of *R. fredii* USDA191, which induced pseudonodules, have been restored to the wild-type phenotype by introduction of a pVK102 cosmid clone of strain USDA201 via pRK2013 triparental matings.[180]

pVK100 has also proved effective in the genetic analysis of *Azospirillum brasilense:* a cosmid bank, constructed in this vector, was used for the identification of *glnA* by complementing Gln⁻, Nif⁻ mutants.[181]

Their large size sometimes makes working with pRK290 and its derivatives difficult; the cosmid derivatives especially have a rather low cloning capacity. Therefore, smaller derivatives have been constructed. One RK2-derived vehicle of only 11.1 kb (pRK2501, Tc, Km)[183] has been modified by insertion of the λ cos site to create cosmid pHK17.[184]

However, these replicons are Mob⁻ and Tra⁻. To make them more suitable, the *oriT* site of RK2 has been cloned into pRK2501 and its progenitor pRK248 to yield pGS65[185] and pRK252,[176] respectively. pRK252 has been altered by inserting a multiple cloning site and the *lacZ* gene, resulting in the derivative plasmid pRK404 (10.6 kb). This vector was used, for example, in the study of the genes *psi* and *psr* of *R. phaseoli*,[77] both involved in the regulation of polysaccharide synthesis.

b. IncQ Derivatives

Because of their small size, efficient mobilization by conjugative plasmids and high copy number, IncQ plasmids have attracted much interest as the basis for the development of broad host-range vectors for general and specific purposes. However, the replicon shows some undesirable characteristics such as limited unique restriction sites suitable for cloning experiments and inconvenient selectable markers (Sm, Su). In order to provide IncQ plasmids with more usable restriction sites and selectable phenotypes, an extensive series of derivatives have been constructed (see, e.g., References 142 and 186-190). Some of these derivatives appear very versatile and may also be useful to the research in nitrogen fixation.

The most commonly used constructs are the series of pKT vectors, based on RSF1010 and developed by Bagdasarian and co-workers (for an overview see References 182, 191 and 192). Again, only a few derivatives, which have found application in nitrogen-fixation research, will be described in detail here.

Plasmid pKT230[192] is a double replicon obtained by fusing pACYC177C with RSF1010 via their *Pst*I sites. It encodes Sm and Km resistance and offers cloning sites for *Hind*III, *Xho*I, *Xma*I, *Bst*EII, *Bam*HI, *Eco*RI, and *Sst*I. Much of the genetic analysis of symbiotic genes in *R. leguminosarum* has been carried out using this vector plasmid.[193,194,197]

Another useful application of pKT230 has been demonstrated by Kennedy and Robson.[195] The *K. pneumoniae nifA* gene and flanking portions of the *nifB* and *nifL* genes were cloned into pKT230 to produce plasmid pCK1. Its introduction into regulatory mutants of *Azotobacter* resulted in restoration of the Nif phenotype. Elimination of pCK1 by introducing pKT248 (Sm,Cm) proved that the complementation was actually due to a *trans* effect of the cloned *nifA* gene product, indicating the existence of a *nifA*-like gene in *Azotobacter*. A similar construct was able to activate the *R. meliloti* p1 symbiotic promoter in *E. coli*.[196]

Cosmid derivatives belonging to the set of pKT cloning vectors, e.g., pKT247,[192] are also available. The most useful cosmids of this series are the almost identical vectors pMMB33 and pMMB34 (both 13.75 kb) constructed by Frey et al.[198] They allow selective cloning of DNA fragments generated by *Bam*HI, *Sau*3A, or *Mbo*I into the single *Bam*HI site. Their major advantage compared to pLAFR1, is not only their larger cloning capacity, but also that polycosmid formation can be avoided during the cloning procedure without dephosphorylation by the generation of cosmid "arms". pMMB33 has been especially successful in the construction of genomic libraries. Dowling et al.[199] described the construction of a gene bank in pMMB33 of *R. leguminosarum* strain PF2, which yielded 1200 clones containing inserts of 30 to 35 kb. The availability of a *B. japonicum* cosmid clone bank, carrying large DNA segments, allowed the localization of *nifH* 17 kb away from the *nifDK* operon in this strain.[200]

Another approach in the construction of broad host-range vectors based on RSF1010 replication functions was undertaken by our group. The idea was to extend the host range of the well-characterized *E. coli* cloning vectors by the incorporation of RSF1010 regions specifying broad host-range replication and maintenance, rather than to manipulate and improve the RSF1010 replicon by inserting additional genes or restriction sites. These constructions resulted in the set of replicative pSUP vectors (also called class II vectors).[201] These vector derivatives are not only being used in our laboratory, but also by other groups, e.g., Djordjevic et al.[202] used the cosmid pSUP106 (Cm, Tc), derived from pACYC184, for complementation studies with *R. trifolii hsn* genes, and Ronson et al.[203] employed it in the analysis of the *R. meliloti ntrA* gene.

c. IncW Derivatives.

The IncW plasmid pSa has also been the basis for the construction of broad host-range cloning vectors. Since the replication functions are limited to a short fragment, the construction of a mini-Sa replicon has proved straightforward and has been achieved by recircularization of a 8.4-kb *Bgl*II fragment.[204] The resulting plasmid, pGV1106 and its derivatives (pGV1113, pGV1120, pGV1122, and pGV1124, however, had simultaneously lost their conjugative properties, so that they are only of limited application for the genetic analysis of symbiotic bacteria.

By circularization of a 13.1-kb fragment encoding Sp and Km resistance, the derivatives pSa151 and pSa727 (carrying an additional Cm resistance) were obtained.[205] They, too, have lost the self-transferability, but they can be mobilized by a separate plasmid containing the pSa transfer functions cloned into pBR322 (see also Section II). This helper plasmid mobilizes pSa151 and pSa727 into a variety of Gram-negative bacterial species, including *Rhizobium*.

Based on pSa151, a set of pUCD[206] vectors has been constructed by fusion with pBR322, including high and low copy number cosmid derivatives[207] which have been introduced into *A. tumefaciens* and *R. meliloti* with the help of pRK2013.

All these IncW-derived vectors are theoretically applicable in a variety of Gram-negative bacteria. They have found application in *Agrobacterium* and *Pseudomonas* and are potentially also valuable in genetic analysis of nitrogen-fixing bacteria.

3. Specific-Purpose Cloning Vectors

It is often not sufficient merely to insert the DNA under study, with its own transcriptional and translational initiation signals, into a suitable vector in order to obtain and assay gene expression and regulation. Genes, especially those involved in symbiotic nitrogen fixation, are subject to complex regulatory circuits; consequently, specific constructions are necessary to obtain information about changes in gene expression that lead to nodule formation and nitrogen fixation.

This section will briefly summarize the broad host-range vectors available and employed for studying the expression and regulation of symbiotic nitrogen-fixation genes.

a. Expression Vectors: Vectors That Provide Improved Expression of Inserted DNA

The principle of this type of vector is that the gene under study is placed under the control of a foreign, vector-encoded promoter and synthesized either from its own N terminus or as a fusion protein.

Some of the general-purpose cloning vectors described in the previous section may be used in this context. For example, since the *Eco*RI and *Sst*I cloning sites of pKT230 are within the Sm gene, inserts into these sites can be expressed under the control of the Sm promoter. Hence, by cloning a given gene, one can have either constitutive expression from the vector promoter or expression from a promoter within the cloned fragment, depending on the orientation. This strategy has been applied in studies of the symbiotic genes of *R. leguminosarum*, e.g., for constitutive expression of the *nodD* gene.[208]

Similarly, the Km promoter of pKT230 was used to express constitutively the *K. pneu-*

moniae nifA gene in *Azotobacter*[195] and in *R. meliloti*.[196] Additionally, DNA cloned into the *Bg*lII site of pRK290 might be subject to transcriptional read-through originating from the Tc promoter.[176] However, since it is much more desirable to have vectors allowing a controlled gene expression, specific expression vectors have been constructed which contain promoters, the transcriptional activity of which is controllable so that expression can be turned on or off.

Besides specific promoters which most probably function only in a limited number of species (e.g., References 209 and 210) the well-characterized *trp-*, *tac-*, and λ-promoters have chiefly been used for the construction of broad host-range expression vectors.

The pRK290 derivative pTE3 (20.7 kb), constructed to permit overexpression of the *R. meliloti nodABC* genes,[211] contains the *S. typhimurium trp*-promoter and should be applicable for the overproduction of other gene products, not only in *R. meliloti,* but also in other nitrogen-fixing bacteria.

A very strong promoter, frequently used in *E. coli* expression vectors, is the *tac*-promoter, which is a fusion between the −35 region of the *trp-* and the −10 region of the *lacUV5*-promoter. This promoter was cloned into a derivative of RSF1010, together with the *lacI*q gene to yield the expression vectors pMMB22 and pMMB24.[212] They both have a length of 12.7 kb, specify an additional Ap resistance as a selectable marker, and differ only in the orientation of the cloned promoter fragment. Inserts cloned into the *Eco*RI site (pMMB22) or the *Hin*dIII site (pMMB24) are linked to the *tac*-promoter, and transcriptional activity is subject to either repression by the *lacI*q gene product or induction in the presence of IPTG. These vectors have proved to be useful for the expression of genes in *Pseudomonas;* they have also been used as cloning vectors in *Rhizobium.*[213]

A derivative of the IncQ plasmid pKT240,[212] constructed by inserting the strong λ p$_L$-promoter as well as the λ cI857 repressor gene, was shown to function in *Erwinia* and *Serratia* species[214] and is also potentially applicable in other Gram-negative bacteria, provided the p$_L$-promoter is recognized and regulatable in these hosts. Recognition of λ-promoters *in vitro* and transcriptional inhibition by the cI repressor has been observed in *Cyanobacteria,*[215] and has been used to accomplish regulated expression of cloned genes in *Anacystis nidulans.*[216]

b. Promoter Probe Vectors: Vectors That Allow the Identification of Promoter Activities

This type of vector generally involves the joining of DNA sequences that regulate the gene of interest to another gene, for which sensitive assays are available, facilitating easy identification of promoter sequences and controlling elements as well as permitting the investigation of regulatory functions. This is especially desirable for the study of the activation or repression of regulated symbiotic promoters, since their original gene products are difficult to assay directly. Moreover, variations in the concentration of the original gene product cannot be readily monitored. The general principle of this technique is described in more detail in Section II.

Some of the normal broad host-range cloning vectors in use may be suitable for monitoring the activity of promoters. For instance, the *R. meliloti* fragment carrying the *ntrA* gene was cloned into the *Hin*dIII site within the Tc gene of the vector pSUP106, thus disrupting the Tc promoter but leaving the coding region intact. Using this construct, it could be established that the *ntrA* promoter is not active in *E. coli* but constitutively expressed in *R. meliloti.*[203]

Similarly, the pKT230 encoded Km resistance has been placed under the control of the *nifA*-dependent *nifB* promoter of *K. pneumoniae*[195] This construction, designated pCK1, helped to identify a *ntrA* gene in *Azotobacter* by monitoring the activity of the *nifB* promoter by the expression of the Km phosphotransferase gene.[217]

Additionally, specific broad host-range promoter probe vectors have been constructed, both based on IncP and IncQ plasmids. Apart from promoterless resistance genes, the *E. coli* Lac operon (see also Section II) has found widespread use, although the presence of endogenous β-galactosidase activity in many strains might be problematical.

pRK248,[218] a 9.6-kb, Tc-resistant derivative of RK2, was modified to contain an additional Cm resistance gene and the promoterless Km phosphotransferase gene from Tn5, to yield the transcriptional fusion vector pMK341; similarly, fusion vectors with the Lac operon have been constructed.[219] One of these translational fusion vehicles, pMK353 (22.2 kb) was used to clone promoter sequences by random insertion of *R. meliloti* total DNA. One disadvantage of these vectors might be their lack of transmissibility by conjugation. However, they have the phage P4 cos site incorporated and are thus capable of being introduced into *R. meliloti* 104A14 by bacteriophage P2 transduction.[219]

Based on pRK290 and using *lacZ* as the assay gene, Ditta et al.[176] constructed two transcriptional (pGD499, pGD500) and one translational fusion vector (pGD926). pGD926 (28.1 kb) which allows cloning of a *Hind*III-*Bam*HI fragment, was used for an extensive analysis of *Bal*31-generated deletions of the *R. meliloti* symbiotic promoters p1 and p2.[196] *LacZ* protein fusion vectors have also been extremely useful in the analysis of *nod* gene expression. Vector pIJ1363, constructed by Rossen et al.[208] is similar to pGD926: it is based on pRK290 and contains the *lacZY* genes immediately downstream from a unique *Bam*HI site, so that genes cloned into this site can be fused in frame with the eighth codon of *lacZ*. Additionally, it contains a transcriptional terminator upstream of the insertion site to exclude any external transcription. Using this vector, it was established that the *R. leguminosarum nodD* gene is autoregulatory while the *nodABC* promoter is induced by the *nodD* gene product in the presence of plant exudates. Similarly, the regulation of the *nodE* and *nodF* promoter has been determined.[220] The same fusions have been used to establish that other plant-secreted compounds may inhibit *nod* gene induction.[221]

IncQ plasmids have also been modified toward assaying promoter activity and regulation. One example is pKT240[212] (12.9 kb) which resulted from replacement of part of RSF1010 with a fragment encoding Ap and Km resistances. This manipulation led to the deletion of the Su/Sm promoter and the Su coding sequence. Consequently, pKT240 specifies resistance to Ap and Km and offers a promoterless Sm gene which can be fused with *Eco*RI-generated fragments. In *Rhizobium*, pKT240 has been used as a normal cloning vector,[202,222] but has found limited use for monitoring expression and regulation of nitrogen-fixation genes, probably because Sm resistance is not a very convenient assay gene in many circumstances. Similar constructions have been made by Tsygankov and Chistoserdov.[223] These derivatives (pAYC36, pAYC37, 9.4 kb) are somewhat smaller and provide more cloning sites, but again carry a promoterless Sm gene as indicator. Consequently, IncQ derivatives containing the more versatile β-galactosidase test system have been developed, e.g., the *lacZ* gene devoid of its transcriptional and translational signals has been cloned into RSF1010 to yield the expression vectors pUI108 and pUI109.[224] The transcriptional fusion vector pMP190 is a 15-kb derivative of the IncQ plasmid pKT214[225] and contains a multiple cloning site directly upstream from a promoterless *lacZ* gene and Sm and Cm as selection markers.[226] This vector has been used to monitor the effect of different substances present in *Vicia sativa* root exudates on the induction of the *R. leguminosarum nodA* promoter.[226]

Many more constructions for monitoring gene expression in nitrogen-fixing bacteria have been made and successfully applied. This short overview summarizes only the principle of their application and demonstrates their usefulness in the analysis of the regulation of symbiotic genes.

4. Potential Problems and Drawbacks

Although there is no doubt about the usefulness of broad host-range vectors, they also exhibit some undesirable characteristics which might be problematic in routine work.

a. Selectable Markers

It must be noted that many of the constructs described above appear very versatile, but are, in fact, not applicable in many organisms simply because of their selection markers. Ap and Cm resistance genes are especially inconvenient selectable phenotypes in many nitrogen-fixing species, and the identification of clones harboring the vector plasmid may be problematic.

Km resistance might cause selection problems as well, especially in complementation assays of Tn5-induced mutants. However, as described by Rossen et al.,[197] the aminoglycoside phosphotransferase encoded on pKT230, which originates from pACYC177, is different from that encoded on Tn5 and also confers resistance to lividomycin. So, transfer of recombinant plasmids can be selected by lividomycin, at least in *R. leguminosarum*.

b. Stability

Replication and maintenance of broad host-range plasmids are dependent on a number of functions, which are often dispersed around the plasmid genome and not clustered as in narrow host-range plasmids. Consequently, despite the efforts to generate small derivatives, the commonly used broad host-range cloning vehicles are on the whole not as small and well characterized as the *E. coli* vectors.

In fact, reduction in size often results in pronounced plasmid instability. Many of the small IncP derivatives described above exhibit a relatively high instability in the absence of selective pressure; plasmid pRK2501 is unstably maintained not only in *E. coli*, but also in *Acinetobacter, Rhizobium*, and *Azotobacter* strains.[227] This is in contrast to the larger pRK290 derivatives. For pRK290, the frequency of loss in *E. coli* (Rec⁻) has been reported to be lower than 1% per generation under nonselective conditions,[160] and a similar rate was found for *R. meliloti*, whereas in the case of *B. japonicum* segregation was more pronounced.[228]

It should also be noted that many IncQ-derived constructs, mainly those containing the complete or part of the pBR replicon, were reported to be unstable or reduced in host range.[142,189,225] One such construct was modified by various manipulations, including removal of the *oriV* of pBR322, and was then found to be stably inherited.[188] We, too, have observed a similar increase in stability, when a fragment containing the pBR325 origin was removed from pSUP204. It appears that the pBR and the IncQ replication functions are not compatible with each other.

Additionally, the complex regulatory mechanisms which govern broad host-range replication are currently not well understood, and alterations, such as removing part of the genome or inserting foreign DNA, may influence this balanced situation and result in reduction of plasmid stability or replication host range.[229] This has been found, e.g., for pLAFR1 and pVK100 cosmid clones which are reported to be somewhat unstable in nodule bacteria.[168,170,178,230,231] In some cases, even significant loss has been observed.[169,173] Also, the pRK290-derived expression vector pTE3,[211] carrying the *R. meliloti nodABC* genes under the control of the *trp*-promoter, appeared to be unstable in *R. meliloti* 1021 in the absence of selection, although it was stably inherited in *E. coli*.

pKT230 was reported to be relatively stable in *Rhizobium*: pKT230 derivatives were retained in more than 90% of *R. leguminosarum* cells isolated from *Vicia hirsuta* nodules.[232,233] However, there are also reports that pMMB33 cosmid clones are rather unstable and subject to deletion even in *E. coli recA* hosts.[199] We have also experienced a similar instability of cosmid derivatives of pSUP106, although the vector itself proved relatively stable in *E. coli* as well as in a series of *Rhizobium* strains.[201] Cloning of smaller fragments obviously did not affect plasmid stability.

The factors that cause unstable inheritance can be manifold, and stability can vary greatly within different species. Whatever the reasons, the instability of hybrid plasmids, not only in terms of their complete loss, but also the generation of rearrangements, can be problematical.

c. Copy Number

The vectors derived from RK2 retain the low copy number typical for IncP plasmids. In contrast, the IncQ-derived cloning vectors, described above, are characterized by a high copy number, which might cause unexpected effects.

A multicopy effect has been observed, for example, when the *nodABC* genes of *R. leguminosarum*, cloned into pKT230, were introduced into a normally nodulating strain. Nodulation was drastically reduced, probably due to overexpression of these genes rather than to a regulatory effect.[232] Similarly, it was found that gene *psi* inhibits polysaccharide synthesis in *R. phaseoli* and that this effect was much more pronounced when *psi* was cloned in IncQ vectors as compared to pLAFR1.[233] Also, fragments of *B. japonicum* carrying Hac genes, subcloned in pKT230, gave more frequent curling of root hairs than they did in the single copy vector pVK102, when introduced into *Pseudomonas putida*.[234] *Lignobacter* transconjugants, carrying *R. trifolii nod* genes cloned into pKT240, were able to nodulate white and subterranean clover, whereas transconjugants carrying the same genes in a low copy IncP derivative, nodulated only white clover plants.[222]

Although the high copy number of IncQ derivatives might not necessarily be undesirable, it still has to be considered, especially when regulatory or regulated genes are under study.

C. ALTERNATIVE SYSTEMS

Introduction and establishment of cloned DNA in a wide range of bacterial hosts is not necessarily dependent on the availability of a suitable broad host-range cloning vehicle, but there are also possibilities which allow the use of the sophisticated *E. coli* technology and its application in nonenteric strains. The two principles briefly described below may avoid recloning steps or the construction of broad host-range vectors for specific experimental purposes and may also circumvent problems concerning vector stability and the effects of high copy number.

1. Cointegrate Formation

As previously mentioned, *E. coli*-specific cloning vectors are of limited use in the genetic analysis of nitrogen-fixing organisms, mainly because they display only a very restricted replication host range. An alternative to the construction of broad host-range vectors is to extend the host range of *E. coli* vectors by cointegrate formation with RP4 or suitable derivatives (lacking one or more resistance markers). This methodology is schematically shown in Figure 8. The DNA fragment of interest, cloned into a narrow host-range *E. coli* vector, is integrated *in vivo* into RP4, a process which occurs spontaneously and at high frequencies. The mechanism of cointegrate formation is not clearly understood. It is assumed that homologous regions, present on either replicon (e.g., the Ap resistance in the case of pBR derivatives or the Mob region in the case of pSUP vectors) are responsible for the recombination process.

We have applied this technique successfully in *R. meliloti* as well as in *R. leguminosarum*,[107] for example, to select nodulation genes directly from a cosmid bank constructed in the narrow host-range cosmid pSUP205. A similar protocol was used to establish pJB8 derivatives in *Rhizobium loti*.[213] However, in some cases, we have observed that the presence of the RP4 plasmid might influence the symbiotic proficiency of the host, especially in strains of *R. leguminosarum*.

The cosmid clone identified can also be rescued by transferring the cointegrates back into *E. coli* where spontaneous resolution occurs; superinfection with λ then separates and purifies the cosmid of interest from the RP4 component.

In vivo fusion with RP4 is, of course, not restricted to cosmid derivatives, but occurs with any replicon provided it contains regions homologous to RP4. Using mobilizable *E. coli* vectors, this method combines two different types of experiments in one cloning step:

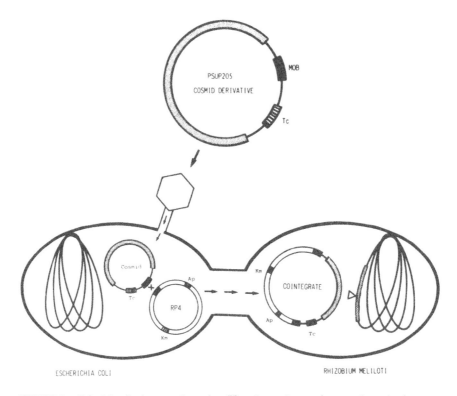

FIGURE 8.   Principle of cointegrate formation. The scheme shows cointegrate formation between RP4 and the cosmid vector pSUP205[26] which carries *R. meliloti* DNA. After introduction (e.g., by λ infection) of the cosmid derivative into an *E. coli* strain harboring a Tc[s] RP4 plasmid, recombination between the two replicons can occur spontaneously via homologous regions, such as the Mob-site present on both molecules. Upon transfer into *R. meliloti* this cointegrate, which can be selected for by the vector-encoded Tc resistance, is stably established due to the wide host-range replication functions of RP4. This system has been successfully used, for example, to directly select cosmid clones able to complement *R. meliloti* Nod::Tn5 mutants.[107]

the cloned DNA can, for example, be loaded with Tn5 and — based on the suicide effect in non-*E. coli* species — recombined into the original genome by homogenotization (see Section II), and the same recombinant molecule can be fused with RP4, transferred back into its original (or any other) host where the effect of the Tn5 insertion can be monitored in *trans*. Additionally, other genetic constructs, such as protein fusions generated in narrow host-range vectors, can be established and assayed in a broad range of bacteria without the need to reclone.

2. Stable Integration into the Host Genome

An alternative to the introduction on a separate autonomous replicon is the incorporation of the genetic material under investigation into the host genome. This can be readily achieved by cloning the DNA fragment under study into a narrow host-range vector and inserting the complete recombinant molecule into the host genome via homologous recombination between the cloned and the corresponding endogenous DNA, provided the vector is transmissible by conjugation. If necessary, introduction of the Mob-site, e.g., by cloning a Mob-cassette[235] or by introducing Tn5-Mob (see Section II) may convert the vector into a mobilizable derivative. Again, this technique allows different experiments with the same recombinant molecule, as demonstrated, e.g., by the work of Aguilar et al.[236] who used the same narrow host-range recombinant plasmid for Tn5 mutagenesis and regulation studies in *E. coli* as well as in *R. meliloti*. However, since integration depends on DNA homology, this method is restricted to the original host or very close relatives. Theoretically, this limitation can be partly overcome by providing the vector with a sequence which confers homology to a variety of bacterial strains.

Furthermore, bacterial transposons may be exploited as insertional cloning vectors. Such a transposable broad host-range cloning system, based on two compatible broad host-range plasmids, has been devised by Grinter[237] and modified by Barry.[238] The carrier replicon is derived from RP4 and retains its self-transmissibility and Tc resistance, but is unstably inherited without selective pressure. It carries a variant of Tn7 which possesses a selection marker, cloning sites, and the Tn7 termini, but has lost its *trans*-acting transposition functions. The second component, the Km-resistant helper plasmid, is based on the IncQ plasmid R300B and provides the Tn7 transposition functions missing from the carrier transposon.

This system has been successfully tested in *E. coli, Methylophilus methylotrophus,* and *P. aeruginosa,* but it is expected to be applicable also in other organisms. Theoretically, both components can be conjugally transmitted into any Gram-negative bacterium. Since the carrier is not stably replicated, selection for the Tn7-encoded resistance will identify those cells in which the defective transposon along with the cloned DNA has transposed into the host genome with the aid of the transposition functions provided in *trans* by the helper plasmid. Upon curing of the helper plasmid, e.g., by introduction of another IncQ plasmid, the stability of the inserted sequences is ensured.

Since integration of foreign DNA via transposons does not depend on DNA homology and occurs at different sites in the bacterial chromosome, these or similar systems might become extremely useful for *in vivo* and *in vitro* genetic engineering in nitrogen-fixing bacteria, because they allow the stable introduction of genetic constructs, including gene or operon fusions, into any host genome of choice.

D. CONCLUDING REMARKS

The unique features of broad host-range plasmids, i.e., their high-frequent transfer into and stable propagation in most Gram-negative bacteria, have made these plasmids not only very important as a basis for the construction of cloning vectors, but the plasmids themselves are extremely valuable tools for intergeneric transfer and expression of the DNA under study. Although IncP-mediated intra- and interspecific mobilization of chromosomal portions or plasmids, which are *per se* not self-transmissible (as is true for many symbiotic plasmids), has contributed greatly to the analysis of symbiotic nitrogen fixation, it has not been considered in this section. Instead, the focus of this review was to summarize the vector systems and methodologies available in nitrogen-fixation research.

The list of broad host-range vectors and cloning systems, outlined above, certainly does not represent all the potential possibilities, but illustrates an overview of the currently available technology. Each system has its advantages and disadvantages, and depending on the organism and on the type of experiment, one or another is more suitable. In any case, it is of value and importance to have cloning vectors that belong to different incompatibility groups, since complementation and gene regulation assays often require the simultaneous introduction and presence of more than one cloned gene in the host cell. Combinations between IncP- and IncQ-cloned genes were, for example, instrumental in the identification of gene *psr* (polysaccharide restoration) which negatively regulates *psi* expression in *R. phaseoli*[77] and in establishing that the *R. trifolii hsn* regions (cloned into an IncQ vector) can interact with the *R. meliloti* common *nod* genes (on pLAFR1) to confer nodulation ability to an *R. trifolii* strain cured of its symbiotic plasmid.[202] Activation studies of symbiotic promoters especially are largely dependent on the availability of different vectors, e.g., having the promoter under study cloned in one Inc-type vector which allows an easy assay for promoter activity, and the activating gene products (such as *nifA* or *nodD*) constitutively expressed and provided in *trans* by a second type of vector (see, for example, References 196 and 208).

Initially, broad host-range vectors, mainly the IncQ derivatives, were developed for *Pseudomonas* and other industrially important soil bacteria. However, with the increasing interest in nitrogen-fixing bacteria, not only those vectors that are available have been adapted, but also new systems have been worked out especially for use in these strains. They have contributed immensely to the progress achieved in the analysis of symbiotic genes in fast- and slow-growing *Rhizobia* and will, without doubt, also be extremely valuable research tools in other symbiotic nitrogen-fixing organisms.

ACKNOWLEDGMENTS

We wish to thank V. Krishnapillai, W. Klipp, and B. Kosier for critically reading and correcting the manuscript.

REFERENCES

1. **Kleckner, N., Roth, J., and Botstein, D.**, Genetic engineering *in vivo* using translocatable drug resistance elements: new methods in bacterial genetics, *J. Mol Biol.*, 116, 125, 1977.
2. **Kleckner, N.**, Transposable elements in prokaryotes, *Annu. Rev. Genet.*, 15, 341, 1981.
3. **Campbell, A., Berg, D. E., Botstein, D., Lederberg, E. M., Novick, R. P., et al.**, Nomenclature of transposable elements in bacteria, *Gene*, 5, 197, 1979.
4. **Calos, M. P. and Miller, J. H.**, Transposable elements, *Cell*, 20, 579, 1980.
5. **Starlinger, P.**, IS elements and transposons, *Plasmid*, 3, 242, 1980.
6. **Shapiro, J. A., Ed.**, *Mobile Genetic Elements*, Academic Press, New York, 1983.
7. **Berg, D. E. and Berg, C. M.**, The prokaryotic transposable element Tn5,, *Biotechnology*, 1, 417, 1983.
8. **Beringer, J. E., Beynon, J. L., Buchanan-Wollaston, A. V., and Johnston, A. W. B.**, Transfer of the drug-resistance transposon Tn5 to *Rhizobium*, *Nature (London)*, 276, 633, 1978.
9. **De Bruijn, F. J. and Lupski, J. R.**, The use of transposon Tn5 mutagenesis in the rapid generation of correlated physical and genetic maps of DNA segments cloned into multicopy plasmids — a review, *Gene*, 27, 131, 1984.
10. **Mills, D.**, Transposon mutagenesis and its potential for studying virulence genes in plant pathogens, *Annu. Rev. Phytopathol.*, 23, 297, 1985.
11. **Berg, D. E., Schmandt, M. A., and Lowe, J. B.**, Specificity of transposon Tn5 insertion, *Genetics*, 105, 813, 1983.
12. **Singer, J. T. and Finnerty, W. R.**, Insertional specificity of transposon Tn5 in *Acinetobacter* sp., *J. Bacteriol.*, 157, 607, 1984.
13. **Berg, D. E., Davies, J., Allet, B., and Rochaix, J. D.**, Transposition of R-factor genes to bacteriophage λ, *Proc. Natl. Acad. Sci. U.S.A.*, 72, 3628, 1975.
14. **Jorgensen, R. A., Rothstein, S. J., and Reznikoff, W. S.**, A restriction enzyme cleavage map of Tn5 and localization of a region encoding neomycin resistance, *Mol. Gen. Genet.*, 177, 65, 1979.
15. **Genilloud, O., Garrido, M. C., and Moreno, F.**, The transposon Tn5 carries a bleomycin-resistance determinant, *Gene*, 32, 225, 1984.
16. **Mazodier, P., Cossart, P., Giraud, E., and Gasser, F.**, Completion of the nucleotide sequence of the central region of Tn5 confirms the presence of three resistance genes, *Nucleic Acids Res.* 13, 195, 1985.
17. **Collins, C. M. and Hall, R. M.**, Identification of a Tn5 determinant conferring resistance to phleomycins, bleomycins, and tallysomycin, *Plasmid*, 4, 143, 1985.
18. **Mazodier, P., Giraud, E., and Gasser, F.**, Genetic analysis of the streptomycin resistance encoded by Tn5, *Mol. Gen. Genet.*, 192, 155, 1983.
19. **Putnoky, P., Kiss, G. B., Ott, I., and Kondorosi, A.**, Tn5 carries a streptomycin resistance determinant downstream from the kanamycin resistance gene, *Mol. Gen. Genet.*, 191, 288, 1983.
20. **Selveraj, G. and Iyer, V. N.**, Transposon Tn5 specifies streptomycin resistance in *Rhizobium* spp., *J. Bacteriol.*, 158, 580, 1984.
21. **De Vos, G. F., Finan, T. M., Signer, E. R., and Walker, G. C.**, Host dependent transposon Tn5-mediated streptomycin resistance, *J. Bacteriol.*, 159, 395, 1984.
22. **O'Niell, E. A., Kiely, G. M., and Bender, R. A.**, Transposon Tn5 encodes streptomycin resistance in nonenteric bacteria, *J. Bacteriol.*, 159, 388, 1984.

23. **Don, R. H., Weightman, A. J., Knackmus, H. J., and Timmis, K. N.,** Transposon mutagenesis and cloning analysis of the pathways for degradation of 2,4-dichlorophenoxyacetic acid and 3-chlorobenzoate in *Alcaligenes eutrophus* JMP134 (pJP4), *J. Bacteriol.,* 161, 85, 1985.

24. **Auerswald, E. A., Ludwig, G., and Schaller, H.,** Structural analysis of Tn5, *Cold Spring Harbor Symp. Quant. Biol.,* 45, 107, 1980.

25. **Beck, E., Ludwig, G., Auerswald, E. A., Reiss, B., and Schaller, H.,** Nucleotide sequence and exact location of the neomycin phosphotransferase gene from transposon Tn5, *Gene,* 19, 327, 1982.

26. **Simon, R., Priefer, U., and Pühler, A.,** A broad host range mobilization system for *in vivo* genetic engineering: transposon mutagenesis in Gram-negative bacteria, *Biotechnology,* 1, 784, 1983.

27. **Zsebo, K. M., Wu, F., and Hearst, J. E.,** Tn5.7 construction and physical mapping of pRPS404 containing photosynthetic genes from *Rhodopseudomonas capsulata, Plasmid,* 11, 182, 1984.

28. **Hirsch, P. R., Wang, C. L., and Woodward, M. J.,** Construction of a Tn5 derivative determining resistance to gentamycin and spectinomycin using a fragment cloned from R1033, *Gene,* 4, 203, 1986.

29. **De Vos, G. F., Walker, G. C., and Signer, E. R.,** Genetic manipulations in *Rhizobium meliloti* utilizing two new transposon Tn5 derivatives, *Mol. Gen. Genet.,* 204, 485, 1986.

30. **Rella, M., Mercenier, A., and Haas, D.,** Transposon insertion mutagenesis of *Pseudomonas aeruginosa* with a Tn5 derivative: application to physical mapping of the arc gene cluster, *Gene,* 33, 293, 1985.

31. **Casadesus, J., Janez, E., and Olivares, J.,** Transposition of Tn1 to the *Rhizobium meliloti* genome, *Mol. Gen. Genet.,* 180, 405, 1980.

32. **Krishnapillai, V., Royle, P., and Lehrer, J.,** Insertion of the transposon Tn1 into the *Pseudomonas aeruginosa* chromosome, *Genetics,* 97, 295, 1981.

33. **Bolton, E., Glynn, P., and O'Gara, F.,** Site specific transposition of Tn7 into a *Rhizobium meliloti* megaplasmid, *Mol. Gen. Genet.,* 193, 153, 1984.

34. **Caruso, M. and Shapiro. J. A.,** Interaction of Tn7 and temperate phage F116L of *Pseudomonas aeruginosa, Mol. Gen. Genet.,* 188, 292, 1982.

35. **Ely, B. and Croft, R. H.,** Transposon mutagenesis in *Caulobacter crescentus, J. Bacteriol.,* 149, 620, 1977.

36. **Ely, B.,** Vectors for transposon mutagenesis of non-enteric bacteria, *Mol. Gen. Genet.,* 200, 302, 1985.

37. **Thomson, J. A., Hendson, M., and Magnes, R. M.,** Mutagenesis by insertion of drug resistance transposon Tn7 into *Vibrio* species, *J. Bacteriol.,* 148, 374, 1981.

38. **Turner, P., Barber, C., and Daniels, M.,** Behaviour of the transposons Tn5 and Tn7 in *Xanthomonas campestris* pv. *campestris, Mol. Gen. Genet.,* 195, 101, 1984.

39. **Way, J. C., Davis, M. A., Morisato, D., Roberts, D. E., and Kleckner, N.,** New Tn10 derivatives for transposon mutagenesis and for construction of *lacZ* operon fusions by transposition, *Gene,* 32, 369, 1984.

40. **Selveraj, G. and Iyer, V. N.,** Suicide plasmid vehicles for insertion mutagenesis in *Rhizobium meliloti* and related bacteria, *J. Bacteriol.,* 156, 1292, 1983.

41. **Berg, D. E., Weiss, A., and Crossland, L.,** Polarity of Tn5 insertion mutation in *E. coli, J. Bacteriol.,* 142, 439, 1980.

42. **Poole, K. and Hancock, R. E. W.,** Isolation of a Tn501 insertion mutant lacking porin protein P of *Pseudomonas aeruginosa* PAO, *Mol. Gen. Genet.,* 202, 403, 1986.

43. **Tsuda, M., Harayama, S., and Iino, T.,** Tn501 insertion mutagenesis in *Pseudomonas aeruginosa* PAO, *Mol. Gen. Genet.,* 196, 494, 1984.

44. **Yakobson, E. A. and Guiney, D. G.,** Conjugal transfer of bacterial chromosomes mediated by the RK2 plasmid transfer origin cloned into transposon Tn5, *J. Bacteriol.,* 160, 451, 1984.

45. **Simon, R., Priefer, U., and Pühler, A.,** Vector plasmids for *in vivo* and *in vitro* manipulations of Gram-negative bacteria, in *Molecular Genetics of the Bacteria-Plant Interaction,* Pühler, A., Ed., Springer-Verlag, Heidelberg, 1983, 98.

46. **Simon R.,** High frequency mobilization of Gram-negative bacterial replicons by the *in vitro* constructed Tn5-Mob transposon, *Mol. Gen. Genet.,* 196, 413, 1984.

47. **Hynes, M. F., Simon, R., and Pühler, A.,** The development of plasmid-free strains of *Agrobacterium tumefaciens* by using incompatibility with a *Rhizobium meliloti* plasmid to eliminate pAtC58, *Plasmid,* 13, 99, 1985.

48. **Hynes, M. F., Simon, R., Müller, P., Niehaus, K., Labes, M., and Pühler, A.,** The two megaplasmids of *Rhizobium meliloti* are involved in the effective nodulation of alfalfa, *Mol. Gen. Genet.,* 202, 356, 1986.

49. **Martinez, E., Palacios, R., and Sanchez, F.,** Nitrogen-fixing nodules induced by *Agrobacterium tumefaciens* harboring *Rhizobium phaseoli* plasmids, *J. Bacteriol.,* 169, 2828, 1987.

50. **Broughton, W. J., Wong, C. H., Lewin, A., Samrey, U., Myint, H., Meyer, Z. A. H., Dowling D. N., and Simon, R.,** Identification of *Rhizobium* plasmid sequences involved in recognition of *Psophocarpus, Vigna,* and other legumes, *J. Cell Biol.,* 102, 1173, 1986.

51. **Silhavy, T. J. and Beckwith, J. R.,** Uses of *lac* fusions for the study of biological problems, *Microbiol. Rev.,* 49, 398, 1985.

52. **Silhavy, T. J., Berman, M. L., and Enquist, L. W.,** *Experiments with Gene Fusions,* Cold Spring Harbor Laboratory, Cold Spring Harbor, NY, 1984.

53. **Shapira, S. K., Chou, J., Richaud, F. V., and Casadaban, M. J.,** New versatile plasmid vectors for the expression of hybrid proteins coded by a cloned gene fused to *lac* Z gene sequences encoding an enzymatically active carboxy-terminal protein of β-galactosidase, *Gene,* 25, 71, 1983.

54. **Casadaban, M., Martinez-Arias, A., Shapira, S. K., and Chou, J.,** β-galactosidase gene fusions for analyzing gene expression in *E. coli* and yeast, in *Methods in Enzymology,* Vol 100-B, Wu, R., Grossman, L., and Moldave, K., Eds., Academic Press, New York, 1983, 293.

55. **Guan, C., Wanner, B., and Inouye, H.,** Analysis of regulation of *pho* B expression using a *pho* B-*cat* fusion *J. Bacteriol.,* 156, 710, 1983.

56. **Close, T. J. and Rodriguez, R. L.,** Construction and characterization of the chloramphenicol-resistance gene cartridge: a new approach to the transcriptional mapping of extrachromosomal elements, *Gene,* 20, 305, 1982.

57. **McKenney, K., Shimatake, H., Court, D., Schmeissner, U., Brady, C., and Rosenberg, M.,** A system to study promoter and terminator signals recognized by *Escherichia coli* RNA polymerase, in *Gene, Amplification and Analysis,* Chirikjian, J. S. and Papas, T. S., Eds., Elsevier Science, New York, 1981, 384.

58. **Van den Broeck, G., Timko, M. P., Kausch, A. P., Cashmore, A. R., Van Montagu, M., and Herrera-Estrella, L.,** Targeting of a foreign protein to chloroplasts by fusion to the transit peptide from the small subunit of ribulose 1,5-bisphosphate carboxylase, *Nature (London),* 313, 358, 1985.

59. **Hoffmann, C. and Wright, A.,** Fusions of secreted proteins to alkaline phosphatase: a new approach for studying protein secretion, *Proc. Natl. Acad. Sci. U.S.A.* 82, 5107, 1985.

60. **Engebrecht, J., Simon, M., and Silverman, M.,** Measuring gene expression with light, *Science,* 227, 1345, 1985.

61. **Shaw, J. J. and Kado, C. I.,** Development of a *Vibrio* bioluminescence gene-set to monitor phytopathogenic bacteria during the ongoing disease process in a non-disruptive manner, *Biotechnology,* 4, 560, 1986.

61a. **Shaw, J. J. and Kado, C. I.,** Direct analysis of the invasiveness of *Xanthomonas campestris* mutants generated by Tn4431, a transposon containing a promoterless luciferase cassette for monitoring gene expression, in *Molecular Genetics of Plant-Microbe Interaction,* Verma, D. P. S. and Brisson, N., Eds., Martinus Nijhoff, The Hague, 1987, 57.

62. **MacNeil, D., Zhu, J., and Brill, W. J.,** Regulation of nitrogen fixation in *Klebsiella pneumoniae*: isolation and characterization of strains with *nif-lac* fusions, *J. Bacteriol.,* 145, 348, 1981.

63. **Dixon, R., Eady, R. R., Espin, G., Hill, S., Iaccarino, M., Kahn, D., and Merrick, M.,** Analysis of regulation of *Klebsiella pneumoniae* nitrogen fixation *(nif)* gene cluster with gene fusions, *Nature (London),* 286, 128, 1980.

64. **Baker, T. A., Howe, M. M., and Gross, C. A.,** MudX, a derivative of Mudl(*lac* Ap) which makes stable *lacZ* fusions at high temperature, *J. Bacteriol.,* 156, 960, 1983.

65. **Casadaban, M. J. and Chou, J.,** In vivo formation of gene fusions encoding hybrid β-galactosidase proteins in one step with transposable Mu-*lac* transducing phage, *Proc. Natl. Acad. Sci. U.S.A.,* 81, 535, 1984.

66. **Castilho, B. A., Olfson, P., and Casadaban, M. J.,** Plasmid insertion mutagenesis and *lac* gene fusion with mini-Mu bacteriophage transposons, *J. Bacteriol.,* 158, 488, 1984.

67. **Huges, K. T. and Roth, J. R.,** Conditional transposition defective derivatives of Mudl(Amp *lac*), *J. Bacteriol.,* 159, 130, 1984.

68. **Bremer, E., Thomas, J. S., and Weinstock, G. M.,** Transposable λ p*lac* Mu bacteriophages for creating *lacZ* operon fusions and kanamycin resistance insertion in *Escherichia coli, J. Bacteriol.,* 162, 1092, 1985.

69. **Groisman, E. A. and Casadaban, M. J.,** Mini-Mu bacteriophage with plasmid replicons for in vivo cloning and *lac* gene fusing, *J. Bacteriol.,* 168, 357, 1986.

70. **Olson, E. R., Sadowsky, M. J., and Verma, D. P. S.,** Identification of genes involved in the *Rhizobium*-legume symbiosis by Mu-dl (kan,*lac*)-generated transcription fusions, *Biotechnology,* 3, 143, 1985.

71. **Innes, R. W., Kuempel, P. L., Plazinski, J., Canter-Cremers, H., Rolfe, B. G., and Djordjevic, M. A.,** Plant factors induce expression of nodulation and host range genes in *Rhizobium trifolii, Mol. Gen. Genet.,* 201, 426, 1985.

72. **Redmond, J. W., Batley, M., Djordievic, M. A., Innes, R. W., Kuempel, P. L., and Rolfe, B. G.,** Flavones induce expression of nodulation genes in *Rhizobium, Nature (London),* 323, 632, 1986.

73. **Djordjevic, M. A., Redmond, J. W., Batley, M., and Rolfe, B. G.,** Clover secrete specific phenolic compounds which either stimulate or repress *nod* gene expression in *Rhizobium trifolii, EMBO J.,* 6, 1173, 1987.

74. **Toussaint, A. and Resibois, A.,** Phage Mu: transposition as a life style, in *Mobile Genetic Elements,* Shapiro, J. A., Ed., Academic Press, New York, 1983, 105.

75. **Kroos, L. and Kaiser, D.,** Construction of Tn5 *lac,* a transposon that fuses *lacZ* expression to exogenous promoters, and its introduction into *Myxococcus xanthus, Proc. Natl. Acad. Sci., U.S.A.,* 81, 581, 1984.

76. **Stachel, S. E., Gynheung, A., Flores, C., and Nester, E. W.,** A Tn3 *lacZ* transposon for the random generation of β-galactosidase gene fusions: application to the analysis of gene expression in *Agrobacterium, EMBO J.,* 4, 891, 1985.

77. **Borthakur, D. and Johnston, A. W. B.,** Sequence of *psi*, a gene on the symbiotic plasmid of *Rhizobium phaseoli* which inhibits exopolysaccharide synthesis and nodulation and demonstration that its transcription is inhibited by *psr*, another gene on the symbiotic plasmid, *Mol. Gen. Genet.*, 207, 149, 1987.

78. **Borthakur, D., Lamb, J. W., and Johnston, A. W. B.,** Identification of two classes of *Rhizobium phaseoli* genes required for melanin synthesis, one of which is required for nitrogen fixation and activates the transcription of the other, *Mol. Gen. Genet.*, 207, 155, 1987.

79. **Bellofatto, V., Shapiro. L., and Hodgson, D. A.,** Generation of a Tn5 promoter probe and its use in the study of gene expression in *Caulobacter crescentus*, *Proc. Natl. Acad. Sci. U.S.A.* 81, 1035, 1984.

80. **Manoil, C. and Beckwith, J.,** Tn*phoA*: a transposon probe for protein export signals, *Proc. Natl. Acad. Sci. U.S.A.* 82, 8129, 1985.

81. **Boulnois, G. J., Varley, J. M., Sharpe, G. S., and Franklin, F. C. H.,** Transposon donor plasmids, based on ColIb-P9 for use in *Pseudomonas putida* and a variety of other Gram-negative bacteria, *Mol. Gen. Genet.*, 200, 65, 1985.

82. **Eaton, R. W. and Timmis, K. N.,** Characterization of a plasmid-specified pathway for catabolism of isopropylbenzene in *Pseudomonas putida* RE204, *J. Bacteriol.*, 168, 123, 1986.

83. **Whitta, S., Sinclair, M. I., and Holloway, B. W.,** Transposon mutagenesis in *Methylobacterium* AM1 (*Pseudomonas* AM1), *J. Gen. Microbiol.*, 131, 1547, 1985.

84. **Boucher, C., Bergeron, B., Barate de Bertalmio, M., and Denarie, J.,** Introduction of bacteriophage Mu into *Pseudomonas solanacearum* and *Rhizobium meliloti* using the R factor RP4, *J. Gen. Microbiol.*, 98, 253, 1977.

85. **Denarie, J., Rosenberg, C., Bergeron, B., Boucher, C., Michel, M., and Borate de Bertalmio, M.,** Potential of RP4∷Mu plasmids for in vivo genetic engineering of Gram-negative bacteria, in *DNA Insertion Elements, Plasmids and Episomes*, Bukhari, A. I., Shapiro, J. A., and Adhya, S. L., Eds., Cold Spring Harbor Laboratory, Cold Spring Harbor, NY, 1977, 507.

86. **Van Vliet, F., Silva, B., Van Montagu, M., and Schell, J.,** Transfer of RP4∷Mu plasmids to *Agrobacterium tumefaciens*, *Plasmid*, 1, 446, 1978.

87. **Hirsch, P. R. and Beringer, J. E.,** A physical map of pPHIJI and pJB4JI, *Plasmid*, 12, 133, 1984.

88. **Boucher, C., Message, B., Debieu, D., and Zischek, C.,** Use of P-1 incompatibility group plasmids to introduce transposons into *Pseudomonas solanacearum*, *Phytopathology*, 71, 639, 1981.

89. **Morales, V. M. and Sequeira, L.,** Suicide vector for transposon mutagenesis in *Pseudomonas solanacearum*, *J. Bacteriol.*, 163, 1263, 1985.

90. **Kaufmann, N., Hüdig, H., and Drews, G.,** Transposon Tn5 mutagenesis of genes for the photosynthetic apparatus in *Rhodopseudomonas capsulata*, *Mol. Gen. Genet.*, 198, 153, 1984.

91. **Singh, M. and Klingmüller, W.,** Transposon mutagenesis in *Azospirillum brasilense*: isolation of auxotrophic and Nif⁻ mutants and molecular cloning of the mutagenized *nif* DNA, *Mol. Gen. Genet.*, 202, 136, 1986.

92. **Banfalvi, Z., Sakanyan, V., Koncz, C., Kiss, A., Dusha, I., and Kondorosi, A.,** Location of nitrogen fixation genes on a high molecular weight plasmid of *R. meliloti*, *Mol. Gen. Genet.*, 184, 129, 1981.

93. **Casey, C., Bolton, E., and O'Gara, F.,** Behaviour of bacteriophage Mu-based IncP suicide vector plasmids in *Rhizobium* spp., *FEMS Microbiol. Lett.*, 20, 217, 1983.

94. **Meade, H. M., Long, S. R., Ruvkun, G. B., Brown, S. E., and Ausubel, F. M.,** Physical and genetic characterization of symbiotic and auxotrophic mutants of *Rhizobium meliloti* induced by transposon Tn5 mutagenesis, *J. Bacteriol.*, 149, 114, 1982.

95. **Noel, K. D., Sanchez, A., Fernandez, L., Leemans, J., and Cevallos, M.,** *Rhizobium phaseoli* symbiotic mutants with transposon Tn5 insertions, *J. Bacteriol.*, 158, 148, 1984.

96. **Forrai, T., Vincze, E., Banfalvi, Z., Kiss, G. B., Randhawa, G. S., and Kondorosi, A.,** Localization of symbiotic mutations in *Rhizobium meliloti*, *J. Bacteriol.*, 153, 635, 1983.

97. **Figurski, D. H. and Helinski, D. R.,** Replication of an origin-containing derivative of plasmid RK2 dependent on a plasmid function provided in trans, *Proc. Natl. Acad. Sci. U.S.A.*, 76, 1648, 1979.

98. **Lam, S. T., Lam, B. S., and Strobel, G.,** A vehicle for the introduction of transposons into plant-associated pseudomonads, *Plasmid*, 13, 200, 1985.

99. **Forster, T. J., Davis, M. A., Roberts, D. E., Takerhita, K., and Kleckner, N.,** Genetic organization of transposon Tn10, *Cell*, 23, 201, 1981.

100. **Weiss, A., Hewlett, E. L., Myers, G. A., and Falkow, S.,** Transposon Tn5-induced mutations affecting virulence factors of *Bordetella pertussis*, *Infect. Immunol.*, 42, 33, 1983.

101. **Bullerjahn, G. S. and Benzinger, R. H.,** Introduction of the mercury transposon Tn501 into *Rhizobium japonicum* strains 31 and 110, *FEMS Microbiol. Lett.*, 22, 183, 1984.

102. **Bolivar, F.,** Construction and characterization of new cloning vehicles. III. Derivatives of plasmid pBR322 carrying unique EcoRI sites for selection of EcoRI generated recombinant molecules, *Gene*, 4, 121, 1978.

103. **Chang, A. C. Y. and Cohen, S. N.,** Construction and characterization of amplifiable multicopy DNA cloning vehicles derived from the pA15 cryptic miniplasmid, *J. Bacteriol.*, 134, 1141, 1978.

104. **Rostas, K., Sista, P. R., Stanley, J., and Verma, D. P. S.,** Transposon mutagenesis of *Rhizobium japonicum, Mol. Gen. Genet.,* 197, 230, 1984.

105. **Kennedy, C., Gamal, R., Humphrey, R., Ramos, I., Brigle, K., and Dean, D.,** The *nif*H, *nif*M and *nif*N genes of *Azotobacter vinelandii:* characterization by Tn5 mutagenesis and isolation from pLAFR1 gene banks, *Mol. Gen. Genet.,* 205, 318, 1986.

106. **Guiney, D. G. and Helinski, D. R.,** The DNA-protein relaxation complex of the plasmid RK2: location of the site-specific nick in the region of the proposed origin of transfer, *Mol. Gen. Genet.,* 176, 183, 1979.

107. **Simon, R., O'Connell, M., Labes, M., and Pühler, A.,** Plasmid vectors for the genetic analysis and manipulation of Rhizobia and other gram-negative bacteria in *Methods in Enzymology,* Vol 118, Weisbach, A. and Weisbach, H., Eds., Academic Press, New York, 1986, 640.

108. **Sasakawa, C., Lowe, J. B., McDivitt, L., and Berg, D. E.,** Control of transposon Tn5 transposition in *Escherichia coli, Proc. Natl. Acad. Sci. U.S.A.* 79, 7450, 1982.

109. **Berg, D. E., Johnsrud, L., McDivitt, L., Ramabhadran, R., and Hirschel, B. J.,** The inverted repeats of Tn5 are transposable elements, *Proc. Natl. Acad. Sci. U.S.A.* 79, 2632, 1984.

110. **Hirschel, B. J. and Berg, D. E.,** A derivative of Tn5 with direct terminal repeats can transpose, *J. Mol. Biol.,* 155, 105, 1982.

111. **Hirschel, B. J., Galas, D. J., Berg, D. E., and Chandler, M.,** Structure and stability of transposon 5-mediated cointegrates, *J. Mol. Biol.,* 159, 557, 1982.

112. **Isberg, R. R. and Syvanen, M.,** Replicon fusions promoted by the inverted repeats of Tn5: the right repeat is an insertion sequence, *J. Mol. Biol.,* 150, 15, 1981.

113. **Anderson, D. M. and Mills, D.,** The use of transposon mutagenesis in the isolation of nutritional and virulence mutants in two pathovars of *Pseudomonas syringae, Phytopathology,* 75, 104, 1985.

114. **So, J. S., Hodgson, A. L. M., Haugland, R., Leavitt, M., Banfalvi, Z., Nieuwkoop, A. A., and Stacey, G.,** Transposon-induced symbiotic mutants of *Bradyrhizobium japonicum:* isolation of two gene regions essential for nodulation, *Mol. Gen. Genet.,* 207, 15, 1987.

115. **Donald, R. G., Raymond, C. K., and Ludwig, R. A.,** Vector insertion mutagenesis of *Rhizobium* sp. strain ORS571: direct cloning of mutagenized DNA sequences, *J. Bacteriol.,* 162, 317, 1985.

116. **Ruvkun, G. B. and Ausubel, F. M.,** A general method for site-directed mutagenesis in prokaryotes, *Nature (London),* 289, 85, 1981.

117. **Ruvkun, G. B., Sundaresan, V., and Ausubel, F. M.,** Directed transposon mutagenesis and complementation analysis of *Rhizobium meliloti* symbiotic nitrogen fixation genes, *Cell,* 29, 551, 1982.

118. **Ditta, G.,** Tn5 mapping of Rhizobium nitrogen fixation genes, in *Methods in Enzymology,* Vol. 118, Weisbach, A. and Weisbach, H., Eds., Academic Press, New York, 1986, 519.

119. **Messing, J. and Vierra, J.,** A new pair of M13 vectors for selecting either strand of a double digest restriction fragment, *Gene,* 19, 269, 1982.

120. **Müller, W., Keppner, W., and Rasched, I.,** Versatile kanamycin-resistance cartridges for vector construction in *Escherichia coli, Gene,* 46, 131, 1986.

121. **Prentki, P. and Kirsch, H. M.,** In vitro insertional mutagenesis with a selectable DNA fragment, *Gene,* 29, 303, 1984.

122. **Frey, J. and Krisch, H. M.,** Ω mutagenesis in Gram-negative bacteria: a selectable interposon which is strongly polar in a wide range of bacterial species, *Gene,* 36, 143, 1985.

123. **Mulligan, J. T. and Long, S. R.,** Induction of *Rhizobium meliloti nod* C expression by plant exudate required *nod* D, *Proc. Natl. Acad. Sci. U.S.A.,* 82, 6609, 1985.

124. **Casadaban, M. J., Chou, J., and Cohen, S. N.,** In vitro gene fusions that join an enzymatically active β-galactosidase segment to amino-terminal fragments of exogenous proteins: *Escherichia coli* plasmid vectors for the detection of translational initiation signals, *J. Bacteriol.,* 143, 971, 1980.

125. **Jagadish, M. N., Bookner, S. D., and Szalay, A. A.,** A method for site-directed transplacement of in vitro altered DNA sequences in *Rhizobium, Mol. Gen. Genet.,* 199, 249, 1985.

126. **Lowbury, E. J. L., Kidson, A., Lilly, H. A., Ayliffe, G. A., and Jones, R. J.,** Sensitivity of *Pseudomonas aeruginosa* to antibiotics: emergence of strains highly resistant to carbenicillin, *Lancet,* 2, 448, 1969.

127. **Datta, N., Hedges, R. W., Shaw, E. J., Sykes, R. B., and Richmond, M. H.,** Properties of an R factor from *Pseudomonas aeruginosa, J. Bacteriol.,* 108, 1244, 1971.

128. **Stanisich, V. A. and Holloway, B. W.,** Chromosome transfer in *Pseudomonas aeruginosa* mediated by R factors, *Genet. Res.,* 17, 169, 1971.

129. **Ingram. L. C., Richmond, M. H., and Sykes, R. B.,** Molecular characterization of the R-factors implicated in the carbenicillin resistance of a sequence of *Pseudomonas aeruginosa* strains isolated from burns, *Antimicrob. Agents Chemother.,* 3, 279, 1973.

130. **Burkardt, H.-J., Riess, G., and Pühler, A.,** Relationship of group P1 plasmids revealed by heteroduplex experiments: RP1, RP4, R68 and RK2 are identical, *J. Gen. Microbiol.,* 114, 341, 1979.

131. **Thomas, C. M., Smith, C. A., Shingler, V., Cross, M. A., Hussain, A. A. K., and Pinkney, M.,** Regulation of replication and maintenance functions of broad host range plasmid RK2, in *Plasmids in Bacteria,* Helinski, D. R., Cohen, S. N., Clewell, D. B., Jackson, D. A., and Hollaender, A., Eds., Plenum Press, New York, 1985, 261.

132. **Figurski, D. H., Meyer, R. J., and Helinski, D. R.,** Suppression of ColE1 replication properties by the IncP-1 plasmid RK2 in hybrid plasmids constructed *in-vitro, J. Mol. Biol.,* 133, 295, 1979.

133. **Meyer, R. and Helinski, D. R.,** Unidirectional replication of the P-group plasmid RK2, *Biochim. Biophys. Acta,* 478, 109, 1977.

134. **Thomas, C. M., Stalker, D. M., and Helinski, D. R.,** Replication and incompatibility properties of segments of the origin region of replication of the broad host range plasmid RK2, *Mol. Gen. Genet.,* 181, 1, 1981.

135. **Schmidhauser, T. J. and Helinski, D. R.,** Regions of broad host range plasmid RK2 involved in replication and stable maintenance in nine species of Gram-negative bacteria, *J. Bacteriol.,* 164, 446, 1985.

136. **Guiney, D. G., Chikami, G., Deiss, C., and Yakobson, E.,** The origin of plasmid DNA transfer during bacterial conjugation, in *Plasmids in Bacteria,* Helinski, D. R., Cohen, S. N., Clewell, D. B., Jackson, D. A., and Hollaender, A., Eds., Plenum Press, New York, 1985, 521.

137. **Barth, P. R. and Grinter, N. J.,** Comparison of the deoxyribonucleic acid molecular weights and homologies of plasmids conferring linked resistance to streptomycin and sulfonamides, *J. Bacteriol.,* 120, 618, 1974.

138. **Meyer, R., Hinds, M., and Brasch, M.,** Properties of R1162, a broad-host range, high-copy-number plasmid, *J. Bacteriol.,* 150, 552, 1982.

139. **Bryan, L. E., Van den Elzen, H. M., and Tseng, J. T.,** Transferable drug resistance in *Pseudomonas aeruginosa, Antimicrob. Agents Chemother.,* 1, 22, 1972.

140. **Guerry, P., Van Embden, J., and Falkow, S.,** Molecular nature of two non-conjugative plasmids carrying drug resistance genes, *J. Bacteriol.,* 117, 619, 1974.

141. **Willets, N. and Crowther, C.,** Mobilization of the non-conjugative IncQ plasmid RSF1010, *Genet. Res. Camb.,* 37, 311, 1981.

142. **Gautier, F. and Bonewald, R.,** The use of plasmid R1162 and derivatives for gene cloning in the methanol-utilizing *Pseudomonas* Am1, *Mol. Gen. Genet.,* 178, 375, 1980.

143. **De Graaff, J., Crosa, J. H., Heffron, F., and Falkow, S.,** Replication of the nonconjugative plasmid RSF1010, in *Escherichia coli* K-12, *J. Bacteriol.,* 134, 1117, 1978.

144. **Scholz, P., Haring, V., Scherzinger, E., Lurz, R., Bagdasarian, M. M., Schuster, H., and Bagdasarian, M.,** Replication determinants of the broad host range plasmid RSF1010, in *Plasmids in Bacteria,* Helinski, D. R., Cohen, S. N., Clewell, D. B., Jackson, D. A., and Hollaender, A., Eds., Plenum Press, New York, 1984, 243.

145. **Scherzinger, E., Bagdasarian, M. M., Scholz, P., Lurz, R., Rückert, B., and Bagdasarian, M.,** Replication of the broad host range plasmid RSF1010: requirement for three plasmid-encoded proteins, *Proc. Natl. Acad. Sci. U.S.A.,* 81, 654, 1984.

146. **Haring, V., Scholz, P., Scherzinger, E., Frey, J., Derbyshire, K., Hatfull, G., Willets, N. S., and Bagdasarian, M.,** Protein RepC is involved in copy number control of the broad host range plasmid RSF1010, *Proc. Natl. Acad. Sci. U.S.A.,* 82, 6090, 1985.

147. **Brasch, M. A. and Meyer, R. J.,** Genetic organization of plasmid R1162 DNA involved in conjugative mobilization, *J. Bacteriol.,* 167, 703, 1986.

148. **Watanabe, T., Furuse, C., and Sakaizumi, S.,** Transduction of various R-factors by phage P1 in *Escherichia coli* and by phage P22 in *Salmonella typhimurium, J. Bacteriol.,* 96, 1791, 1968.

149. **Datta, N. and Hedges, R. W.,** Trimethoprim resistance conferred by W plasmids in Enterobacteriaceae, *J. Gen. Microbiol.,* 72, 349, 1972.

150. **Coetzee, J. N., Datta, N., and Hedges, R. W.,** R factors from *Proteus rettgeri, J. Gen. Microbiol.,* 72, 534, 1972.

151. **Gorai, A. P., Heffron, F., Falkow, S., Hedges, R. W., and Datta, N.,** Electron microscope heteroduplex studies of sequence relationships among plasmids of the W incompatibility group, *Plasmid,* 2, 485, 1979.

152. **Ward, J. M. and Grinsted, J.,** Physical and genetic analysis of the IncW-group plasmids R388, Sa and R7K, *Plasmid,* 7, 239, 1982.

153. **Tait, R. C., Lundquist, R. C., and Kado, C. I.,** Genetic map of the crown-gall suppressive IncW plasmid pSa, *Mol. Gen. Genet.,* 186, 10, 1982.

154. **Tait, R. C., Close, T. J., Rodriguez, R. L., and Kado, C. I.,** Isolation of the origin of replication of the IncW group plasmid pSa, *Gene,* 20, 39, 1982.

155. **Farrand, S. K., Kado, C. I., and Ireland, C. R.,** Suppression of tumorigenicity by the IncW R plasmid pSa in *Agrobacterium tumefaciens, Mol. Gen. Genetic.,* 181, 44, 1981.

156. **Leong, S. A., Ditta, G. S., and Helinski, D. R.,** Heme biosynthesis in *Rhizobium:* identification of a cloned gene coding for delta-aminolevulinic acid synthesis from *Rhizobium meliloti, J. Biol. Chem.,* 257, 8724, 1982.

157. **Julliot, J. S. and Boistard, P.,** Use of RP4-prime plasmids constructed *in vitro* to promote a polarized transfer of the chromosome in *Escherichia coli* and *Rhizobium meliloti, Mol. Gen. Genet.,* 173, 289, 1979.

158. **Vincze, E., Koncz, C., and Kondorosi, A.,** Construction *in vitro* of R-prime plasmids and their use for transfer of chromosomal genes and plasmids of *Rhizobium meliloti, Acta Biol. Acad. Sci. Hung.,* 32(3-4), 195, 1981.

159. **Scott, K. F., Hughes, J. E., Gresshoff, P. M., Beringer, J. E., Rolfe, B. G., and Shine, J.,** Molecular cloning of *Rhizobium trifolii* genes involved in symbiotic nitrogen fixation, *J. Mol. Appl. Genet.,* 1, 315, 1982.

160. **Ditta, G., Stanfield, S., Corbin, D., and Helinski, D. R.,** Broad host range DNA cloning system for Gram-negative bacteria. Construction of a gene bank of *Rhizobium meliloti, Proc. Natl. Acad. Sci. U.S.A.,* 77, 7347, 1980.

161. **Corbin, D., Ditta, G., and Helinski, D.,** Clustering of nitrogen fixation *(nif)* genes in *Rhizobium meliloti, J. Bacteriol.,* 149, 221, 1982.

162. **Corbin, D., Barran, L., and Ditta, G.,** Organization and expression of *Rhizobium meliloti* nitrogen fixation genes, *Proc. Natl. Acad. Sci. U.S.A.,* 80, 3005, 1983.

163. **Ruvkun, G. B., Long, S. R., Meade, H. M., Van den Bos, R. C., and Ausubel, F. M.,** ISRml: a *Rhizobium meliloti* insertion sequence which preferentially transposes into nitrogen fixation *(nif)* genes, *J. Mol. Appl. Genet.,* 1, 405, 1982.

164. **Scott, K. F.,** Conserved nodulation genes from the non-legume symbiont *Bradyrhizobium* sp. *(Parasponia), Nucleic Acids Res.,* 14, 2905, 1986.

165. **Bachem, C. W. B., Kondorosi, E., Banfalvi, Z., Horvath, B., Kondorosi, A., and Schell, J.,** Identification and cloning of nodulation genes from the wide host range *Rhizobium* strain MPIK3030, *Mol. Gen. Genet.,* 199, 271, 1985.

166. **Jara, P., Quiviger, B., Laurent, P., and Elmerich, C.,** Isolation and genetic analysis of *Azospirillum brasilense* Nif⁻ mutants, *Can. J. Microbiol.,* 29, 968, 1983.

167. **Friedman, A. M., Long, S. R., Brown, S. E., Buikema, W. J., and Ausubel, F.,** Construction of a broad host range cosmid cloning vector and its use in the genetic analysis of *Rhizobium* mutants, *Gene,* 18, 289, 1982.

168. **Long, S. R., Buikema, W. J., and Ausubel, F. M.,** Cloning of *Rhizobium meliloti* nodulation genes by direct complementation of Nod⁻ mutants, *Nature (London),* 298, 485, 1982.

168a. **Staskawicz, B. J., Dahlbeck, D., and Napoli, C.,** Molecular genetics of race-specific avirulence in *Pseudomonas syringae* pv. *glyeinea,* in *Current Communication in Molecular Biology: Plant Cell/Cell Interactions,* Sussex, I., Ellingboe, A., Crouch, M., and Malmberg, R., Eds., Cold Spring Harbor Laboratory, Cold Spring Harbor, N.Y., 1986, 109.

169. **Downie, J. A., Ma, Q. S., Knight, C. D., Hombrecher, G., and Johnston, A. W. B.,** Cloning of the symbiotic region of *Rhizobium leguminosarum;* the nodulation genes are between the nitrogenase genes and a *nif*A like gene, *EMBO J.,* 2, 947, 1983.

170. **Hombrecher, G., Götz, R., Dibb, N. J., Downie, J. A., Johnston, A. W. B., and Brewin, N. J.,** Cloning and mutagenesis of nodulation genes from *Rhizobium leguminosarum* TOM, a strain with extended host range, *Mol. Gen. Genet.,* 194, 293, 1984.

171. **Putnoky, P. and Kondorosi, A.,** Two gene clusters of *Rhizobium meliloti* code for early essential nodulation functions and a third influences nodulation efficiency, *J. Bacteriol.,* 167, 881, 1986.

172. **Adams, T. H., McClung, C. R., and Chelm, B. K.,** Physical organization of the *Bradyrhizobium japonicum* nitrogenase gene region, *J. Bacteriol.,* 159, 857, 1984.

173. **Cantrell, M. A., Haugland, R. A., and Evans, H. J.,** Construction of a *Rhizobium japonicum* gene bank and use in isolation of a hydrogen uptake gene, *Proc. Natl. Acad. Sci. U.S.A.,* 80, 181, 1983.

174. **Lamb, J. W., Downie, J. A., and Johnston, A. W. B.,** Cloning of the nodulation *(nod)* genes of *Rhizobium plaseoli* and their homology to *R. leguminosarum nod* genes, *Gene,* 4, 235, 1985.

175. **Pawlowski, K., Ratet, P., Schell, J., and de Bruijn, F. J.,** Cloning and characterization of *nif*A and *ntr*C genes of the stem nodulating bacterium ORS751, the nitrogen fixing symbiont of *Sesbania rostrata:* regulation of nitrogen fixation *(nif)* genes in the free-living versus symbiotic state, *Mol. Gen. Genet.,* 206, 207, 1987.

176. **Ditta, G., Schmidhauser, T., Yakobson, E., Lu, P., Liang, Y.-W., Finlay, D. R., Guiney, D., and Helinski, D. R.,** Plasmids related to the broad host range vector pRK290, useful for gene cloning and for monitoring gene expression, *Plasmid,* 13, 149, 1985.

177. **Knauf, V. C. and Nester, E. W.,** Wide host range cloning vectors: a cosmid clone bank of an *Agrobacterium* Ti plasmid, *Plasmid,* 8, 45, 1982.

178. **Jagadish, M. N. and Szalay, A. A.,** Directed transposon Tn5 mutagenesis and complementation in slow-growing, broad-host range cowpea *Rhizobium, Mol. Gen. Genet.,* 196, 290, 1984.

179. **Masterson, R. V. and Atherly, A. G.,** The presence of repeated DNA sequences and a partial restriction map of the pSym of *Rhizobium fredii* USDA193, *Plasmid,* 16, 37, 1986.

180. **Stanley, J., Longtin, D., Madrazak, C., and Verma, D. P. S.,** Genetic locus in *Rhizobium japonicum* (fredii) affecting soybean root nodule differentiation, *J. Bacteriol.,* 166, 628, 1986.

181. **Elmerich, C., Fogher, C., Bozouklian, H., Perroud, B., and Dusha, I.,** Advances in the genetics of *Azospirillum,* in *Nitrogen Fixation Research Progress,* Evans, H. J., Bottomley, P. J., and Newton, W. E., Eds., Martinus Nijhoff, The Hague, 1985, 477.

182. **Bagdasarian, M., Bagdasarian, M. M., Lurz, R., Nordheim, A., Frey, J., and Timmis, K. N.,** Molecular and functional analysis of the broad host range plasmid RSF1010 and construction of vectors for gene cloning in Gram-negative bacteria, in *Drug Resistance in Bacteria*, Mitshuhashi, S., Ed., Japan Scientific Societies Press, Tokyo, 1982, 183.

183. **Kahn, M., Kolter, R., Thomas, C., Figurski, D., Meyer, R., Remaut, E., and Helinski, D. R.,** Plasmid cloning vehicles derived from plasmids ColEl, F, R6K, and RK2, *Methods Enzymol.*, 68, 268, 1979.

184. **Klee, H. J., Gordon, M. P., and Nester, E. W.,** Complementation analysis of *Agrobacterium tumefaciens* Ti plasmid mutations affecting oncogenicity, *J. Bacteriol.*, 150, 327, 1982.

185. **Selvaraj, G. and Iyer, V. N.,** A small mobilizable IncP group plasmid vector packageable into bacteriophage lambda capsids *in vitro*, *Plasmid*, 13, 70, 1985.

186. **Sakaguchi, K.,** Vectors for gene cloning in *Pseudomonas* and their applications, in *Curr. Top. Microbiol. Immunol.*, 96, 3, 1982.

187. **Chistoserdov, A. Y. and Tsygankov, Y. D.,** Broad host range vectors derived from an RSF1010::Tnl plasmid, *Plasmid*, 16, 161, 1986.

188. **Sharpe, G. S.,** Broad host range cloning vectors for Gram-negative bacteria, *Gene*, 29, 93, 1984.

189. **Wood, D. O., Hollinger, M. F., and Tindol, M. B.,** Versatile cloning vectors for *Pseudomonas aeruginosa*, *J. Bacteriol.*, 14, 1448, 1981.

190. **Davison, J. D., Heusterspreute, M., Chevalier, N., Ha-Thi, V., and Brunel, F.,** Vectors with restriction site banks. V. pJRD215, a wide-host-range cosmid vector with multiple cloning sites, *Gene*, 51, 275, 1987.

191. **Bagdasarian, M. and Timmis, K. N.,** Host:vector systems for gene cloning in *Pseudomonas*, *Curr. Top. Microbiol. Immunol.*, 96, 47, 1982.

192. **Bagdasarian, M., Lurz, R., Rückert, B., Franklin, F. C. H., Bagdasarian, M. M., Frey, J., and Timmis, K. N.,** Specific purpose plasmid cloning vectors. II. Broad host range, high copy number, RSF1010-derived vectors and host-vector system for gene cloning in *Pseudomonas*, *Gene*, 6, 237, 1981.

193. **Rossen, L., Johnston, A. W. B., and Downie, J. A.,** DNA sequence of the *Rhizobium leguminosarum* genes nodAB and C required for root hair curling, *Nucleic Acids Res.*, 12, 9497, 1984.

194. **Downie, J. A., Knight, C. D., Johnston, A. W. B., and Rossen, L.,** Identification of genes and gene products involved in the nodulation of peas by *Rhizobium leguminosarum*, *Mol. Gen. Genet.*, 198, 255, 1985.

195. **Kennedy, C. and Robson, R. L.,** Activation of nif gene expression in *Azotobacter* by the nifA gene product of *Klebsiella pneumoniae*, *Nature (London)*, 301, 626, 1983.

196. **Better, M., Ditta, G., and Helinski, D. R.,** Deletion analysis of *Rhizobium meliloti* symbiotic promoters, *EMBO J.*, 4, 2419, 1985.

197. **Rossen, L., Ma, Q.-S., Mudd, E. A., Johnston, A. W. B., and Downie, J. A.,** Identification and DNA sequence of fixZ, and nifB-like gene from *Rhizobium leguminosarum*, *Nucleic Acids Res.*, 12, 7123, 1984.

198. **Frey, J., Bagdasarian, M., Feiss, D., Franklin, F. C. H., and Deshusses, J.,** Stable cosmid vectors that enable the introduction of cloned fragments into a wide range of Gram-negative bacteria, *Gene*, 24, 299, 1983.

199. **Dowling, D. N., Samrey, U., Stanley, J., and Broughton, W. J.,** Cloning of *Rhizobium leguminosarum* genes for competitive nodulation blocking on peas, *J. Bacteriol.*, 169, 1345, 1987.

200. **Fischer, H.-M. and Hennecke, H.,** Linkage map of the *Rhizobium japonicum* nifH and nifDK operons encoding the polypeptides of the nitrogenase enzyme complex, *Mol. Gen. Genet.*, 196, 537, 1984.

201. **Priefer, U. B., Simon, R., and Pühler, A.,** Extension of the host range of *Escherichia coli* vectors by incorporation of RSF1010 replication and mobilization functions, *J. Bacteriol.*, 163, 324, 1985.

202. **Djordjevic, M. A., Innes, R. W., Wijffelman, C. A., Schofield, P. R., and Rolfe, B. G.,** Nodulation of specific legumes is controlled by several distinct loci in *Rhizobium trifolii*, *Plant Mol. Biol.*, 6, 389, 1986.

203. **Ronson, C. W., Nixon, B. T., Albright, L. M., and Ausubel, F. M.,** The *Rhizobium meliloti* ntrA gene is required for diverse metabolic functions, *J. Bacteriol.*, 169, 424, 1987.

204. **Leemans, J., Langenakens, J., DeGreve, H., Deblaere, R., Van Montagu, M., and Schell, J.,** Broad-host-range cloning vectors derived from the W-plasmid Sa, *Gene*, 19, 361, 1982.

205. **Tait, R. C., Close, T. J., Lundquist, R. C., Hagiya, M., Rodriguez, R. L., and Kado, C. I.,** Construction and characterization of a versatile broad host range DNA cloning system for Gram-negative bacteria, *Biotechnology*, 1, 269, 1983.

206. **Close, T. J., Zaitlin, D., and Kado, C. I.,** Design and development of amplifiable broad-host-range cloning vectors: analysis of the vir region of *Agrobacterium tumefaciens* plasmid pTiC58, *Plasmid*, 12, 111, 1984.

207. **Gallie, D. R., Novak, S., and Kado, C. I.,** Novel high- and low-copy stable cosmids for use in *Agrobacterium* and *Rhizobium*, *Plasmid*, 14, 171, 1985.

208. **Rossen, L., Shearman, C. A., Johnston, A. W. B., and Downie, J. A.,** The nodD gene of *Rhizobium leguminosarum* is autoregulatory and in the presence of plant exudate induces the nodA,B,C genes, *EMBO J.*, 4, 3369, 1985.

209. **Johnson, J. A., Wong, W. K. R., and Beatty, J. T.,** Expression of cellulase genes in *Rhodobacter capsulatus* by use of plasmid expression vectors, *J. Bacteriol.,* 167, 604, 1986.

210. **Mermod, N., Ramos, J. L., Lehrbach, P. R., and Timmis, K. N.,** Vector for regulated expression of cloned genes in a wide range of Gram-negative bacteria, *J. Bacteriol.,* 167, 447, 1986.

211. **Egelhoff, R. R. and Long, S. R.,** *Rhizobium meliloti* nodulation genes: identification of *nod*DABC gene products, purification of *nod*A protein, and expression of *nod*A in *Rhizobium meliloti, J. Bacteriol.,* 164, 591, 1985.

212. **Bagdasarian, M. M., Amann, E., Lurz, R., Rückert, B., and Bagdasarian, M.,** Activity of the hybrid *trp-lac (tac)* promoter of *Escherichia coli* in *Pseudomonas putida.* Construction of broad-host-range, controlled-expression vectors, *Gene,* 931, 273, 1983.

213. **Lewin, A., Rosenberg, C., Meyer, H., Wong, C. H., Hirtz, R. -D., Nelson, L., Stanley, J., Manem, J. -F., Dowling, D. N., Denarie, J., and Broughton, W. J.,** Multiple host-specificity regions of the widely compatible *Rhizobium* NGR234 selected using the promiscuous legume *Vigna unguiculata, Plant Mol. Biol.,* 8, 447, 1987.

214. **Leemans, R., Remaut, E., and Fiers, W.,** A broad-host-range expression vector based on the p_L promoter of coliphage λ: regulated synthesis of human interleukin 2 in *Erwinia* and *Serratia* species, *J. Bacteriol.,* 169, 1899, 1987.

215. **Miller, S. S., Ausubel, F. M., and Bogorad, L.,** Cyanobacterial ribonucleic acid polymerase recognize Lambda promoters, *J. Bacteriol.,* 140, 246, 1979.

216. **Friedberg, D. and Seijffers, J.,** Controlled gene expression utilising lambda phage regulatory signals in a cyanobacterium host, *Mol. Gen. Genet.,* 203, 505, 1986.

217. **Santero, E., Luque, F., Medina, J. R., and Tortolero, M.,** Isolation of *ntr*A-like mutants of *Azotobacter vinelandii, J. Bacteriol.,* 166, 541, 1986.

218. **Thomas, C. M., Meyer, R., and Helinski, D. R.,** Regions of broad-host-range plasmid RK2 which are essential for replication and maintenance, *J. Bacteriol.,* 141, 213, 1980.

219. **Kahn, M. L. and Timblin, C. R.,** Gene fusion vehicles for the analysis of gene expression in *Rhizobium meliloti, J. Bacteriol.,* 158, 1070, 1984.

220. **Shearman, C. A., Rossen, L., Johnston, A. W. B., and Downie, J. A.,** The *Rhizobium leguminosarum* nodulation gene *nod*F encodes a polypeptide similar to acyl-carrier protein and is regulated by *nod*D plus a factor in pea root exudate, *EMBO J.,* 5, 647, 1986.

221. **Firmin, J. L., Wilson, K. E., Rossen, L., and Johnston, A. W. B.,** Flavonoid activation of nodulation genes in *Rhizobium* reversed by other compounds present in plants, *Nature (London),* 324, 90, 1986.

222. **Plazinski, J. and Rolfe, B. G.,** Sym plasmid genes of *Rhizobium trifolii* expressed in *Lignobacter* and *Pseudomonas* strains, *J. Bacteriol.,* 162, 1261, 1985.

223. **Tsygankov, Y. D. and Chistoserdov, A. Y.,** Specific-purpose broad-host-range vectors, *Plasmid,* 14, 118, 1985.

224. **Nano, F. E., Shepherd, W. D., Watkins, M. M., Kuhl, S. A., and Kaplan, S.,** Broad-host-range plasmid vector for the *in vitro* construction of transcriptional/translational *lac* fusions, *Gene,* 34, 219, 1984.

225. **Bagdasarian, M., Bagdasarian, M. M., Coleman, S., and Timmis, K. N.,** New vector plasmids for cloning in *Pseudomonas,* in *Plasmids of Medical, Environmental and Commercial Importance,* Timmis, K. N. and Pühler, A., Eds., Elsevier/North-Holland, Amsterdam, 1979, 411.

226. **Zaat, S. A. J., Wijffelman, C. A., Spaink, H. P., Van Brussel, A. N., Okker, R. J. H., and Lugtenberg, B. J. J.,** Induction of the *nod*A promoter of *Rhizobium leguminosarum* sym plasmid pRL1JI by plant flavanones and flavones, *J. Bacteriol.,* 169, 198, 1987.

227. **Schmidhauser, T., Filutowicz, M., and Helinski, D. R.,** Replication of derivatives of the broad host range plasmid RK2 in two distantly related bacteria, *Plasmid,* 9, 325, 1983.

228. **O'Gara, F. and Donnelly, D.,** Stability of vector plasmids in *Rhizobium* spp., in Plasmid Instability, EEC Meeting in Heraklion, Abstract book, October 16 to 18, 1985.

229. **Meyer, R., Laux, R., Boch, G., Hinds, M., Bayly, R., and Shapiro, J. A.,** Broad-host-range IncP-4 plasmid R1162: effects of deletions and insertions on plasmid maintenance and host range, *J. Bacteriol.,* 152, 140, 1982.

230. **Downie, J. A., Hombrecher, G., Ma, Q. S., Knight, C. D., Wells, B., and Johnston, A. W. B.,** Cloned nodulation genes of *Rhizobium leguminosarum* determine host-range specificity, *Mol. Gen. Genet.,* 190, 359, 1983.

231. **Lambert, G. R., Cantrell, M. A., Hanus, F. J., Russell, S. A., Haddad, K. R., and Evans, H. J.,** Intra- and interspecies transfer and expression of *Rhizobium japonicum* hydrogen uptake genes and autotrophic growth capability, *Proc. Natl. Acad. Sci. U.S.A.,* 82, 3232, 1985.

232. **Knight, C. D., Rossen, L., Robertson, J. G., Wells, B., and Downie, J. A.,** Nodulation inhibition by *Rhizobium leguminosarum* multicopy *nod*ABC genes and analysis of early stages of plant infection, *J. Bacteriol.,* 166, 552, 1986.

233. **Borthakur, D., Downie, J. A., Johnston, A. W. B., and Lamb, J. W.,** Psi, a plasmid-linked *Rhizobium phaseoli* gene that inhibits exopolysaccharide production and which is required for symbiotic nitrogen fixation, *Mol. Gen. Genet.,* 200, 278, 1985.

234. **Sutton, B. C., Stanley, J., Zelechowska, M. G., and Verma, D.-P. S.,** Isolation and expression of *Rhizobium japonicum* cloned DNA encoding an early soybean nodulation function, *J. Bacteriol.*, 158, 920, 1984.

235. **Selvaraj, G., Fong, Y. C., and Iyer, V. N.,** A portable DNA sequence carrying the cohesive site *(cos)* of bacteriophage λ and the *mob* (mobilization) region of the broad-host-range plasmid RK2: a module for the construction of new cosmids, *Gene*, 32, 235, 1984.

236. **Aguilar, O. M., Kapp, D., and Pühler, A.,** Characterization of a *Rhizobium meliloti* fixation gene (*fixF*) located near the common nodulation region, *J. Bacteriol.*, 164, 245, 1985.

237. **Grinter, N. J.,** A broad-host-range cloning vector transposable to various replicons, *Gene*, 21, 133, 1983.

238. **Barry, G. F.,** Permanent insertion of foreign genes into the chromosomes of soil bacteria, *Biotechnology*, 4, 446, 1986.

Chapter 3

THE *AZOLLA-ANABAENA* SYMBIOSIS

Jacek Plazinski

TABLE OF CONTENTS

I. INTRODUCTION

The heterosporous aquatic ferns in the genus *Azolla* grow on the surfaces of freshwater ponds, lakes, streams, swamps, and other small bodies of water, or on moist shorelines. The genus was established by Lamarck in 1783,[1] and its members are native to Asia (e.g., China, Vietnam), Africa (e.g., Senegal, Zaire, Sierra Leone), the Americas (e.g., southern South America to Alaska), and the Antipodes.[2] Individual species have been distributed by man and natural means to temperate as well as tropical and subtropical areas.[3]

The most remarkable characteristic of *Azolla* is its symbiotic relationship with the nitrogen-fixing blue-green alga (cyanobacterium), *Anabaena azollae*.[4] The *Azolla* fern provides nutrients including fixed carbon and a protective cavity in each leaf for *Anabaena* colonies and in exchange receives fixed atmospheric nitrogen and possibly other growth-promoting substances.[5,6]

The agronomic importance of *Azolla* is related to its ability to grow very successfully in habitats where little or no combined nitrogen is available. Moreover, through natural senescence and artificial incorporation (green manuring) *Azolla-Anabaena* can contribute a significant amount of combined nitrogen into its habitats. The peoples of Asia have long recognized the benefits of growing *Azolla* for fish feeding, weed control, and, particularly, in rice paddies as an alternative source of nitrogen fertilizer.[4]

The intent of this chapter is to provide a current view on the *Azolla-Anabaena* symbiosis, focusing on recent genetic and molecular analyses of its endosymbiotic partner, *Anabaena azollae*.

II. HISTORICAL OUTLINE

Fossilized *Azolla* specimens have been dated back to around 2 to 65 million years ago.[7] Fossil records show evidence of at least 25 extinct species,[8,9] although Fowler[10] indicated that the number may be as high as 48.

Vietnam and China are the only countries known to have a long history of *Azolla* cultivation. *Azolla* has been used by farmers for centuries in these countries. Unfortunately, much of the information on its management until recently has not been available outside Vietnam and China. *Azolla* was used as a fodder for domesticated animals such as pigs, ducks, and fish, as well as for weed suppression in flooded crops.[3] Expanding the use of *Azolla* began in China and Vietnam in the early 1960s, and by 1980 its cultivation was being practiced in many Asian and African countries.[3] Existing species of *Azolla* are now widely distributed all over the world except in polar and subpolar regions.[2]

III. REPRODUCTION OF *AZOLLA*

A. SEXUAL REPRODUCTION

Azolla reproduces both asexually and sexually. Sexual reproduction of *Azolla* is a most complex and improbable process that presents many fascinating challenges to developmental biologists. The fern is the sporophyte generation. The male and female gametophytes are small, different from the other, and arise in microspores and megaspores, respectively. The two kinds of spore develop in micro- and megasporangia, borne in micro- and megasporocarps. Sporulation is sporadic in nature. Ferns of some species sporulate at regular intervals, while others very rarely, if ever, produce spores. The factors controlling the induction of sporulation are not understood; however, in *Azolla filiculoides* it is associated with mat formation[11] and occurs in summer months in temperate regions.[12]

Sporocarps are subtended on short stalks on the first ventral lobe of the first leaf of a lateral branch and occur in pairs, except in *A. nilotica* in which they occur in tetrads.[13] These pairs may be of the same or opposite sex. Microsporocarps are large and globular relative to the small and ovoid megasporocarps. While immature, each pair is covered by a protective hood[14] or involucre[15] which is formed by a modification of the lamina of the dorsal lobe.[15,16]

Megasporocarps and microsporocarps share the same early stages of development.[17] The ventral leaf lobe initial produces a pair of megasporangial initial cells. The central cell of the young sporangium undergoes many divisions, resulting in a primary sporocyte surrounded by a layer of tapetum. The sporocyte is then cleaved into eight megaspore mother cells. In parallel, the cell walls of the tapetal cells dissolve and their cytoplasms fuse, forming the tapetal syncytium or periplasmodium. The walls of the megaspore mother cells dissolve, and all eight cell nuclei undergo meiosis, producing 32 haploid megaspore nuclei.[2] Only one megaspore nucleus survives to form a megaspore, and the other 31 abort into the syncytium. While the megaspore matures, the surrounding tapetal syncytium gives rise to the megaspore apparatus. The periplasmodium becomes highly vacuolate and divides into four (section *Euazolla*) or ten (section *Rhizosperma*) parts. The part immediately surrounding the developing megaspore forms the gula, girdle, and perine, all parts of the megaspore apparatus. The perine eventually becomes very elaborate and may be decorated with filamentous processes referred to as the capture mechanism[18] or filosum.[19] The other parts of the periplasmodium form "Schwimmapparat",[17] or floats. The number and arrangement of floats on the apparatus have been very useful along with perine characteristics in establishing taxonomic relationships among extant and fossil species of *Azolla*.[20]

If all 32 megaspore nuclei abort, the megasporangium senesces and numerous stalked microsporangia arise from initials at the base of the columella.[21] Development in each microsporangium produces 32 to 64 microspore nuclei in a periplasmodium formed by the tapetum much as in megasporangia. The periplasmodium divides into three or more parts which form the pseudocellular masses called massulae.[1] Microspores are segregated equally into the massulae and come to occupy alveolate near their surfaces. In all extant species except *A. nilotica* the surface of these alveolate acellular structures bears special elongate barbed processes called glochidia. Species in section Rhizosperma either lack glochidia (for instance, *A. nilotica*) or their glochidia lack the barbed tips (*A. pinnata*) which are typical of all species in section Euazolla.

Fertilization takes place after the megasporocarps and massulae fall to the bottom of the pond, where, through the agency of water movements, the glochidia anchor the massulae to megaspore perine entanglements.[22] Afterward, microspores germinate and release spermatozoids which escape through the gelatinized massula to fertilize the egg cell, which has formed in the female gametophyte, still contained within the megasporocarp. Figure 1 represents in a simplifed form the life cycle of *Azolla*.

The embryo develops root and foot initials from the lower product of the first division of the zygote, and shoot and cotyledon initials from the upper cell. As the cotyledon and first leaf emerge, the sporeling floats to the surface.[23] The cotyledon lacks a cavity for the symbiont, but the dorsal leaf lobes and shoot apex entrap *Anabaena* short filaments (hormogonia) that were carried throughout megasporogenesis and retained under the indusium cap of the megasporocarp. *Anabaena* hormogonia persist at the shoot apex and enter each successive leaf cavity, where their development is geared to that of the leaf (see later).

B. VEGETATIVE REPRODUCTION

Vegetative reproduction of *Azolla* is an extremely efficient process. The sporophyte of the fern reproduces by fragmentation via an abscission layer that forms at the base of each branch. Since roots develop in fixed sequence along with leaves on each branch, the separating fragments are autonomous. They then drift away from the parent plant and become independent. The plant is almost fully grown in 15 to 20 d.[3] Under favorable conditions populations of some species can double in weight in 3 to 5 d.

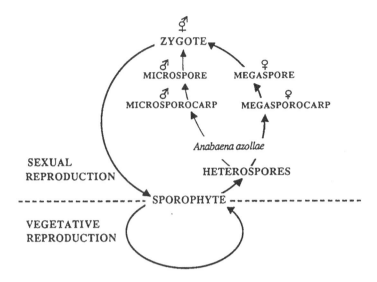

FIGURE 1. Schematic illustration of the *Azolla* life cycle showing the continuation of the association with *Anabaena azollae* (modified from Ashton and Walmsley[6]).

IV. TAXONOMY OF *AZOLLA*

The genus *Azolla*, a conjugation of two Greek words *azo* (to dry) and *allyo* (to kill), was established by Lamarck in 1783 with the description of *Azolla filiculoides*. In 1810, Robert Brown put the genus on a firmer scientific footing by presenting illustrations of reproductive structures in his descriptions of several new species, including *Azolla pinnata* and *A. rubra*.

In 1944 Svenson[1] published a critical account of the New World species, relying primarily on features of reproductive structures for species delineation.

Azolla is an unusual fern, easily set apart from other ferns by several features. Ferns are regarded as true land plants, but *Azolla* is one of the few water ferns. Most ferns produce only one kind of spore (homosporous), but *Azolla* produces two kinds (heterosporous). Based on these characteristics *Azolla* has been grouped with another heterosporous water fern, *Salvinia*, into the order Salviniales, but in its reproductive structures *Azolla* differs sufficiently from *Salvinia* to be placed into the monotypic family Azollaceae.[10]

All recent studies agree that features of sporulating structures are the most useful characters for species recognition in *Azolla*.[19,20,24] On the basis of the structure of the megaspore apparatus and the nature of the glochidia of the massulae, over 50 fossil and extant species have been described and organized into some seven subgeneric sections.[25] Most taxonomic accounts have relied on float number and arrangement together with sporoderm zonation and ornamentation for species definition.

Perkins et al.[26] found that scanning electron microscopy reveals distinctive perine organizations which are constant within taxa and consistently different between taxa. Martin[20] suggests that other features such as the nature of the collar, the column, float retention mechanism, and distribution and type of leaf hairs (see Figure 2) might also be important taxonomic characteristics. The presence or location of glochidia on the massulae can also be a distinguishing characteristic.[15] The species belonging to *Euazolla* have glochidia positioned on the total surface, while those of *Rhizosperma* are located on the inner surface of *A. pinnata* and are absent in *A. nilotica*.[15]

There have also been a few reports suggesting that the chromosome number observed during meiosis might be a useful taxonomic criterion for distinguishing *Azolla* species.[27,28]

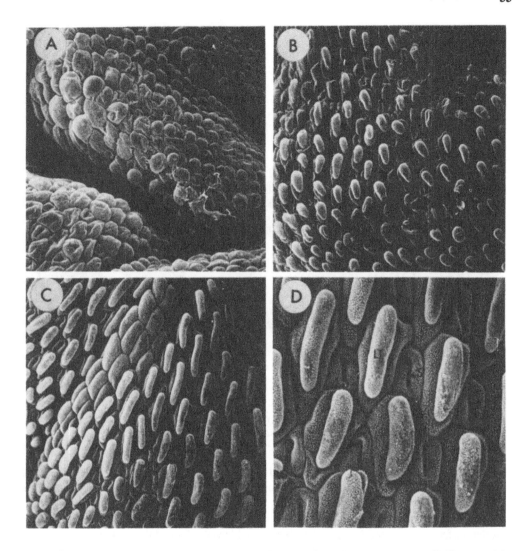

FIGURE 2. Scanning electron micrographs of the *Azolla* leaf surface. Pictures show structural differences of the epidermal trichomes (hairs). Such differences can be used as taxonomic criteria. (A) Micrograph of the *A. filiculoides* leaf surface (magnification × 235); (B) leaf trichomes of *A. mexicana* (magnification × 135); (C) leaf trichomes of *A. pinnata;* (D) close up of (C) showing epidermis of a leaf with trichomes and stomata. LT = leaf trichome, S = stomata.

Most present taxonomic treatments recognize seven distinct species of *Azolla* on the basis of reproductive and morphological features. They are grouped into two sections:[3] section *Euazolla* includes *Azolla caroliniana* Willdenow (1810), *A. filiculoides* Lamarck (1783), *A. mexicana* Presl (1845), *A. microphylla* Kalfuss (1824), *A. rubra* R. Brown (1810); and section *Rhizosperma* includes *A. pinnata* R. Brown (1810) and *A. nilotica* De Caisne (1867). Figure 3 shows examples of some *Azolla* species.

V. MORPHOLOGICAL ASPECTS

A. PLANT HOST

Azolla plants are triangular or polygonal in shape and bear deeply bilobed leaves and adventitious roots. Fronds float on the water surface individually or in mats. In bulk they give the appearance of a dark-green to reddish carpet. Each of the small, alternately arranged

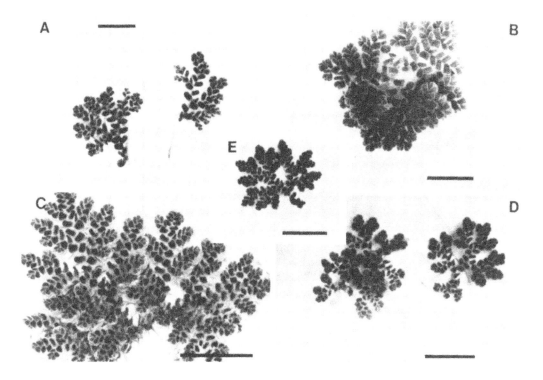

FIGURE 3.   General appearance of *Azolla* fronds. (A) *A. mexicana;* (B) *A. pinnata* (China); (C) *A. filiculoides (China); (D) A. microphylla;* and (E) *A. caroliniana.* Bars = 1 cm.

cuticulate leaves which cover the stem and branches is composed of a thick aerial dorsal and a thin floating ventral lobe.[2] The papillose dorsal lobes are chlorophyllous except in the colorless margin and contain the symbiotic *Anabaena azollae* within an ovoid cavity connected to the atmosphere by a pore (Figure 4). The translucent ventral lobes resting on the water surface support the frond and are nearly achlorophyllous. Adventitious roots hang in the water or, when in shallow water, may penetrate into mud. The roots develop in an acropetal fashion at branch points along the ventral surface of the stem and exhibit a root cap and numerous root hairs.[29] Root development appears to be at least indirectly controlled by the symbiont since *Anabaena*-free plants often exhibit increased root production.[30]

As the leaf primordium differentiates at the growing point, a slight depression is formed near the base on the adaxial side of the dorsal lobe (Figure 5). A portion of the *Anabaena* colony at the shoot apex above the dorsal lobe primordia is scooped into the enlarging depression by a glove-shaped transfer hair that differentiates very close to the apical cell of the stem.[31] Multicellular hairs of two or more cells emerge from epidermal cells within the depression as soon as the cavity begins to form.[18] During maturation of the cavity the hairs undergo a change in ultrastructure, indicating a change in function. The hairs demonstrate the transfer cell morphology of a cell wall labyrinth and dense cytoplasm containing abundant endoplasmic reticulum and numerous mitochondria.[18,32] In the cavities the transfer hairs are composed of several branched or unbranched cells. The hair originating from the axil of a leaf primordium on the apical meristem is termed the primary branched hair (PBH).[31] As a leaf cavity is closing, additional epidermal hairs are formed at the back of the cavity depression and are termed the secondary branched hairs (SBH). All the other hairs are comprised of only two cells — a stalk cell and a terminal cell — and are termed simple

57

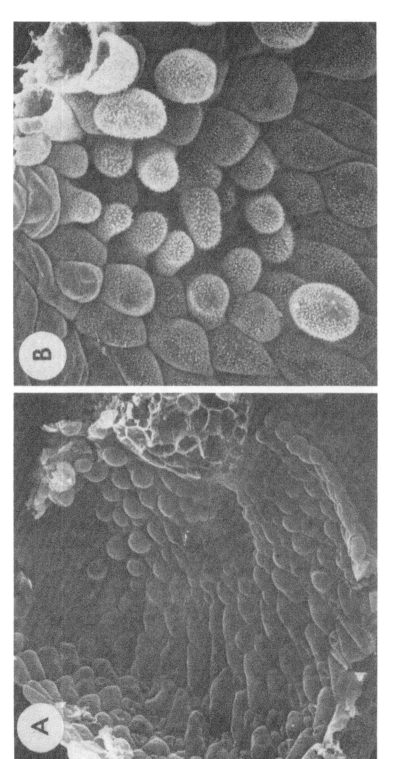

FIGURE 4.   SEM micrographs showing the adaxial face of the dorsal leaf lobe of *A. pinnata*. (A) The cavity in this young leaf has not yet closed, and some filaments of *Anabaena* can be seen through a large pore (Magnification x 450); (B) in a mature leaf the pore is occluded by proliferous outgrowths from the margins of the pore (Magnification x 950).

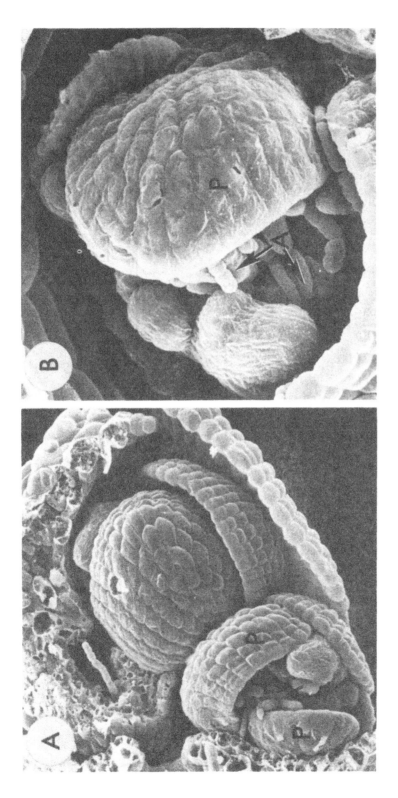

FIGURE 5.   Micrographs of the *A. pinnata* stem apex. (A) Cyanobacterial filaments are entrapped between leaf primordia (P) where they divide (Magnification x 450); (B) close up of (A) showing *Anabaena* filaments packed in the stem apex. At this stage only vegetative cells can be seen (Magnification x 670). P = leaf primordia; A = *Anabaena* filaments.

hairs.[31] Transfer cells are believed to be involved in the exchange of metabolites and products of nitrogen fixation[2] (e.g., glutamate dehydrogenase activity is associated with the hairs),[33] and, in general, are secretory and/or absorptive in function. Meristematic activity at the margin of the cavity pore, followed by hair development, effectively closes the cavity and prevents the contained *Anabaena* from escaping, and free-living algae and cyanobacteria from entering (Figure 4 A and B).

B. *ANABAENA AZOLLAE*, THE ENDOPHYTE

1. Taxonomy

Anabaena azollae Strasburger is the only species mentioned in symbiotic association with *Azolla*. Fjerdingstad[34] claimed that the alga is actually an ecoform of *Anabaena variabilis* and should, therefore, be called *A. variabilis* status *azollae*. Taxonomists place *A. azollae* within the Cyanobacteria, order Nostocales, family Nostocaceae.[35] The evidence that there are, in fact, different strains of *A. azollae* is very limited, since the isolates obtained from different fern species are morphologically similar.[2,36] In a recent serological survey using polyclonal antibodies the similarities among 32 isolates far outweighed the differences.[37] Qualitative differences were recently demonstrated in antigenic structures between fresh and cultured isolates of *A. azollae*.[38,39] From these reports, two conclusions could be drawn: (1) use of polyclonal antibodies can differentiate the *A. azollae* of *Euazolla* from that of *Rhizosperma*, and (2) cultured isolates of *A. azollae* have different antigenic properties from those isolated freshly from the fern.

Recently, in the People's Republic of China, 13 hybridoma cell lines secreting monoclonal antibody against *A. azollae* isolates have been established.[40] Among the 13 monoclonal antibodies (MAbs), three groups were established: (1) a subgroup-specific MAb (reacting only with *A. azollae* from Euazolla), (2) a species-specific MAb (reacting only with all *A. azollae* from *Azolla pinnata*), and (3) an *A. azollae*-specific MAb (reacting with all *A. azollae* representing seven species of *Azolla*). None of the MAbs reacted with two free-living N_2-fixing cyanobacteria, *Anabaena azotica* and *Tolypothrix*. The authors concluded that there are at least four subgroups of *A. azollae* in *Azolla* species. This was the first report showing differences between cyanobacterial isolates from different species of *Azolla*. New data on the taxonomic status of *A. azollae* obtained by means of molecular techniques are presented in a later section of this chapter.

2. Properties of the *Anabaena* Cells

Anabaena azollae grows as a filament or trichome (thread) of cells. Growth leads to elongation of the trichome and to an increase, through intercalary cell division, in the number of its constituent cells. Reproduction occurs by breakage of the trichome into shorter chains of cells called hormogonia. The trichome is unbranched and uniseriate (one cell in width), since intercalary cell division is confined to a plane at right angles to its long axis.[35] There are three types of cells: vegetative cells (primary site of photosynthesis), heterocysts (site of N_2-fixation),[41] and akinetes (thick-walled resting spores formed from vegetative cells).

In the vegetative cell the protoplast is always surrounded by a bilayered wall. The cell membrane (plasmalemma) is usually simple in structure.[35] Much of the protoplast is occupied by the components of the photosynthetic apparatus, which is located in a series of membranous sacs or thylakoids (the sites of the lipophilic pigments such as chlorophyll *a* and carotenoids as well as centers of the photochemical reaction).[41] The cyanobacterial phososynthetic apparatus includes phycobiliproteins as major light-harvesting pigments (as granules on the cytoplasmic face of the thylakoids). *Anabaena*, like other cyanobacteria, synthesizes glycogen as a principal nonnitrogenous organic reserve material.[41] Cyanobacteria also synthesize cyanophycin, a highly branched polypeptide, composed of only two amino acids; a polyaspartic acid core bearing arginyl residues attached to all free carboxyl groups of the core.[42]

Besides their characteristic shape and size, heterocysts can readily be distinguished from the vegetative cells by three structural properties: (1) they are weakly pigmented; (2) they have a thick outer envelope; and (3) near each junction with a vegetative cell they contain a polar granule.[35] The ability to form heterocysts is associated with a physiological property, the capacity of nitrogen fixation under aerobic conditions.[43] The heterocyst envelope is composed of three layers and a laminated inner layer.[44] As the two inner layers are formed, the initially rounded cell pole in contact with the adjacent vegetative cell changes into a short, narrow tubular neck around which the two inner envelope layers are thickened. This modification reduces the contact with the adjacent vegetative cell to a small flattened area at the end of the neck, in which microplasmodesmata develop.[44]

The heterocysts do not contain the glycolipids that are the major neutral lipids of the membranes of vegetative cells.[45] Heterocyst differentiation is subject to inductive control; however, not every cell in the trichome is competent to give rise to heterocysts.[46] A mature heterocyst is a terminal state, i.e., it has lost the ability to de-differentiate.

Differentiation of akinetes usually occurs adjacent to heterocysts and is accompanied by the formation of thick, often pigmented envelopes around the walls of vegetative cells.[47] The cells, as a rule, enlarge considerably and may become considerably longer and wider than vegetative cells.[47]

C. MORPHOLOGY OF THE SYMBIOSIS

The endophyte is protected from the natural aquatic and atmospheric environments in the dorsal leaf lobe of the *Azolla*. Mineral nutrients, carbon sources, and water must pass through tissues of the host plant to reach the *A. azollae* cells. Even the light energy incident on the endophyte has been filtered by the fern leaf tissue . The endophyte thus receives physical protection in a highly specialized environment where required minerals, other nutrients, and adequate moisture are provided (Figure 6).

A colony of cyanobacteria is always present on the apical meristem, in the dorsal lobes (see Figure 5), and under the developing indusium (cap) of both the microsporocarps and megasporocarps.[48] *A. azollae* persists in the mature megasporocarps, and infects sporelings, unlike the colonies entrapped within the microsporocarps. The role of akinetes is difficult to assess. When an akinete germinates, its contents divide and give rise to a hormogonium. The spore membrane becomes mucilaginous, swells, and then ruptures releasing its contents.[48,49] Whether *A. azollae* can persist in a free-living state after akinetes are released when *Azolla* plants decay is not clear, nor whether infection of plants can occur under field conditions.

Anabaena in the cavity develops in synchrony with cavity formation in the *Azolla* leaves. The filaments are undifferentiated in the plant apex, but their cells divide as leaf primordia develop.[50] The depression in the adaxial surface becomes occupied by *Anabaena* from the apical colony, and after cavities have been formed the hair cells develop.[50,51] Filaments of the apical colony are unable to fix nitrogen, but those isolated into leaf cavities progressively differentiate heterocysts and begin to fix nitrogen.[52,53] The frequency of the heterocysts increases from zero at the stem apex to as high as 33% in the 15th leaf.[52] Recently, Kaplan et al.[54] have presented information on the total phycobiliprotein content (found also in heterocysts) as a function of leaf cavity and endophyte age. They have also determined adsorption and fluorescence spectra of the phycobiliproteins in individual vegetative cells and heterocysts from leaf cavities of different developmental stages in *A. caroliniana* and *A. pinnata*. Nitrogen fixation by the microsymbiont is highest in mature leaves and declines as the leaves begin senescence.

Peters et al.[50] showed that the contents of the cavity are enclosed by a thin membrane and can be released when *Azolla* leaves are treated with cellulolytic enzymes. The structure released in this way was called an "algal packet". Algal packets contain *Anabaena* filaments,

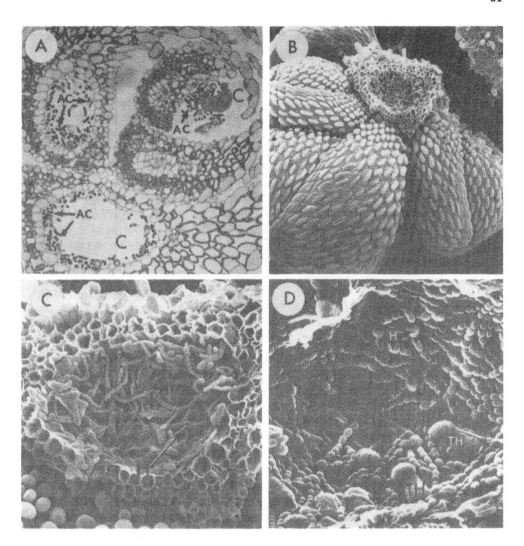

FIGURE 6.   (A) Longitudinal section through the dorsal lobes of successive leaves (parallel to the water surface) of *A. filiculoides*. *Anabaena azollae* cells are present inside cavities (magnification × 168); (B) SEM micrograph of the outer leaf surface and section through a cavity of *A. mexicana* (magnification × 120); (C) close up of (B); tightly packed filaments of *A. azollae* within cavity (Magnification × 235); (D) internal part of mature cavity of *A. pinnata*. Simple transfer hairs are already developed. Many of the *Anabaena* cells have differentiated into heterocysts (magnification × 600). C = cavity, AC = *Anabaena* cells, TH = simple transfer cells, H = heterocyst.

hair cells, and an envelope membrane. The authors also noted that algal packets could not be obtained from *Anabaena*-free *Azolla;* however, Uheda[55] has succeeded in isolation of empty packets from *Anabaena*-free *Azolla pinnata* var. *imbricata* by means of hydrolytic enzymes. This result suggests that the envelope is of plant origin and that the existence of *Anabaena* filaments does not affect its structural integrity. Moreover, there were no differences in structural features[18] and developmental changes of branched and unbranched hair distribution[31] between *Anabaena*-free and *Anabaena*-containing *Azolla* (Figure 7). Thus, the algal packet appears to be a constitutive structure of *Azolla* which has evolved to accommodate its endophyte, whether the latter is present or not.

Robins et al.[56] have observed that the surface of cells of *A. azollae* (isolated from *Azolla filiculoides*) was covered in a layer of hydrated mucilage. This mucilage was observed in the *A. azollae* cells adhering to plant cells in the cavity of the frond. The authors proposed

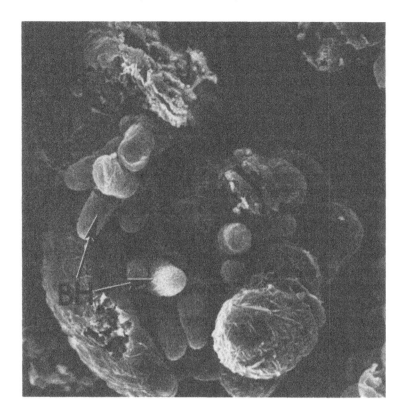

FIGURE 7.   SEM micrograph of a shoot apex of *Anabaena*-free *Azolla pinnata*. Despite the absence of *Anabaena* filaments, glove-shaped, branched transfer hairs are still being formed very close to the apical cell (magnification × 930). BH = branched transfer hairs.

that the cyanobacterial cells were held attached to the surface on which they were growing by the layer of hydrated mucilage which adhered to both the cell surfaces and a variety of support matrices. The mucilage, in the authors' opinion, allows a firmer, long-term attachment of the endosymbiont. Since the mucilage production occurred on inert surfaces, it seemed to be produced by the endophyte, and most probably the adhesion did not depend on specific interactions involving lectins as previously proposed.[57,58]

Little is known about the control of symbiotic development, but it is reasonable to assume that the process is coordinated by a sequence of interorganism communications. Certainly, the exchange of fixed carbon and fixed nitrogen has been demonstrated between *Azolla* and *Anabaena*.[33,53] The host receives and utilizes nitrogen fixed by its endophyte, and the *Anabaena* receives fixed carbon from *Azolla*.[53]

Several workers have speculated that special epidermal trichomes may be involved in metabolite exchange between *Azolla* and *Anabaena*.[18,50] Anatomic studies have proved that two types of cavity trichomes (simple and branched hairs) become intimately associated with each cavity's *Anabaena* filaments.[31,59] Every cavity contains only two branched hairs. They are associated with the *Anabaena* from the onset of cavity formation.[18] Simple hairs appear later but increase in number (about 20 to 25 per cavity) as the cavity matures.[31]

Calvert et al.[60] have recently suggested that during the symbiotic interaction branched hairs are involved in exchange of fixed nitrogen between the symbionts, while simple hairs may participate in exchange of fixed carbon from *Azolla* to *Anabaena*.

VI. PHYSIOLOGICAL ASPECTS

A. PHYSIOLOGICAL ECOLOGY

The physiology of *Azolla* has been studied extensively by numerous workers (for reviews see References 2, 4, 30, and 59) in the last decade. Experiments utilizing constant environment chambers have revealed a number of interesting features concerning the growth requirements and nutrition of *Azolla* ferns. Because the fern floats on water, it can be influenced by physical and chemical factors from both the air and water phases. In general, however, the fern behaves in much the same way as most green plants in that it has optimal temperature, light, pH, salt concentration, and nutrient requirements. *Azolla*, as a plant which reproduces vegetatively through fragmentation, has potential for maintaining an exponential growth rate under optimal conditions.

The geographic distribution of *Azolla* clearly indicates that the genus is adapted to very wide-ranging temperature conditions. Generally, the optimum is between 20 and 30°C. Upper and lower limits vary with the species and strains. Some strains will tolerate frost when others will grow when the temperature exceeds 45°C (one Australian isolate of *A. pinnata* is known to survive the temperature of 55°C[61]).[2,6,62]

Peters et al.[63] have shown that *A. mexicana* is the strain most tolerant to elevated temperature whereas *A. filiculoides* is the least tolerant one. Other workers[62] have found that the growth rate of *Azolla* slows down severely at high temperature, yielding greater biomass at 22°C than at 33°C in *A. pinnata*, *A. filiculoides*, and *A. mexicana* strains. They also showed that *A. filiculoides* requires lower temperature than other species for its growth. *A. pinnata* was found to have the best tolerance to high temperature of all species. Talley et al.[11] reported that *A. filiculoides* could withstand temperatures as low as −5°C without apparent harm. Ashton[12] concluded that cold tolerance increases with pH and is highest in the pH range of 8 to 10.

Azolla is largely tolerant to pH variations; it survives in a range of pH from 3.5 to 10 and grows well from 4.5 to 8.[12,64] Ashton[12] found that relative growth rate is influenced by a direct relationship between light intensity and pH; high light intensity (60 klx) with high pH (9 to 10) and low light intensity (15 klx) with low pH (5 to 6) allowed maximum relative growth rates. In contrast, Peters et al.[63] did not observe any differences in the temperature optima at pH 6 and 8 as a function of light intensity. Ashton[12] and Watanabe[65] reported that nitrogen fixation decreased at neutral pH.

Azolla is very sensitive to drought and dies in a few hours if its substratum becomes dry. Growth is promoted by a fairly shallow depth of water (~5 cm) in which there can be little turbulence. Such a situation favors mineral nutrition, since roots are near the soil, and reduces the negative effect on growth of water turbulence; on the other hand, it does not allow rooting in the soil, which in effect creates a premature state of overpopulation.

The growth of *Azolla* saturates at approximately 25 to 50% of full sunlight (40 to 57.5 klx).[12,66] Also, nitrogenase activity declines more rapidly when light intensity increases in comparison to when light intensity decreases. Peters et al.[67] reported that CO_2 fixation saturated at 8000 lx and nitrogen fixation saturated at 5000 lx in *A. caroliniana*. Lu et al.[68] reported that 47 klx resulted in the highest rates of growth and nitrogen fixation for *A. pinnata*. They also reported that *A. pinnata* can survive in the range from 3.5 to 120 klx.

Salinity has rarely been mentioned as a problem in the cultivation of *Azolla*; however, it is a factor which should be considered wherever the introduction of *Azolla* is being considered. Haller et al.[69] reported that the growth of *A. caroliniana* ceases in a solution containing 1.3% salt (33% of seawater). Ge Shi-an et al.[70] found that an increase in salt concentration from 0.3 to 0.9% caused a decrease in N percentage and frond size, and an increase in root number.

Like most plants, *Azolla* is sensitive to changes and deficiencies in the supply of plant nutrients. For optimal growth, the fern requires all the macro- and micronutrients which are essential for normal plant growth.[4] Macronutrients such as phosphorus (the most common factor limiting the growth of *Azolla*),[2,71] nitrogen, potassium,[72] calcium,[71] and magnesium[72] are especially important and produce marked effects on growth of the fern if present in too high or too low concentrations. Since nitrogen fixation by the cyanobiont also plays a dominant role in regulating the growth of the fern, micronutrients such as iron,[73] cobalt,[74] and molybdenum,[75] which have been shown to be essential for the nitrogen-fixing process,[4] must also be present in adequate supply in waters where the fern grows.

B. NITROGEN FIXATION

Although *Azolla* also extracts nitrogen from its aquatic environment, the cyanobiont is capable of meeting the entire nitrogen requirement of the association. The nitrogen-fixation capacity of the *Azolla-Anabaena* association has been demonstrated indirectly (acetylene reduction-gas chromatography and assay of H_2 production) and directly (by use of $^{15}N_2$).[4] In comparison to the legume-*Rhizobium* symbiosis, nitrogenase activity by *Anabaena* does not cease when the association is grown for long periods with adequate or excessive amounts of combined N in the medium.[50] Okoronkwo and Van Hove[76] have studied the *Azolla-Anabaena* nitrogenase activity in the presence of combined nitrogen and conclude that in comparison to other symbioses, the *Azolla-Anabaena* relationship is tolerant to repression by combined N. The results implied that nitrogenase in *Anabaena azollae* is not inactivated, but only diluted out, in the presence of NH_4^+ nitrogen.

It has been shown that *Anabaena,* immediately after isolation from *Azolla,* fixes nitrogen at a rate at least equal to free-living cyanobacteria.[77,78] The conclusion was drawn that ammonium was released by the symbiont and assimilated by the host tissue in the association.

Recently, Meeks et al.[79] have confirmed that symbiotic *Anabaena* releases a substantial portion of its newly fixed nitrogen as ammonium. The authors showed that the nitrogen assimilation is conducted by the glutamine synthetase-glutamate synthase (GS/GOGAT) pathway with little or no synthesis of glutamate by glutamate dehydrogenase (GDH). The little biosynthetic activity of GDH in the isolated *Anabaena* indicated that all of GDH activity is associated with the hair cells of the cavity. This suggestion has been confirmed by other workers.[33,80] It was also shown that N_2-fixation and NH_4^+ assimilation were not tightly coupled as they are in free-living cyanobacteria.[81,82] The release of ammonium supports the idea that GS, the initial enzyme of the primary ammonium assimilatory pathway, is repressed in symbiotic *Anabaena.*[83,84]

Recently, Yoneyama et al.[85] have shown that *Anabaena* has higher ^{15}N abundance than the host *Azolla* plants, and that the contribution of N_2-fixation to the N nutrition of *Azolla* is as high as 99%.

Azolla and its symbiotic *Anabaena* are both photosynthetic organisms and their pigmentation is complementary. The association and individual partners exhibit Calvin cycle intermediates of photosynthetic CO_2 fixation. Sucrose is a major fixation product in the *Azolla,* but it is not detectable as a labeled product in the isolated endophyte.[86] The *Anabaena* endosymbionts contribute as little as 2% of the association's total photosynthetic capability; however, it is known that there is a cross-feeding of fixed carbon from *Azolla* to the *Anabaena.*[86]

Enhancement of nitrogenase activity by sugar supply has been demonstrated in the cultured cyanobiont isolated from *A. caroliniana.*[87] Relatively low concentration of fructose (up to 1 m*M*) and higher concentration of sucrose (up to 20 m*M*) were found to support nitrogenase activity.[88] In contrast, Rozen et al.[89] have found that the *A. azollae* isolated from *A. filiculoides* takes up and utilizes fructose in the light for mixotrophic growth. Fructose was favored by the endophyte as a substrate over sucrose and glucose. The *A. azollae* cells grown in the presence of 8 m*M* fructose accumulated glycogen within 2 to 3 d followed by reduction of glycogen content during the fourth day. In fructose-grown cells the glucose-6-phosphate dehydrogenase (G6PDH) activity was increased five- to sixfold, the frequency of heterocysts was increased from 6 to 18%, and the acetylene reduction activity was increased threefold as compared with control cells. The authors suggested a link between sugar uptake, the turnover of its stored derivative glycogen, and the events leading to heterocyst differentiation. Probably sudden changes in the C/N ratio and/or presence of specific metabolites of carbohydrate metabolism trigger the differentiation.[89] Peters[86] postulated that there is a transition from photoautotrophic metabolism in generative *Anabaena* filaments to a photoheterotrophic or mixotrophic mode of metabolism with increasing differentiation of heterocysts and an exogenous carbon source, sucrose being utilized to maintain levels of reducing power.

Watanabe[65] has established the potential capacity of *Azolla* nitrogen fixation in the field as 1.1 kg N per ha per day — one of very many valuable studies of N_2-fixation in field conditions which cannot be reviewed here because of space limitations.

The close relationship between photosynthesis and nitrogen fixation in *Azolla-Anabaena* symbiosis is well documented.[78,90] Photosynthesis is the ultimate source of ATP, reductant, and carbon skeletons used in N_2 fixation. Dark, aerobic nitrogenase-catalyzed reduction of substrates is dependent upon endogenous reserves of photosynthate, and dark rates are always less than one half of those obtained in the light.[86] A number of studies strongly indicate that reductant from endogenous reserves of photosynthate and photophosphorylation are the primary driving forces for nitrogenase activity in the light; and that dark, respiratory-driven activities may be ATP limited.[78,90,91] Interactions between photosynthesis and nitrogenase activity have been demonstrated by comparing action spectra for nitrogenase catalyzed C_2H_2 reduction in the *A. caroliniana-Anabaena* association and isolated endophyte.[92]

It has been well documented that the *Azolla-Anabaena* system has a nitrogenase-catalyzed hydrogen evolution activity.[30,67] Ruschel et al.[93] concluded that there is a certain independence of hydrogenase from nitrogenase or photosynthetic activity in the symbiosis and that the hydrogen evolved through nitrogenase is recycled. They postulated that the *Anabaena azollae* may use three sources of energy: CO_2 (from photosynthesis), H_2 from air, and H_2 evolved via nitrogenase. This versatility may account for the effectiveness of N_2 fixation in the *Azolla-Anabaena* system.

VII. NEW COMBINATIONS OF *AZOLLA-ANABAENA*

In studies of symbioses between *Rhizobium* and legumes, it is possible to separate the partners and recombine them in new associations and to engineer new genotypes. By contrast, research on *Azolla-Anabaena* combinations and recombinations lags behind. All of the extant strains presumably represent the outcome of natural selection operating on naturally occurring mutations in either the host or the symbiont, but two major problems have so far prevented substantial progress from being made in attempts to make new combinations in the laboratory. The first problem is that the two partners normally remain together throughout both vegetative and sexual reproduction. The current diversity of strains may, therefore, have been generated by natural selection following mutation *in situ*. Artificial recombination requires: (1) preparation of *Anabaena*-free *Azolla* for use as recipients (this is relatively easy) and (2) isolation of *Anabaena* cells, followed by reinsertion of algal filaments into the compartment around the stem apex of the fern.

The second problem is taxonomic. The seven *Azolla* species are fairly easy to identify, but as already stated, very little is known about the taxonomy of *Anabaena*. This is a fundamental gap in knowledge. Thus, when recombination procedures are attempted, there is no unambiguous means of checking that new associations really have been established and that they are subsequently maintained in field conditions. If the contribution of *Anabaena* to strain variation is to be studied, a reliable taxonomic system must be devised.

The difficulty of culturing *A. azollae* in isolation has been a major obstacle in elucidating taxonomic aspects of the symbiosis. Numerous authors[6,87,94-96] claimed to have grown *A. azollae* in isolation, but these reports were disputed by others who were unable to culture the symbiotic alga in isolation.[52,75,90] Several attempts to recombine isolated *A. azollae* with alga-free *Azolla* were reported to be unsuccessful.[75,97]

In 1984 Liu[98] reported successful recombination of algae-free *Azolla* (obtained by culturing the tip of the frond) and freshly isolated *A. azollae*. The frequency of recombination was 43%, and the introduced cyanobiont showed nitrogen-fixing capacity. This result was controversial, but Liu and co-workers have now developed a method which gives consistent recombinants.[99] Laboratory-grown *A. azollae* was introduced into germinating sporocarps obtained from *Anabaena*-free *Azolla*, or sporocarps from which *Anabaena* had been removed. Because these results represent a major breakthrough, they deserve more detailed explanation. In one set of experiments, *Anabaena* cells were eliminated from megaspores before germination of the sporocarps. The authors observed that the removal of the indusium and apical membrane of megaspore apparatus could eliminate symbiotic *Anabaena* and give rise to a new generation of *Anabaena*-free *Azolla* fern (unable to grow in N-free medium). Apical membrane and indusium were removed from the megaspore by cutting horizontally at the top of the float. The decapitated megasporocarps were placed vertically on a sloped filter in petri dishes and left to germinate. During embryogeny, indusia from other fern megasporocarps, containing *Anabaena* filaments on the inner surface, could be grafted onto the original position of the megaspore apparatus of the same or another species of fern (*A. microphylla* to *A. filiculoides* and vice versa). In about 50% of cases, *Anabaena* cells from the grafted indusia were successfully transferred to the surface of the pericolumellar area. During the first few stages of fern development *Anabaena* cells migrated into the shoot apex region, probably by crossing the cotyledon and the first leaf. Some *Anabaena* cells entered into the leaf cavities and others stayed at the apical site or were attached to the hair cells. About 10% of the seedlings could grow in N-free solution and showed acetylene reduction activity. Also, two lines of laboratory-grown *A. azollae* isolated by Newton and Herman[87] were inoculated through a capillary tube onto the apical portion of the decapitated megaspore apparatus at its embryogenic stage. The authors observed that some of the *Anabaena* entered cavities whereas others stayed outside the cavities forming filamentous connections from one leaf to another. The resultant *Azolla* failed to grow on N-free medium, confirming other work indicating that Newton's isolates are not, in fact, *A. azollae*.[38,39]

In order to confirm the identity of the newly established symbiotic *Anabaena*, the monoclonal antibodies described earlier were used. Using fluorescent antibodies, the authors proved that they produced new combinations of *Azolla* and *Anabaena* and that effective symbioses could be established. In general, these findings will obviously open new ways of studying the *Azolla-Anabaena* symbiosis.

VIII. GENETIC AND MOLECULAR STUDIES

A. STUDIES OF THE *ANABAENA AZOLLAE* ISOLATES

The difficulty of culturing *A. azollae in vitro* has been a major obstacle to the genetic study of this important cyanobacterium. Thus, there is no convenient system of genetic analysis in symbiotic *Anabaena* with which, for example, a mutant screening or complementation test can be performed.

Nevertheless, in recent years several laboratories have begun to use techniques of recombinant DNA research in order to characterize *A. azollae* strains.[100-103]

In nitrogen-fixing prokaryotes, including *A. azollae*, biological nitrogen fixation is catalyzed by the nitrogenase enzyme complex which is composed of two components: the nitrogenase (called MoFe protein) and the nitrogenase reductase (called Fe protein).[104] In the free-living *Anabaena* sp. PCC7120 these components are genetically determined by three genes: *nifH*, *nifD*, and *nifK*.[105,106] Another *nif* gene (*nifS*) is required for maturation of the nitrogenase complex.[106] In vegetative cells of *Anabaena* sp. PCC7120, *nifK* is separated from *nifDH* by an 11-kilobase (kb) DNA fragment.[106] Golden et al.[107] have recently demonstrated that this intervening region is excised in subsequent linking of the *nifK* and *nifDH* genes. Lammers et al.[108] have shown that the 11-kb DNA element is excised by site-specific recombination events between directly repeated 11-base pair (bp) sequences at each of its ends, and encodes a gene, *xisA*, required for its own excision. A second rearrangement which leads to the excision of a circular DNA element from vegetative-cell DNA is also observed near *nifS*.[107] All these findings, including organization of nitrogen-fixation genes and a developmentally regulated *nifK* /*nif D* gene rearrangement phenomenon, were shown only in the free-living *Anabaena* sp.[107,109]

Less is known about *nif* gene organization in symbiotic *Anabaena*. However, using DNA probes from the free-living *Anabaena* sp. PCC7120, Franche and Cohen-Bazire[100] have reported that the restriction sites within *nifK*, *nifD*, and *nifH* genes of four freshly isolated symbiotic *A. azollae* are strongly conserved, although some differences were recorded in the *nifH* genes. In the same paper the presence of a second copy of *nifH* in *A. azollae* was also reported. Recently, the same authors have compared the restriction sites in the region of the nitrogenase structural genes of five *Rhizosperma* endosymbionts to those of the four *Euazolla*.[103] The restriction fragments which hybridized to the *nifH* probe were identical in the five endosymbionts of *Rhizosperma* but differed from those observed in the four *A. azollae* extracted from *Euazolla* species. They presented evidence that symbiotic *Anabaena* strains derive from a common ancestral *A. azollae* and belong to two slightly divergent evolutionary lines. However, the *nif* genes that were investigated were not satisfactory for taxonomic purposes since the nitrogenase structural genes have been strongly conserved during evolution, presumably because of stringent structural requirements in the protein.[110] No hybridization was found between DNA from the endosymbionts and a probe carrying 1.8-kb DNA fragment of the region separating *nifK* and *nifDH* in the vegetative cells of *Anabaena* sp. PCC7120, whereas the same probe hybridized to the DNA extracted from seven free-living *Anabaena-Nostoc* strains. Recently, it has been confirmed that the *nifHDK* genes are contiguous in the vegetative cells of *A. azollae* (extracted from *A. pinnata* var. *imbricata* IND and *A. microphylla*)[111] as was suggested by mRNA studies.[112] This finding implies different organization of the nitrogen-fixation genes in symbiotic *A. azollae* strains in comparison to the free-living cyanobacteria.

We have studied 15 different *Azolla* isolates and their symbionts, in conjunction with a wide variety of heterologous gene clones as the DNA probes.[101,102] The probes fell into two groups, one including DNA fragments which contain nitrogenase genes isolated from free-living *Anabaena* sp. PCC7120 (*nifH* and *nifS* clones), and the second containing non-nitrogenase genes including the rRNA genes and the ribulose-1,5-biphosphate carboxylase/oxygenase (RuBisCo) genes isolated from *Anacystis nidulans*[113,114] as well as DNA fragments representing three of the four indigenous plasmids isolated from *Anabaena* sp. PCC7120.[102] The susceptibility of *A. azollae* DNA to 25 different endonucleases was first determined and four restriction enzymes (i.e., *Eco*RI, *Hind*III, *Bgl*II, and *Kpn*I) were selected for routine use. The DNA hybridization patterns between *A. azollae* isolates extracted from different *Azolla* species representing both sections *Euazolla* and *Rhizosperma*, using single or combined *nif* probes from the free-living *Anabaena* sp. PCC7120, were compared. Most of the restriction sites within and around the *nifH* and *nifS* were different among the endosymbionts of the section *Euazolla*. This permits differentiation between endosymbionts isolated from *A. caroliniana*, *A. filiculoides*, and *A. microphylla* (Figure 8). However, no differences in the hybridization pattern were observed among *A. azollae* isolated from *Azolla* species in section *Rhizosperma*. DNA polymorphism was observed among *A. azollae* isolates when a

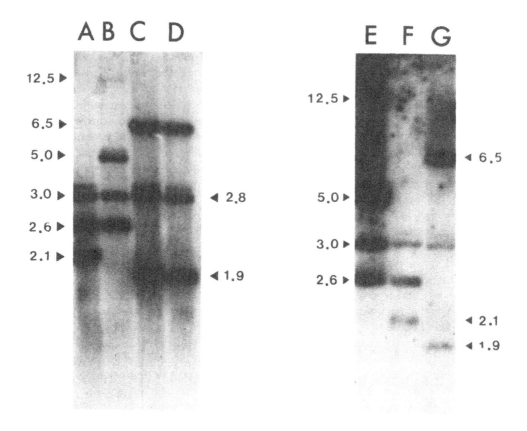

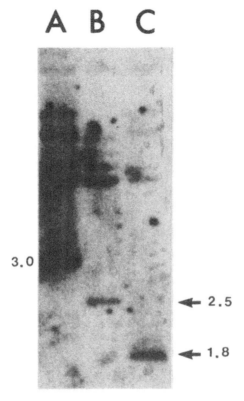

FIGURE 9. Autoradiogram of ^{32}P-labeled Ru-BisCo probe hybridized to *A. azollae* DNA extracted from: (A) *Azolla pinnata* var. *pinnata* (Townsville), (B) *A. filiculoides* (Shepparton), and (C) *A. filiculoides* (E. Germany). The DNA was digested with the endonuclease *Eco*RI. Sizes are in kb. Arrows indicate unique hybridizing bands that distinguish between *A. azollae* isolated from two different biotypes of the same *Azolla filiculoides* species.

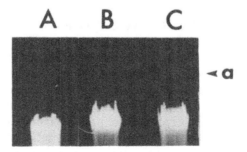

FIGURE 10. Visualization of the small indigenous plasmids in symbiotic *Anabaena* strains. (A) *A. azollae microphylla*, (B) *A. azollae filiculoides* (Snowy River), and (C) *A. azollae pinnata* (Darwin). "a" shows plasmid band migrated into agarose gel. Plasmid size is about 43 kb.

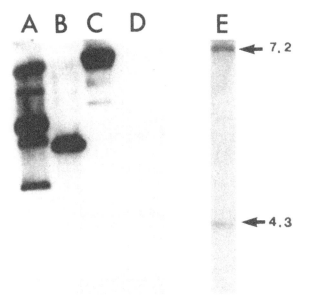

FIGURE 11.   Autoradiogram of ^{32}P-labeled *nifH* (lanes A, B, C, and D) and *nodCBADFE* (lane E) probes hybridized to DNAs extracted from: (A) *Lignobacter* sp. K17, (B) *Azospirillum brasilense* SP245, (C) *Azospirillum brasilense* SP7, (D) *Agrobacterium radiobacter* AFSR-1. All these DNAs were digested with *Eco*RI and probed with *NifH* (7120). (E) *Agrobacterium radiobacter* AFSR-1 DNA was digested with *Hind*III and probed with *Rhizobium trifolii* nodulation genes. *A. radiobacter* strain was isolated from the leaf cavity of *Azolla filiculoides* (Snowy River). Bacterial strains represented in lanes A, B, and C are known to possess the *nifH* gene; therefore, they were used as a control. DNA of AFSR-1 shows homology to the *R. trifolii nod* probe (two hybridizing bands, 7.2 kb and 4.3 kb).

Anabaena associations. Among four isolates of *Azolla filiculoides* there are three different genotypes of *A. azollae,* and among five isolates of *Azolla pinnata* also three different genotypes of the endosymbiont were observed. All Australian *Azolla* strains so far examined appear to contain a symbiont which has phylogenetically diverged from the endosymbionts extracted from the *Azolla* ecotypes collected in the other continents.

B. STUDIES OF BACTERIA IN LEAF CAVITIES

There have been a few reports on isolation of coryneform bacteria from leaf cavities of *Azolla caroliniana* Willd.[87,117-119] These aerobic eubacteria were present in numbers exceeding those of the cyanobacteria.[118] Bacteria such as *Pseudomonas,*[117] *Caulobacter* and *Alcaligenes,*[87] and *Arthrobacter globiformis*[119] have been reported to be present in the leaf cavities of *Azolla*. It has been suggested that the microorganisms in the cavities protect *A. azollae* by diminishing the intracavity NH_4^+ accumulation.[76] In our laboratory we have isolated a bacterial contaminant from the leaf cavity of surface sterilized *Azolla filiculoides* (Snowy River) fern. This bacterial strain grows rather slowly on rich media (3 d on nutrient agar and 5 d on mannitol medium). Using two different identification kits (API 20E and API 20NE) we classify the bacterium as *Agrobacterium radiobacter*.[120] This strain was designated AFSR-1. Most interestingly, the AFSR-1 strain does not show any homology to the *nif* genes, as expected, but when hybridized to the 5.2-kb *Bgl*II DNA fragment of the *Rhizobium trifolii* nodulation genes (*nodCBADFE*),[121] two weakly hybridizing bands were detected (7.2-kb and 4.3-kb *Hind*III-cleaved fragments) (Figure 11). Positive hybridization to these DNA sequences has important implications in studying the evolutionary origins of these genes. Furthermore, characterization of this bacterial strain seems to be important because it is still possible that the bacterium plays some role in the *Azolla-Anabaena* symbiosis, since these bacteria were not found in the cavities of *Anabaena*-free *Azolla pinnata* fern.[120]

IX. CONCLUDING REMARKS

In the last decade our knowledge of the *Azolla-Anabaena* symbiosis has dramatically increased. Basic morphological, physiological, and biochemical studies of *Azolla* species or strains have yielded much information on the nature of the association. Biological nitrogen fixation by *Azolla* and *Anabaena azollae* has long been recognized in Southeast Asia as a source of fertilizer for rice cultivation. The association has the property of being able to retain high nitrogenase activity in the presence of combined nitrogen, making the system compatible with inorganic nitrogen supplementation. *Azolla* is also unique in that it is the only agriculturally used symbiosis in which the prokaryotic partner is carried through the sexual cycle of the eukaryote. This property presents problems in developing new host-symbiont combinations that are not met to the same extent in legume-*Rhizobium* systems. However, the successful construction of new *Azolla-Anabaena* combinations now gives us a chance to create associations with specifically desired traits. Temperature-resistant, herbicide-resistant, insecticide-resistant or unusually effective N_2-fixing *Azolla-Anabaena* symbioses may now become available to suit geographical and climatic needs. These prospects would be enhanced if further fundamental advances in knowledge of (1) control of sporulation and (2) *in vitro* culture of *A. azollae* could be made. The latter, in particular, would facilitate the application of genetic techniques in order to generate mutants of symbiotic *Anabaena*. Southern blot hybridization analysis, using heterologous DNA probes, combined with the use of monoclonal antibody techniques provide powerful tools for characterization of symbiotic *Anabaena* strains. Now that reliable systems for classifying *A. azollae* are available, it is relatively simple to verify successful construction of new *Azolla-Anabaena* combinations.

ACKNOWLEDGMENTS

I wish to thank Professor Brian Gunning for reviewing the manuscript and Dr. Barry Rolfe for his encouragement. I am greatly indebted to all the researchers and colleagues who sent me strains containing DNA probes. Dr. C. Franche is thanked for her collaboration during tenure of a Visiting Fellowship in the Research School of Biological Sciences at Australian National University, and for many valuable suggestions which have been incorporated into this review. I also thank Mrs. R. Taylor for her skillful technical assistance and Mrs. Cathy Busby for taking SEM pictures. The author's research was supported in part by the Australian Centre for International Agricultural Research Grant 8501, and I wish to acknowledge the support of Dr. Eric Craswell and our Chinese collaborators, particularly Mr. Liu Chung-Chu, in the National Azolla Research Centre, Fujian Academy of Agricultural Science, Fuzhou, P.R.C.

REFERENCES

1. **Svenson, H. K.**, The new world species of *Azolla, Am. Fern. J.*, 34, 69, 1944.
2. **Lumpkin, T. A. and Plucknett, D. L.**, *Azolla* botany, physiology and use as a green manure, *Econ. Bot.*, 34, 111, 1980.
3. **Lumpkin, T. A. and Plucknett, D. L.**, *Azolla* as a Green Manure: Use and Management in Crop Production, West View Press, Boulder, CO, 1982.

4. **Moore, A. W.,** *Azolla:* biology and agronomic significance, *Bot. Rev.,* 35, 17, 1969.
5. **Schaede, R.,** Untersuchungen über *Azolla* und ihre Symbiose mit Blaualgen, *Planta,* 35, 319, 1947.
6. **Ashton, P. J. and Walmsley, R. D.,** The aquatic fern *Azolla* and its *Anabaena* symbiont, *Endeavour,* 35, 39, 1976.
7. **West, R. G.,** The occurrence of *Azolla* in the British interglacial deposits, *New Phytol.,* 52, 267, 1953.
8. **Hills, L. V. and Gopal, B.,** *Azolla primaeva* and its phylogenetic significance, *Can. J. Bot.,* 45, 1179, 1967.
9. **Reed, C. F.,** Index marsileata et salviniata, *Bol. Soc. Bot. Ser. 2a,* 28, 5, 1954.
10. **Fowler, K.,** An escape mechanism for spermatozoids in *Azolla* massulae, *Fern. J.,* 65, 7, 1975.
11. **Talley, S. N., Talley, B. J., and Rains, D. W.,** Nitrogen fixation by *Azolla* in rice fields, in *Genetic Engineering for Nitrogen Fixation,* Hollaender, A., Ed., Plenum Press, New York, 1977, 259.
12. **Ashton, P. J.,** The effects of some environmental factors on the growth of *Azolla filiculoides* Lam., in The Orange River Progress Report, Institute for Environmental Sciences, University of the O. F. S. Bloemfontein, South Africa, 1974, 123.
13. **Demalsy, P.,** Le sporophyte d'*Azolla nilotica, Cellule,* 56, 5, 1953.
14. **Duncan, R. E.,** The cytology of sporangium development in *Azolla filiculoides, Bull. Torrey Bot. Club,* 67, 391, 1940.
15. **Konar, R. N. and Kapoor, R. K.,** Embryology of *Azolla pinnata, Phytomorphology,* 22, 211, 1974.
16. **Bonnet, A. L. M.,** Contribution a l'étude des Hydropteridées. III. Recherches sur *Azolla filiculoides* Lamk., *Rev. Cytol. Biol. Veg.,* 18, 1, 1957.
17. **Strasburger, E.,** Über *Azolla,* Herman Davis, Jena, Leipzig, 1, 1873.
18. **Duckett, J. G., Toth, R., and Soni, S. L.,** An ultrastructural study of the *Azolla, Anabaena azollae* relationship, *New Phytol.,* 75, 111, 1975.
19. **Fowler, K. and Stennett-Willson, J.,** Sporoderm architecture in modern *Azolla, Fern Gaz.,* 11, 405, 1978.
20. **Martin, A. R. H.,** Some structures in *Azolla* megaspores and an anomalous form, *Rev. Paleobot. Palynol.,* 12, 141, 1976.
21. **Godfrey, R. K., Reinvert, G. W., and Huke, R. D.,** Observations on microsporocarpic materials of *Azolla caroliniana, Am. Fern. J.,* 51, 89, 1961.
22. **Bierhorst, D. W.,** *The Morphology of Vascular Plants, Salviniales,* Macmillan, New York, 1971, 341.
23. **Campbell, D. H.,** On the development of *Azolla filiculoides* Lam., *Ann. Bot. (London),* 7, 155, 1893.
24. **Kempf, E. K.,** Elektronenmikroskopie der Sporodermis von kanozoischen Megasporen der Wasserfarne-Gattung *Azolla, Palaeontol. Z.,* 43, 95, 1969.
25. **Fowler, K.,** Megaspores and massulae of *Azolla prisca* from the Oligocene of the Isle of Wight, *Paleontology,* 18, 483, 1975.
26. **Perkins, S. K., Peters, G. A., Lumpkin, T. A., and Calvert, H. E.,** Scanning electron microscopy of perine architecture as a taxonomic tool in the genus *Azolla* Lamarck, *Scanning Electron Microsc.,* 4, 1719, 1985.
27. **Loyal, D. S.,** Chromosome size and structure in some heterosporous ferns, in *Advancing Frontiers in Cytogenetics and Improvement of Plants,* Kachroo, P. N., Ed., Hindustan Publication, New Delhi, 1972, 293.
28. **Singh, P. K., Patra, R. N., and Nayak, S. K.,** Sporocarp germination, cytology and mineral nutrition of *Azolla,* in *Practical Application of Azolla for Rice Production,* Silver, W. S. and Schroder, E. C., Eds., Martinus Nijhoff/Junk, Dordrecht, 1984, 55.
29. **Gunning, B. E. S., Hughes, J. E., and Hardham, A. R.,** Formative and proliferative cell divisions, cell differentiation, and developmental changes in the meristem of *Azolla* roots, *Planta,* 143, 121, 1978.
30. **Peters, G. A.,** The *Azolla-Anabaena azollae* symbiosis, in *Genetic Engineering for Nitrogen Fixation,* Hollaender, A., Ed., Plenum Press, New York, 1977, 231.
31. **Calvert, H. E. and Peters, G. A.,** The *Azolla-Anabaena* relationship. IX. Morphological analysis of leaf cavity hair populations, *New Phytol.,* 89, 327, 1981.
32. **Gunning, B. E. S. and Pate, J. S.,** "Transfer cells": plant cells with wall ingrowths specialized in relation to short distance transport of solutes — their occurrence, structure, and development, *Protoplasma,* 68, 107, 1969.
33. **Peters, G. A., Kaplan, D., Meeks, J. C., Buzby, K. M., Marsh, B. H., and Corbin, J. L.,** Aspects of carbon and nitrogen exchange in the *Azolla-Anabaena* symbiosis, in *Nitrogen Fixation and CO_2 Metabolism,* Ludden, P. W. and Burris, J. E., Eds., Elsevier Science, New York, 1985, 213.
34. **Fjerdingstad, E.,** *Anabaena variabilis* status azollae, *Arch. Hydrobiol. Suppl. 49, Algol. Studies,* 17, 377, 1976.
35. **Stanier, R. Y. and Cohen-Bazire, G.,** Prototrophic prokaryotes: the Cyanobacteria, *Annu. Rev. Microbiol.,* 31, 225, 1977.
36. **Bozzini, A., De Luca, P., Sabato, S., and Gigliano, C. S.,** Comparative study of six species of *Azolla* in relation to their utilization as green manure for rice, in *Practical Application of Azollae for Rice Production,* Silver, W. S. and Schroder, E. C., Eds., Martinus Nijhoff/Junk, Dordrecht, 1984, 125.

37. **Ladha, J. K. and Watanabe, I.,** Antigenic similarity among *Anabaena azollae* separated from different species of *Azolla, Biochem. Biophys. Res. Commun.,* 109, 675, 1982.
38. **Gates, J. E., Fisher, R. W., Goggin, T. W., and Azrolan, N. I.,** Antigenic differences between *Anabaena azollae* fresh isolates from the *Azolla* fern leaf cavity and the free-living cyanobacteria, *Arch. Microbiol.,* 128, 126, 1980.
39. **Arad, H., Keysari, A., Tel-Or, E., and Kobiler, D.,** A comparison between cell antigens in different isolates of *Anabaena azollae, Symbiosis,* 1, 195, 1985.
40. **Liu, C.-C., Chen, Y., Tang, L., Zheng, Q., Song, T., Chen, M., Li, Y., and Lin, T.,** Studies of monoclonal antibodies against *Anabaena azollae, Acta Bot. Sinica,* in press.
41. **Fogg, G. E., Stewart, W. D. P., and Walsy, A. E.,** *The Blue-Green Algae,* Academic Press, London, 1973.
42. **Simon, R. D.,** Cyanophycin granules from the blue-green alga *Anabaena cylindrica:* a reserve material consisting of copolymers of aspartic acid and arginine, *Proc. Natl. Acad. Sci. U.S.A.,* 68, 265, 1971.
43. **Iyengar, M. O. P. and Desikachary, T. V.,** Occurrence of three-pored heterocysts in *Brachytrichia balani, Curr. Sci.,* 22, 180, 1953.
44. **Lang, N. J.,** Electron microscopic study of heterocyst development in *Anabaena azollae* Strasburger, *J. Phycol.,* 1, 127, 1965.
45. **Walsby, A. E. and Nichols, B. W.,** Lipid composition of heterocysts, *Nature (London),* 221, 673, 1969.
46. **Wilcox, M.,** One-dimensional pattern found in blue-green algae, *Nature (London),* 228, 686, 1970.
47. **Clark, R. L. and Jensen, T. E.,** Ultrastructure of akinete development in a blue-green alga, *Cylindrospermum* sp., *Cytologia,* 34, 439, 1969.
48. **Shen, E. Y.,** *Anabaena azollae* and its host, *Azolla pinnata, Taiwania,* 7, 1, 1960.
49. **Fritsch, F. E.,** Studies on Cyanophyceae. III. Some points in the reproduction of *Anabaena, New Phytol.,* 3, 216, 1904.
50. **Peters, G. A., Toia, R. E., Jr., Raveed, D., and Levine, N. J.,** The *Azolla-Anabaena azollae* relationship. VI. Morphological aspects of the association, *New Phytol.,* 80, 583, 1978.
51. **Neumuller, M. and Bergman, B.,** The ultrastructure of *Anabaena azollae* in *Azolla pinnata, Physiol. Plant,* 51, 69, 1981.
52. **Hill, D. J.,** The pattern of development of *Anabaena* in the *Azolla-Anabaena* symbiosis, *Planta,* 133, 237, 1975.
53. **Kaplan, D. and Peters, G. A.,** The *Azolla-Anabaena azollae* relationship. X. $^{15}N_2$ fixation and transport in main stem axes, *New Phytol.,* 89, 337, 1981.
54. **Kaplan, D., Calvert, H. E., and Peters, G. A.,** Phycobiliprotein in the *Azolla* endophyte as a function of leaf age and cell type, *Plant Physiol.,* 69, 156, 1982.
55. **Uheda, E.,** Isolation of empty packets from *Anabaena*-free *Azolla, Plant Cell Physiol.,* 27, 1187, 1986.
56. **Robins, R. J., Hall, D. O., Shi, D.-J., Turner, R. J., and Rhodes, M. J. C.,** Mucilage acts to adhere cyanobacteria and cultured plant cells to biological and inert surfaces, *FEMS Microbiol. Lett.,* 34, 155, 1986.
57. **Kobiler, D., Cohen-Sharon, A., and Tel-Or, E.,** Recognition between the N_2-fixing *Anabaena* and the water fern *Azolla, FEBS Lett.,* 133, 157, 1981.
58. **Mellor, R. B., Gadd, G. M., Rowell, P., and Stewart, W. D. P.,** A phytohaemaglutinin from the *Azolla-Anabaena azollae* symbiosis, *Biochem. Biophys. Res. Commun.,* 99, 1348, 1981.
59. **Peters, G. A. and Calvert, H. E.,** The *Azolla-Anabaena azollae* symbiosis, in *Algal Symbiosis,* Goff, L. J., Ed., Cambridge University Press, New York, 1983, 109.
60. **Calvert, H. E., Pence, M. K., and Peters, G. A.,** Ultrastructural ontogeny of leaf cavity trichomes in *Azolla* implies a functional role in metabolite exchange, *Protoplasma,* 129, 10, 1985.
61. **Shaw, W.,** personal communication, 1985.
62. **Watanabe, I. and Berja, N. S.,** The growth of four species of *Azolla* as affected by temperature, *Aquat. Bot.,* 15, 175, 1983.
63. **Peters, G. A., Toia, R. E., Jr., Evans, W. R., Crist, D. K., Mayne, B. C., and Poole, R. E.,** Characterization and comparisons of five N_2-fixing *Azolla-Anabaena* associations. I. Optimization of growth conditions for biomass increase and N content in a controlled environment, *Plant Cell Environ.,* 3, 261, 1980.
64. **Nickell, L. G.,** Physiological studies with *Azolla* under aseptic condition. II. Nutritional studies and the effects of the chemicals on growth, *Phyton,* 17, 49, 1967.
65. **Watanabe, I.,** *Azolla* utilization in rice culture, *Int. Rice Res. Newsl.,* 2, 3, 1977.
66. **Talley, S. N. and Rains, D. W.,** *Azolla* as nitrogen source for temperate rice, in *Nitrogen Fixation,* Vol. 2, University Park Press, Baltimore, 1980, 311.
67. **Peters, G. A., Evans, W. R., and Toia, R. E.,** *Azolla-Anabaena azollae* relationship. IV. Photosynthetically driven nitrogenase catalyzed H_2 production, *Plant Physiol.,* 58, 119, 1976.
68. **Lu, S. K., Chen, K., Shen, A., and Ge, S.,** Rice paddy green manure — studies on the biological characteristics of red *Azolla, Zhongguo Nongye Kexue (Chin. Agric. Sci.),* 11, 35, 1963.

69. **Haller, W. T., Sutton, D. L., and Barlowe, W. C.,** Effects of salinity on the growth of several aquatic macrophytes, *Ecology,* 55, 891, 1974.

70. **Ge Shi-an, S., Dai-xing, X., and Zhi-hao, S.,** Salt tolerance of *Azolla filiculoides* and its effects on the growth of paddy in Xinwei Haitu, *Zhejiang Nongye Kexue,* 1, 17, 1980.

71. **Yatazawa, M., Tomomatsu, N., Hosoda, N., and Nunome, K.,** Nitrogen fixation in *Azolla-Anabaena* symbiosis as affected by mineral nutrient status, *Soil Sci. Plant Nutr.,* 26, 415, 1980.

72. **Olsen, C.,** On biological nitrogen fixation in nature, particularly in blue-green algae, *C. R. Trav. Lab. Carlsberg,* 37, 269, 1972.

73. **Watanabe, I. and Espinas, C. R.,** Potentiality of nitrogen fixing *Azolla-Anabaena* complex as fertilizer in paddy soil, Saturday Seminar, International Rice Research Institute, Manila, August 14, 1976.

74. **Johnson, G. V., Mayeux, P. A., and Evans, H. J.,** A cobalt requirement for symbiotic growth of *Azolla filiculoides* in the absence of combined nitrogen, *Plant Physiol.,* 41, 852, 1966.

75. **Bortels, H.,** Über die Bedeutung des Molybdens für stickstoffbindende Nostocaceen, *Arch. Microbiol.,* 11, 155, 1940.

76. **Okoronkwo, N. and Van Hove, C.,** Dynamics of *Azolla-Anabaena* nitrogenase activity in the presence and absence of combined nitrogen, *Microbios,* 49, 39, 1987.

77. **Ray, T. B., Peters, G. A., Toia, R. E., Jr., and Mayne, B. C.,** *Azolla-Anabaena* relationship. VII. Distribution of ammonia-assimilating enzymes, protein, and chlorophyll between host and symbiont, *Plant Physiol.,* 62, 463, 1978.

78. **Peters, G. A., Ray, T. B., Mayne, B. C., and Toia, R. E., Jr.,** *Azolla-Anabaena* association: morphological and physiological studies, in *Nitrogen Fixation,* Newton, W. E. and Orme-Johnson, W. H., Eds., University Park Press, Baltimore, 1980, 293.

79. **Meeks, J. C., Steinberg, N., Joseph, C. M., Enderlin, C. S., Jorgensen, P. A., and Peters, G. A.,** Assimilation of exogenous and dinitrogen-derived $^{13}NH^+_4$ by *Anabaena azollae* separated from *Azolla caroliniana* Willd., *Arch. Microbiol.,* 142, 229, 1985.

80. **Uheda, E.,** Isolation of hair cells from *Azolla filiculoides* var. *japonica* leaves, *Plant. Cell Physiol.,* 27, 1255, 1986.

81. **Stewart, W. D. P. and Rowell, P.,** Effects of L-methionine-DL-sulphoximine on the assimilation of newly fixed NH_3, acetylene reduction and heterocyst production in *Anabaena cylindrica, Biochem. Biophys. Res. Commun.,* 65, 846, 1975.

82. **Walk, C. P., Thomas, J., Shaffer, P. W., Austin, S. M., and Galonsky, A.,** Pathway of nitrogen metabolism after fixation of ^{13}N-labelled nitrogen gas by the cyanobacterium, *Anabaena cylindrica, J. Biol. Chem.,* 251, 5027, 1976.

83. **Stewart, W. D. P., Rowell, P., Rai, A. N.,** Symbiotic nitrogen-fixation cyanobacteria, in *Nitrogen Fixation,* Stewart, W. D. P. and Gallon, J. R., Eds., Academic Press, New York, 1980, 239.

84. **Orr, J. and Haselkorn, R.,** Regulation of glutamine synthetase activity and synthesis in free-living and symbiotic *Anabaena* spp., *J. Bacteriol.,* 152, 626, 1982.

85. **Yoneyama, T., Ladha, J. K., and Watanabe, I.,** Nodule bacteroids and *Anabaena:* Natural ^{15}N enrichment in the legume-*Rhizobium* and *Azolla-Anabaena* symbiotic systems, *J. Plant Physiol.,* 127, 251, 1987.

86. **Peters, G. A.,** Azolla-Anabaena symbioses: basic biology, use, and prospects for the future, in *Practical Application of Azolla for Rice Production,* Silver, W. S. and Schroder, E. C., Eds., Martinus Nijhoff/ Junk, Dordrecht, 1984, 1.

87. **Newton, J. W. and Herman, A. I.,** Isolation of cyanobacteria from the aquatic fern *Azolla, Arch. Microbiol.,* 120, 161, 1979.

88. **Tel-Or, E. and Sadovsky, T.,** The response of the nitrogen fixing cyanobacterium *Anabaena azollae* to combined nitrogen compounds and sugar, *Isr. J. Bot.,* 31, 329, 1982.

89. **Rozen, A., Arad, H., Schonfeld, M., and Tel-Or, E.,** Fructose supports glycogen accumulation, heterocyst differentiation, N_2 fixation and growth of the isolated cyanobiont *Anabaena azollae, Arch. Microbiol.,* 145, 187, 1986.

90. **Peters, G. A.,** Studies on the *Azolla-Anabaena azollae* symbiosis, in *Proc. 1st Int. Symp. Nitrogen Fixation,* Vol. 2, Newton, W. E. and Nyman, C. J., Eds., Washington State University Press, Pullman, 1976, 592.

91. **Peters, G. A.,** The *Azolla-Anabaena azollae* relationship. III. Studies on metabolic capabilities and a further characterization of the symbiont, *Arch. Microbiol.,* 103, 113, 1975.

92. **Tyagi, V. V. S., Ray, T. B., Mayne, B. C., and Peters, G. A.,** The *Azolla-Anabaena azollae* relationship. XI. Phycobiliproteins in the action spectrum for nitrogenase-catalyzed acetylene reduction, *Plant Physiol.,* 68, 1479, 1981.

93. **Ruschel, A. P., de Freitas, J. R., and da Silva, P. M.,** Hydrogen uptake by *Azolla-Anabaena, Plant Soil,* 97, 79, 1986.

94. **Vonk, V. and Wellisch, P.,** Zur Frage der Stickstoffassimiliation einiger symbiontischen Cyanophyceen, *Acta Bot. Inst. Bot. Univ. Zagreb,* 6, 66, 1931.

95. **Tuzimura, K., Ikeda, F., and Tukamoto, K.,** Studies on *Azolla* with reference to its use as a green manure for rice fields, *J. Sci. Soil Manure,* 28, 17, 1957.

96. **Wieringa, K. T.,** A new method for obtaining bacteria-free cultures of blue-green algae, *Antonie van Leeuwenhoek; Ned. Tijdschr. Hyg.,* 34, 54, 1968.

97. **Huneke, A.,** Beitrage zur Kenntnis der Symbiose zwischen *Azolla* und *Anabaena, Beitr. Biol. Pflanz.,* 20, 315, 1933.

98. **Liu, C.-C.,** Recent advances on *Azolla* research, in *Practical Application of Azolla for Rice Production,* Silver, W. S. and Schroder, E. C., Eds., Martinus Nijhoff/Junk, Dordrecht, 1984, 45.

99. **Lin, C., Watanabe, I., Liu, C.-C., and Tang, L.-F.,** Recovery of symbiosis from *Anabaena*-free *Azolla, Acta Bot. Sinica,* in press.

100. **Franche, C. and Cohen-Bazire, G.,** The structural *nif* genes of four symbiotic *Anabaena azollae* show a highly conserved physical arrangement, *Plant Sci.,* 39, 125, 1985.

101. **Franche, C., Shaw, W., Gunning, B. E. S., Rolfe, B. G., and Plazinski, J.,** Genetic evidence of different *Anabaena* strains associated with the *Azolla* fern, in *Proc. 8th Australian Nitrogen Fixation Conf.,* Wallace, W. and Smith, S. E., Eds., Occasional Publication No. 25, Australian Institute of Agricultural Science, Sydney, 1986, 93.

102. **Franche, C., Gunning, B. E. S., Rolfe, B. G., and Plazinski, J.,** Use of heterologous hybridization in the phylogenetic studies of symbiotic *Anabaena* strains, in *Molecular Genetics of Plant-Microbe Interactions,* Verma, D. P. S. and Brisson, N., Eds., Martinus Nijhoff, Dordrecht, 1987, 305.

103. **Franche, C. and Cohen-Bazire, G.,** Evolutionary divergence in the *nif* K, D, H genes region among nine symbiotic *Anabaena azollae* and between *Anabaena azollae* and some free-living heterocystous cyanobacteria, *Symbiosis,* 3, 159, 1987.

104. **Mortenson, L. E. and Thorneley, R. N. F.,** Structure and function of nitrogenase, *Annu. Rev. Biochem.,* 48, 387, 1979.

105. **Mazur, B. J., Rice, D., and Haselkorn, R.,** Identification of blue-green algae nitrogen fixation genes by using heterologous DNA hybridization probes, *Proc. Natl. Acad. Sci. U.S.A.,* 77, 186, 1980.

106. **Rice, D., Mazur, B. J., and Haselkorn, R.,** Isolation and physical mapping of nitrogen fixation genes from the cyanobacterium *Anabaena* PCC7120, *J. Biol. Chem.,* 257, 13157, 1982.

107. **Golden, J. W., Robinson, S. J., and Haselkorn, R.,** Rearrangement of nitrogen fixation genes during heterocyst differentiation in the cyanobacterium *Anabaena, Nature (London),* 314, 419, 1985.

108. **Lammers, P. J., Golden, J. W., and Haselkorn, R.,** Identification and sequence of a gene required for a developmentally regulated DNA excision in *Anabaena, Cell,* 44, 905, 1986.

109. **Haselkorn, R.,** Organization of the genes for nitrogen fixation in photosynthetic bacteria and cyanobacteria, *Annu. Rev. Microbiol.,* 40, 525, 1986.

110. **Hennecke, H., Kaluza, K., Thony, B., Fuhrmann, M., Ludwig, W., and Stackebrandt, E.,** Concurrent evolution of nitrogenase genes and 16S rRNA in *Rhizobium* species and other nitrogen fixing bacteria, *Arch. Microbiol.,* 142, 342, 1985.

111. **Franche, C.,** personal communication, 1987.

112. **Nierzwicki-Bauer, S. A. and Haselkorn, R.,** Regulation of transcription in the *Azolla-Anabaena* symbiosis, *J. Cell Biochem.,* Suppl. 9 (C), 240, 1985.

113. **Tomioka, N. and Sugiura, M.,** The complete nucleotide sequence of a 16S ribosomal RNA gene from a blue-green alga, *Anacystis nidulans, Mol. Gen. Genet.,* 191, 46, 1983.

114. **Shinozaki, K. and Sugiura, M.,** Genes for the large and small subunits of ribulose-1,5-biphosphate carboxylase/oxygenase constitute a single operon in a cyanobacterium *Anacystis nidulans* 6301, *Mol. Gen. Genet.,* 200, 27, 1985.

115. **Taylor, R., Rolfe, B. G., Gunning, B. E. S., and Plazinski, J.,** in preparation.

116. **Plazinski, J., Franche, C., Liu, C.-C., Lin, T., Shaw, W., Gunning, B. E. S., and Rolfe, B. G.,** Taxonomic status of *Anabaena azollae;* an overview, *Plant Soil,* 108, 185, 1988.

117. **Bottomley, W. B.,** The effect of organic matter on the growth of various water plants in culture solution, *Ann. Bot.,* 39, 353, 1920.

118. **Gates, J. E., Fisher, R. W., and Candler, R. A.,** The occurrence of coryneform bacteria in the leaf cavity of *Azolla, Arch. Microbiol.,* 127, 163, 1980.

119. **Wallace, W. H. and Gates, J. E.,** Identification of eubacteria isolated from leaf cavities of four species of the N-fixing *Azolla* fern as *Arthrobacter* Conn and Dimmick, *Appl. Environ. Microbiol.,* 52, 425, 1986.

120. **Plazinski, J. and Taylor, R.,** in preparation.

121. **Djordjevic, M. A., Schofield, P. R., and Rolfe, B. G.,** Tn5 mutagenesis of *Rhizobium trifolii* host-specific nodulation genes results in mutants with altered host-range ability, *Mol. Gen. Genet.,* 200, 463, 1985.

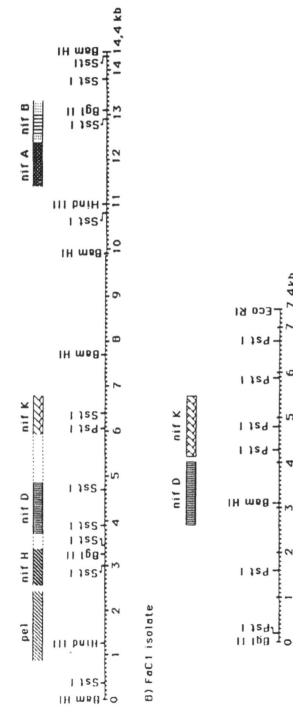

FIGURE 1. Molecular map of the symbiotic region of two *Frankia* strains. (A) Strain ArI3: The three structural *nifHDK* genes have found to be contiguous and are separated from *nifAB* genes by 4.5 kb. Hybridization techniques also detected genes homologous to *pel* genes from *Erwinia chrysantemi*. The *pel Frankia* genes were found to be near the *nif* structural genes. (B) Strain FaCl: the *nifD* and *nifK* genes are shown to be contiguous, but the *nifH* gene was not located linked to *nifD* and would not be clustered with *nifDK*.[56]

Chapter 4

THE GENETICS OF THE *FRANKIA*-ACTINORHIZAL SYMBIOSIS

Pascal Simonet, Philippe Normand, Ann M. Hirsch, and Antoon D. L. Akkermans

TABLE OF CONTENTS

I. INTRODUCTION

The process of biological nitrogen fixation which is carried out by a number of prokaryotic organisms via the nitrogenase enzyme complex remains rate-limiting in many agricultural areas. There has been much discussion about improving biological nitrogen fixation, especially the *Rhizobium*-legume symbiosis which is the best known. However, many other symbiotic associations occur, and these have been recently reviewed.[1,2] Symbiosis between woody, dicotyledonous plants and *Frankia*, a filamentous prokaryote which is classified in the order Actinomycetales,[3] is particularly interesting, and several reviews have been published recently that describe in detail various aspects of the *Frankia*-actinorhizal plant symbiosis. These include the biology of *Frankia*,[4] the efficiency of the nitrogen-fixing activity of the actinorhizal association,[5] the ecology,[6] host plant-endophyte specificity, and the genetics of *Frankia*.[7] The progress in *Frankia*-actinorhizal plant symbiosis has been presented in the proceedings of symposia on this subject during the last decade: Petersham,[8] Corvallis,[9] Madison,[10] Wageningen,[11] Laval,[12] and Umeå.[13]

Study of the *Frankia*-actinorhizal symbiosis is important for several reasons, including the use of actinorhizal shrubs and trees for reforestation and reclamation of depauperate, nitrogen-limiting soils.[14] For example, as more and more land, especially in the tropics, becomes deforested, the utilization of fast-growing, nitrogen-fixing trees for reforestation purposes becomes critical. According to the Office of Technology Assessment (OTA),[15] about 11.3 million hectares of what remains of tropical forests are destroyed annually. In addition, many nonleguminous, nitrogen-fixing plants are colonizers of poor soils. After the Pleistocene glaciation, plants such as *Alnus*, *Dryas*, *Elaeagnus*, *Hippophae*, and *Shepherdia* played an important role in soil reconstruction.[16,17] Currently, their role in nitrogen fixation is recognized as being at least comparable to that of the *Rhizobium*-legume symbiosis[18] and attempts are being made to utilize actinorhizal trees in reforestation and soil reclamation in developing countries.[14,19,20]

Besides the practical application of actinorhizal plants to situations described above, the study of this symbiosis offers much interest to those investigating the basic biology of nitrogen-fixing symbiosis. There are a number of parallels, developmental and genetic, between the *Frankia*-nonlegume symbiosis and the *Rhizobium*-legume association which make the former interesting to study in terms of its molecular biology. Some of these similarities are striking. For example, it has been known for some time that the *nif* genes of *Klebsiella pneumoniae* hybridize with DNA sequences from *Frankia* as well as from *Rhizobium*. Both the *Rhizobium*-legume and the *Frankia*-nonlegume symbiotic associations lead to the formation of root nodules in which the microbes fix atmospheric nitrogen into ammonia. However, we must keep in mind not only the similarities, but also the differences, between the two types of symbiosis. While rhizobia fix nitrogen symbiotically with only the Leguminosae (with the single exception of *Parasponia*, a nonleguminous tree belonging to the Ulmaceae),[21,22,23,24] 21 genera of dicotyledonous plants (including *Allocasuarina* as a new genus) belonging to 8 different families have been reported worldwide to have actinomycete-induced, nitrogen-fixing nodules.[25,26] In addition, unlike the *Rhizobium*-legume system, *Frankia* cells are not released into the host cytoplasm and do not become surrounded by peribacteroid membranes. The actinomycete remains encapsulated by host-derived carbohydrate material. However, recent studies of nodules of woody legumes such as *Andira* indicate that *Rhizobium* in these nodules remain encapsulated within infection threads and are not released into the host cell cytoplasm.[27] Similarly, *Bradyrhizobium* cells are not released from infection threads within *Parasponia* nodules.[28] Hence encapsulation of the nitrogen-fixing partner within host cell material may be a characteristic of nodules of woody plants.

Because of the similarities and differences between the two symbioses, many questions come to mind about the two strategies of making nitrogen-fixing nodules. What symbiotic genes, in addition to *nif* genes, are conserved between *Rhizobium* and *Frankia?* Are the early invasion or *nod* genes conserved? Does *Frankia* produce specific hydrolytic enzymes which facilitate root hair invasion? What factors (genes) allow for a broader host range in the *Frankia*-actinorhizal symbiosis? Do the legume and nonlegume hosts respond to infection by their respective nitrogen-fixing prokaryotic partners differently? Are nodule-specific proteins formed in the various nonlegume hosts in response to *Frankia* infection as they are in legume nodules? If so, are they similar proteins? The *Frankia*-actinorhizal symbiosis, thus, becomes a useful association to study in terms of the similarities to and differences from the *Rhizobium*-legume symbiosis. Investigating various parameters including nodule development, the genes involved in symbiosis and their regulation, host range, etc., may help us come to a greater understanding of how symbiotic associations evolved in general.

Frankia is classified as a "typical" actinomycete, with some strains possessing relatively large genomes (genome molecular weights of ca. 6.0×10^9 and 8.3×10^9 from strains EuI1a and ArI4, respectively).[29,30] Like other actinomycetes, *Frankia* strains are very G-C rich, containing DNA which is about 70% guanine + cytosine.[31,32]

Frankia is far removed taxonomically from *Escherichia coli*, perhaps the prokaryote best genetically characterized, or from *Klebsiella pneumoniae*, the experimental organism of choice for studies on *nif* genes. *K. pneumoniae* is closely related to *E. coli* and fixes nitrogen asymbiotically. Unlike the Gram-negative enterobacteria, *E. coli*, *K. pneumoniae*, and *Rhizobium* spp. *(sensu lato)*, *Frankia* are Gram-positive, filamentous, spore- and vesicle-forming soil bacteria with slow growth rate and filamentous nature. In addition, it has been only since 1978 that *Frankia* strains have been isolated in pure culture from a root nodule.[33,34] Hence, specific tools for analyzing the molecular biology and genetics of *Frankia* have not been developed yet and have had to be adapted from other bacterial methodologies, either from *Rhizobium*, because of its symbiotic relationship with legumes, or from *Streptomyces*, a well-characterized actinomycete.[35,36]

Some problems of *Frankia* molecular biology have been resolved by cloning *Frankia* DNA into *E. coli*. As a consequence, some genes can be identified from a *Frankia* gene library by sequence similarity to genes coding for known functions in other organisms like *Rhizobium* or *Klebsiella*. However, major advances in *Frankia* molecular biology will require the use of endogenous *Frankia* cloning vectors as well as the development of a transformation system. For example, transposon mutagenesis, an extremely useful approach for genetic analysis, cannot be accomplished at this time because there is no efficient method of introducing DNA into *Frankia*. Our goal in this chapter is to examine some of the recent findings on the isolation of microbial and plant genes involved in the *Frankia*-actinorhizal symbiosis and the state of research concerning genetic transformation systems. We will also discuss recent developments that may be helpful for an understanding of the taxonomic classification of *Frankia*.

II. IDENTIFICATION OF SYMBIOTIC GENES IN *FRANKIA*

A. DNA PREPARATION

Early studies on the molecular biology of *Frankia* were hampered by difficulties with growing enough cellular material for DNA extraction and also by the low efficiency of classical lysis techniques of *Frankia* cells. Normand and co-workers[37] used a number of different DNA isolation procedures, exploiting such enzymes as lysozyme, protease, neuraminidase, cellulase, helicase, and mutanase, without much effect on lysing a significant portion of cells. Nevertheless, the use of lysozyme in combination with a drastic extraction method (10% SDS at 90°C and vortexing) allowed the extraction of small amounts of DNA. A similar lysis procedure based mainly on the chemical dissolution of the cell wall by hot lauryl sulfate was used by Dobritsa[38] and An et al.[31] for isolation of *Frankia* DNA.

The use of achromopeptidase improved *Frankia* cell lysis, a critical step in the isolation of high molecular weight DNA. This enzyme, used in combination with lysozyme, resulted in a complete digestion of *Frankia* cell walls, depending on the lot of enzyme.[39] Other purification procedures have been described by An et al.,[31] as a modification[40] of the Marmur[41] procedure, and by Dobritsa,[38] as an application of the method developed for *Streptomyces*.[42] In strains which did not lyse efficiently with enzymatic treatment, a pretreatment with ice-cold acetone[43] has been shown to be successful.[43a]

B. IDENTIFICATION AND ORGANIZATION OF *NIF* GENES

Before we begin a discussion of *nif* genes in *Frankia,* we must define certain terms. The term "sequence similarity" will be used to describe the number of DNA base pair matches, be it the result of total DNA-DNA hybridization (often described as % homology), probe hybridization on Southern blots (which may yield restriction fragment length polymorphisms or RFLPs), or sequence comparison (% positional identify in alignment). In contrast, the term "homology" will be reserved to define common evolutionary origin, following the suggestion of Reeck et al.[44]

The *nif* structural genes encode the nitrogenase enzyme: *nifH* codes for the polypeptides of the Fe protein and *nifDK* codes for the alpha and beta subunits of the MoFe protein. Extensive sequence similarity has been found between the *nifK, D,* and *H* genes from *K. pneumoniae* and those of other nitrogen-fixing bacteria, including *Frankia*.[45] Simonet and co-workers[46] used a *nifKD* probe from *R. meliloti* and, through the use of Southern hybridization techniques, have shown that the three *nif* structural genes (*HDK*) are contiguous in *Frankia* strain ArI3 (Figure 1).

In the fast-growing *Rhizobium* species, these three genes have been found to be closely linked[47-50] and to form a single operon.[51] In addition, they are located on large plasmids known as symbiotic or Sym plasmids (see Reference 52). In contrast, the *nifHDK* genes in the slow-growing *Bradyrhizobium* species are borne on the chromosome. However, *nifH* is separated from *nifDK* by at least 13 kb in *Bradyrhizobium* species[53-55] and is transcribed as a separate operon.

The situation in *Frankia* is still under investigation. Although five out of five *Frankia* strains tested (two *Elaeagnus*-compatible strains, EUN1f and HRN18a; two *Alnus*-compatible strains, AcoN24d and ArI3; and one *Casuarina*-compatible strain, CeD), exhibited close linkage of *nifD* and *H* genes, this does not mean that such a linkage is a general feature of all *Frankia* strains. Ligon and Nakas[56] found that the *nifH* gene was not present on a 49-kb DNA fragment which was isolated from an *Alnus*-compatible strain, FaC1, and which carried both *nifD* and *nifK* genes (Figure 1). Their data suggest that the genes for the Fe protein and MoFe protein may not be as closely linked as in the strains analyzed by Normand et al.[57] Moreover, it is important to keep in mind that the five strains analyzed represent only a small proportion of the actinorhizal endophyte population. *Frankia* isolates from the *Elaeagnus* host compatibility group exhibit considerable genetic diversity and, in addition, no *Frankia* isolates have been described yet from many spore-positive (also known as sp +; i.e. spores and sporangia are abundant within nodules) nodules of *Alnus* nor from nodules developed on species in the families Coriariaceae, Datiscaceae, Rhamnaceae, or Rosaceae.

So far, DNA-DNA hybridization techniques have demonstrated the following:

1. Among *Frankia* strains, the *nifH* gene appears to be more conserved than is the *nifD* gene, and the *nifD* gene is more conserved than the *nifK* gene. The observations agrees with what has been described by Hennecke et al.[58] for other nitrogen-fixing bacteria.

2. The *nifK* gene exhibits some sequence similarity to the *nifD* gene.[58a] Similarity between the two MoFe protein genes has been described previously for *Bradyrhizobium japonicum* by Thony et al.[59] who postulated a common ancestry.

3.   *Frankia* strain ARgP5Ag (ULQ0132105009) has a different organization of *nif* genes compared to the "typical" *Frankia* strains.[60] *Nif* DNA probes were found to hybridize to a 10-kb *EcoRI* fragment on total DNA blots and also to a 5.6-kb *EcoRI* fragment from a large (190 kb) indigenous plasmid.

Thus, in this strain, the *nif* genes appear to be reiterated. Gene reiterations are relatively widespread among nitrogen-fixing microorganisms. They have been described in *R. phaseoli*,[61] *R. fredii*,[62] *Anabaena*,[63,64] *Rhodopseudomonas capsulata*,[65] and *Azotobacter chroococcum*.[66] In *Azotobacter vinelandii*, recent investigations have shown that there are two nitrogenase systems and that these are based on two different *nifH* genes, which are differently expressed, depending on whether high or low molybdenum conditions are present.[67,68] For the other bacteria and particularly for *Frankia*, the biological significance of these gene reiterations is obscure, and it is not known whether the two *nif* reiterations in the DNA of *Frankia* strains ARgP5Ag represent genes with little or a great deal of similarity.

C. LOCALIZATION OF *NIF* GENES

Several genetic and physical studies have now clearly established that in fast-growing *Rhizobium* species the genes involved in symbiosis, particularly the *nif* genes, are located on large plasmids (see review[52]). However, the incidence of plasmid-borne *nif* genes is not restricted to the Rhizobiaceae. *Nif* genes are located on plasmids in both *Lignobacter* K17[69] and *Enterobacter agglomerans*.[70] In some *Frankia* strains, plasmids ranging in size from 8 to 190 kb[39] have been detected, using a modified Eckhardt gel procedure. Large plasmids have also been reported by Dobritsa and co-workers in *Frankia* isolates from alder nodules.[71-74] Using cloned nitrogenase structural genes from *Klebsiella pneumoniae* as DNA probes, Simonet et al.[60] tested the different plasmids detected in the *Alnus*-compatible *Frankia* strains for the presence of *nif* genes. The smaller plasmids (<100 kb) did not hybridize to the *nif* probes. In the *Elaeagnus*-compatible strains, it was found[37] that the plasmid bands could not be visualized routinely and thus, no information has been generated indicating whether *nif* genes are present on the 80-kb, 39-kb and 14-kb plasmids detected in strain EUN1f.[7] However, these small replicons are unlikely candidates for the role of Sym plasmid as found in the fast-growing *Rhizobium* species.

Dobritsa and Tomashevsky[74] demonstrated the occurrence of circular extrachromosomal DNA from nodules of *Alnus glutinosa* and showed that they hybridized with ^{32}P-DNA of the plasmid pSA30 carrying the *K. pneumoniae nifHDK* genes. These authors also discussed possible genome rearrangements inducing multiple circular DNAs in the microsymbiont. So far there is no evidence for major rearrangements of the *nif* genes *HDK* and *AB* during the transition from the free-living stage of *Frankia* ArI3 into the symbiotic (i.e., vesicle cluster) stage. Meesters[75] made a comparative study of the DNA from a culture of strain ArI3 and DNA from root nodules of *A. glutinosa* with ArI3 as microsymbiont. Hybridization of total DNA with *Frankia nif* genes *HDK* and *AB* as probe[46] resulted in similar band patterns.

Simonet et al.[60] have found that a ^{32}P-labeled *K. pneumoniae nif* DNA hybridized to a 5.6-kb *EcoRI* DNA fragment from plasmid pFQ69 (190 kb). Plasmid pFQ69 was detected in the ARgP5Ag strain (ULQ0132105009). This plasmid exhibits a size comparable to that of the symbiotic plasmids of fast-growing *Rhizobium* sp. Unlike most fast-growing *Rhizobium* strains, however, an additional band (10 kb) was found to hybridize to the *nif* gene probe in strain ULQ0132105009 when total DNA was used. This indicates that *nif* sequences in this strain are located both on the chromosome and on the large plasmid pFQ69. What is the significance of the plasmid-borne *nif* genes in this *Frankia* strain, which so far is the only *Frankia* strain of those examined to exhibit this arrangement of *nif* genes? Is plasmid location of symbiotic sequences a widespread or rare characteristic of *Frankia*? What is the biological significance of such a property?

D. REGULATION OF *NIF* GENES

Expression of *nif* genes is regulated by a number of environmental factors, e.g., nitrogen source and oxygen tension, and evidence has been found for the occurrence of regulatory *nif* genes in *Frankia*:

Nitrogen sources — Ammonium salts added to the medium strongly suppress the nitrogenase activity of *Frankia*, as has been found for other nitrogen-fixing organisms. In many *Frankia* strains, but not in all,[75,76] the addition of ammonium salts also suppresses the formation of vesicles, i.e., spherical specialized cells at the end of short hyphae. The simultaneous decrease in nitrogenase activity and disintegration of vesicles has also been observed after addition of ammonium salt to nodulated *Alnus incana* seedlings.[77] There is now good evidence to indicate that both the Fe and the MoFe proteins of nitrogenase are exclusively localized within the vesicles.[75,76,78-80] The genetic basis of the apparent suppression of nitrogenase synthesis in the hyphae of such vesicle-forming *Frankia* strains is still unknown.

Oxygen tension — The process of nitrogen reduction by nitrogenase is inactivated irreversibly by oxygen.[25,81] Unlike the situation observed for *Rhizobium*, pure cultures of *Frankia* are able to reduce dinitrogen to ammonium. Three recent findings have allowed a better understanding of mechanisms developed for protection from O_2 toxicity. Steele and Stowers[82] demonstrated that *Frankia* in well-aerated, pure cultures expressed two superoxide dismutases (SODs) one containing manganese (MnSOD) and one containing iron (FeSOD). These enzymes function by protecting cell components from oxidation brought about by superoxide free radicals (O_2^-) and hydrogen peroxide (H_2O_2). It is tempting to suggest that the vesicles contain one or both forms of SOD. In addition, hemoglobin has been detected in root nodules of nonlegumes, and this hemoglobin is similar to that found in legumes[83,84] (see also Section IV). Hemoglobin may function to facilitate oxygen transport in tissues where the pO_2 is low. This is especially important in nodules of *Casuarina* where no vesicles are present.[85] However, these *Frankia* strains have the genetic potential to and readily synthesize thick-walled, lipid-containing vesicles to protect nitrogenase and fix nitrogen under high pO_2 conditions *in vivo*.[86] Moreover, Berg and McDowell[85] demonstrated the presence of nonseptate, intracellular hyphae in *Casuarina* nodules. Various other aberrant types of nonseptate hyphae have been found in other actinorhizae and might have similar functions as the spherical vesicles.[87,88]

Regulatory *nif* genes — Dixon et al.[89] first demonstrated that the *nifA* gene was involved in *nif* regulation in *K. pneumoniae*. Later it was found that *nif* gene expression was controlled by the products of the *nifLA* operon, which in turn are under control of the general nitrogen regulatory system genes (*ntr* genes).

A number of symbiotic organisms possess a gene functionally or structurally similar to the *K. pneumoniae nifA* gene.[90] For example, hybridization to the *Klebsiella nifA* gene was observed for DNA sequences from *Rhizobium meliloti*,[91,92] *Azospirillum brasilense*, and *A. lipoferum*,[93] and *Bradyrhizobium japonicum*.[94] However, the *Klebsiella nifA* probe failed to hybridize to DNA from *Rhodobacter capsulatus*, but a *nifA* gene was detected in this bacterial species by restoring a Nif$^+$ phenotype to *nifA*-like regulatory mutants.[95]

A *nifA*-like gene in *Frankia* has been detected by hybridization of a 4.5-kb *BamHI* DNA fragment to a *nifA* DNA probe from *R. meliloti*.[46] This *nifA*-like region is situated approximately 4.5 kb away from the *nifHDK* genes. Adjacent to the *nifA*-like region, DNA hybridizations revealed a region exhibiting some sequence similarity to the *R. meliloti nifB* gene. The degree of sequence similarity between *R. meliloti* and *Frankia* DNA is greater for the *nifB* than the *nifA* gene. Thus, *Frankia* represents at least the fifth nitrogen-fixing species which has a *nifB*-like gene directly adjacent to *nifA* in addition to *K. pneumoniae*,[96] *R. leguminosarum*,[97] *R. meliloti*,[98] and *Rhodobacter capsulatus*.[95] Unlike the first four, where the *nifAB* genes are separated by some defined distance from the structural genes from nitrogenase, the *nifA* gene of *R. capsulatus* is adjacent to the *nifHDK* operon. In *Klebsiella*, the region between *nifA* and *nifHDK* contains the nitrogen-fixing genes *nifY E N X Y S V M F L*, and in *R. meliloti*, the *fixABC* genes are located here.[99] Recently, Ligon and Nakas[56] have found that in *Frankia* strain FaC1, a DNA fragment located immediately downstream of *nifDH* hybridized with the *K. pneumoniae nifEN* genes. This opens up interesting possibilities for the localization of other nitrogen-fixation genes in *Frankia*.

E. IDENTIFICATION OF OTHER SYMBIOTIC GENES

Using complementation studies, Long et al.[100] cloned genes involved in nodulation (*nod* genes) from *R. meliloti*. Later, *nod* genes were found in other *Rhizobium* strains by Southern hybridization using *R. meliloti nod* sequences as a probe.[101-103] The genes were analyzed in detail and designated *nodA, B, C,* and *D,* the so-called "common" *nod* genes.[104] Fogher et al.[105] provided an unexpected result when they found that nodulation genes hybridized with *Azospirillum* DNA, suggesting that common mechanisms may exist among bacteria interacting with plants, perhaps in some of the early steps of the recognition between bacteria and plants.

Because of the similarity in the early steps of nodule development between the *Rhizobium*-legume and *Frankia*-actinorhizal associations, i.e., bacterial-plant recognition, root hair curling, or formation of an encapsulation around the penetrating bacteria, it seemed likely that nodulation genes between *Rhizobium* and *Frankia* would be conserved. A number of attempts hybridizing *R. meliloti* common *nod* genes to total *Frankia* DNA have been made, but with little success. Simonet et al.[58a] and Tomashevsky and Dobritsa[73] have not found hybridization between *Rhizobium nod* sequences and *Frankia* DNA. On the other hand, Drake et al.[106] found a low level of hybridization to total DNA isolated from the *Casuarina*-compatible strain HFPCcI3 using a *R. meliloti nodC* probe, but even less when a *nodAB* probe was utilized. Reddy et al.[106a] have looked for complementation of auxotrophic mutations, namely, of Phe⁻, Trp⁻, and Leu⁻ *R. meliloti*, using a HFPCcI3 *Frankia* clone bank made in the broad host range vector mobilizable pLAFR3.[107] The *R. meliloti* Phe⁻ mutants appeared to be complemented by *Frankia* DNA sequences, but none of these auxotrophs elicit symbiotic defects. Using a strategy of genetic complementation of Nod⁻ *R. meliloti* mutants similar to that used by Long et al.[100] and Marvel et al.,[108] Reddy et al.[106a] found that some transconjugants induced nodules on alfalfa. However, after transduction of the pLAFR3/CcI3 genes to a fresh Nod⁻ *R. meliloti*, nodulation was not restored, suggesting that the *Frankia* genes were not solely responsible for the Nod⁺ phenotype of the original transconjugants.[106a]

A necessary step for host penetration by the actinomycete *Frankia* is the ability to hydrolyze the plant cell wall. Once penetration has occurred, the *Frankia* cells become surrounded by a polysaccharide capsule[109-111] that has been postulated to be involved in host-endophyte nutritional interactions.[109] Electron micrographs of infected plant cells in actinorhizas of *Datisca cannabina*[112] and *Elaeagnus angustifolius*[113] show the "cleanness" of the penetration by *Frankia* hypha of plant cell walls. These observations suggested an enzyme-mediated entry of *Frankia* rather than a physically forced penetration. Thus, genes that code for enzymes involved in penetration may be considered *"nod"* genes because they permit entry into the host, as well as *"fix"* genes, which allow *Frankia* to live within the host cell and fix nitrogen.

It is likely that *Frankia* could invade host roots using a biochemical process similar to that used by phytopathogenic bacteria. Indeed, many of the steps used by pathogenic and symbiotic organisms to invade the plant appear similar superficially. Pathogens produce pectate lyase which breaks down the middle lamella and primary cell wall of plant tissues.[114] The genes coding for pectate lyase enzymes from *Erwinia chrysanthemi*, an enterobacterium pathogenic on many plants[115] which produces various pectinases, cellulases, and proteases,[116,117] have been cloned.[118-120] Simonet et al.[58a] have found that *Frankia* DNA sequences exhibited sequence similarity to *Erwinia* genes coding for pectate lyase. They found that a great deal of sequence similarity exists between *Frankia* and *E. chrysanthemi pel* genes on the basis of hybridization intensity. The intensity of the heterologous hybridization band decreased only under conditions of high stringency. *Frankia* DNA exhibited a greater degree of hybridization to a *pelA* clone than to a *pelD* gene.

Several *Frankia* strains grown under a number of culture conditions are able to degrade cellulose or carboxymethylcellulose (CMC) into glucose.[121] Hence, *Frankia* must secrete some type of cellulolytic enzymes which of yet are uncharacterized. Such genes have been cloned from pathogenic organisms, and studies are underway to determine whether there is any sequence similarity between these clones and *Frankia* DNA. Simonet et al.[58a] found that *Clostridium thermocellum cel* genes hybridized with *Frankia* DNA. The *cel*-like sequence has not been localized yet, nor has the extent of sequence similarity between the *C. thermocellum cel* clone and the *Frankia cel*-like gene been established. Reddy et al.[106a] have also detected hybridization between a *cel* clone isolated from *Erwinia chrysanthemi*[122] and HFPCcI3 DNA. Interstingly, a *cel* clone representing an endoglucanase II, an *Erwinia* protein which is not secreted into the medium, did not exhibit much hybridization to the *Frankia* DNA, whereas a clone for endoglucanase I, a protein secreted into the medium, strongly hybridized to the HFPCcI3 DNA sequences.

Other studies have shown that some *Frankia* strains possess an active hydrogen uptake system (*hup* genes) permitting hydrogen to be recycled.[123] In the nitrogen-fixing symbiosis between *Rhizobium* and the legume plant, it has been established that *Rhizobium* strains can be subdivided into Hup+ (possessing an active hydrogen uptake system) and Hup− (lacking an active hydrogen uptake system) strains. Albrecht et al.[124] demonstrated that Hup+ *Rhizobium* strains are more efficient than Hup− ones in fixing nitrogen. The *hup* genes are localized on plasmids in many *Bradyrhizobium* strains.[125,126] Because genes encoding for the same functions are frequently conserved among several microorganisms, cross-hybridization may allow the localization of *hup* genes in *Frankia*. *Bradyrhizobium hup* genes have been found to hybridize to *Frankia* DNA sequences, but so far have not been localized.[126a]

Other genes picked up by cross-hybridization to genes of *Rhizobium* sp. are the *Frankia* glutamine synthetase (GS) genes. Both GS1 and GS2 have been found in *Frankia*.[127] GS2 (subunit molecular weight 43 kilodaltons) is similar to the GS2 enzymes found in all but one member of the Rhizobiaceae analyzed to date and is derepressed during nitrogen starvation. Thus, *Rhizobium* and *Frankia* symbioses have another common feature. It will be interesting to see if additional *Rhizobium* symbiotic genes as well as other genes such as those encoding for IAA and cytokinin biosynthesis from *Agrobacterium tumefaciens* exhibit any sequence similarity to *Frankia* DNA. Such results would raise interesting possibilities not only for the future isolation of symbiotic genes of *Frankia* but also with the respect to the evolution of the two nitrogen-fixing symbioses.

III. CLONING INTO *FRANKIA*

The application of genetic analysis techniques from *E. coli* to studies of *Rhizobium* has provided valuable tools for investigating symbiotic genes and gene regulation in the microsymbiont. Further progress has been marked by the development of endogenous *Rhizobium* cloning vectors. Two groups of broad-host range DNA cloning vectors are now available.[52] In addition, phages detected in *R. meliloti*,[128-131] *R. leguminosarum*, and *R. trifolii*,[132] and in *B. japonicum*[133,134] have been used for genetic studies. The phages permit the cloning of large DNA fragments, especially large plasmids.[52]

A. GENE CLONING SYSTEMS

Several problems remain for the development of *Frankia* endogenous cloning systems. Theoretically DNA can be introduced in recipient cells by four strategies:

Transduction — As of now no *Frankia* phage has been found.[3,7] Detection of actinophages in *Frankia*, however, is seriously hampered by the difficulty of finding plaques, since most *Frankia* strains do not form a lawn on agar plates.[134a]

Conjugation — *Frankia* strains contain plasmids of different sizes and, apart from the nitrogenase genes on plasmid pFQ69, the information coded on these other plasmids is unknown (see Reference 7). Although plasmid profiles have been used for strain identification,[136] so far there is no evidence indicating that a conjugation system exists in any *Frankia* strain. However, for many strains of *Streptomyces*, large segments of chromosome can be transferred from one donor strain to another via conjugation systems.[137] The lack of information concerning conjugative plasmids makes cloning strategies for *Frankia* very limited.

Transformation — As in various other Gram-positive bacteria, protoplast transformation with plasmid vectors seems to be the method of choice for introducing DNA into *Frankia*. Attempts to transform protoplasts of *Frankia* strains EUI, EANIpec, and An2.24 with the *Streptomyces* plasmid pIJ702 have failed.[137a] The possible reasons for this failure will be discussed in Section III.C.

Protoplast fusion — Fusion of protoplasts has been successfully applied in making hybrids of *Streptomyces* spp.[138-141] and attempts were made to fuse protoplasts of unpigmented *Frankia* strains, e.g., EUI, characterized by streptomycin resistance, and the pigmented strains EANIpec and An2.24.[134a] So far no hybrid colonies could be selected after regeneration of the protoplasts. Attempts have been made to introduce the tyrosinase gene and the thiostrepton resistance gene (both located on the high-copy number *Streptomyces* plasmid pIJ702) by direct protoplast fusion of protoplasts from *Streptomyces lividans* (pIJ702) and *Frankia* strains (EANIpec, EUI, An2.24, and CpI1).[134a] No *Frankia* colonies, derived from regenerated protoplasts on complex media,[142,143] were found which showed melanin production and/or thiostrepton resistance. Exciting experiments described by Prakash and Cummings[144] indicated that protoplast fusion of *Frankia* and *Streptomyces griseofuscus* might occur. Among the 20 hybrids isolated, one strain was able to fix nitrogen *in vitro* as a fast-growing *Streptomyces*-like organism that also formed nitrogen-fixing root nodules on *Alnus rubra*. Examination of the root nodules induced by the hybrid revealed only the presence of hyphae-like structures and no vesicles. These results might indicate that protoplast fusion may become a useful way to introduce heterologous DNA in *Frankia* or to introduce *nod* and *nif* genes in biotechnologically important actinomycetes.

B. PRODUCTION OF *FRANKIA* PROTOPLASTS

Three different methods describing the successful isolation of *Frankia* protoplasts have been published. They differ mainly in the enzymatic treatment used. Faure-Raynaud et al.[145] and Normand et al.[146] reported that achromopeptidase as well as lysozyme was necessary to obtain a high level of protoplast production from hyphae. Using 1 gram fresh weight of hyphae, Normand et al.[146] found that 10^5 protoplasts were isolated with lysozyme alone, whereas more than 10^7 protoplasts were produced with the two enzymes in combination. However, a number of difficulties remained which influenced the success of the isolation. For one, cell wall digestion did not proceed synchronously, and considerable heterogeneity was observed in the size of the protoplasts formed, some of them being able to cross a 0.22-μm membrane while others measured greater than 0.5 μm (Figure 2A). Another problem was the difficulty in separating cellular debris from true protoplasts. Because of debris and spores present in the suspension as well as the small size of some of the protoplasts, the

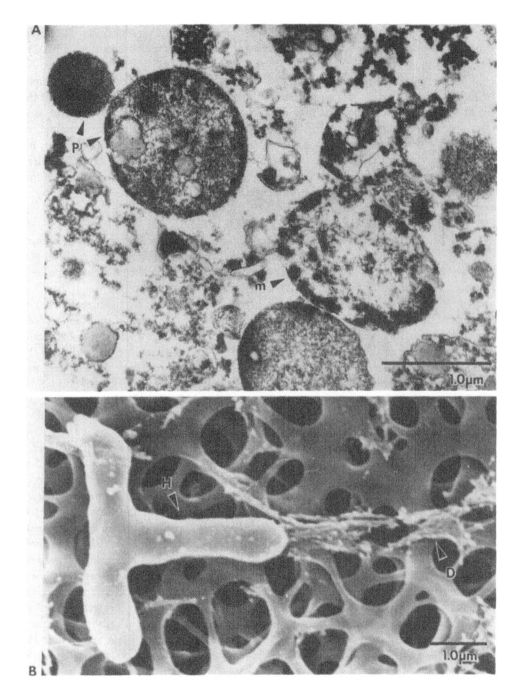

FIGURE 2.   (A) TEM of sections through protoplasts from strain Arl3. Note the marked variation in the size of protoplasts ranging from less than 0.1 μm to more than 0.5 μm and the cell debris present with the protoplasts. (B) SEM of the structure frequently observed during the regeneration process of *Frankia* Arl3 strain protoplasts. Protoplasts appeared to fuse before forming hyphal-like structures.

estimation of the number of protoplasts seen by phase microscopy may be grossly inaccurate. Transmission electron micrographic analysis of such a suspension demonstrates that protoplasts with a continuous cell membrane free of cell wall debris are indeed present (Figure 2A).

Tisa and Ensign,[142] using different *Frankia* strains, have reported that lysozyme alone was sufficient for total digestion of the cell wall and that subsequent separation of protoplasts from cell debris and intact cells was unnecessary. However, the concentration of lysozyme was found to be critical; the optimum was 250 μg per ml.

All the published methods report that an optimal conversion of hyphae to protoplasts requires young cultures, previous growth in medium containing glycine, and the presence of $MgCl_2$ and $CaCl_2$ in the lysis mixture. However, it must be pointed out that the strains and growth conditions used in the different published reports are quite different and may result in the discrepancies noted above regarding lysozyme efficiency. Some strains, e.g., Cc1.17, are difficult to convert into protoplasts. Such strains generally were also difficult to lyse, even under conditions for DNA extraction.[142a] This supports the importance of initial strain selection for the development of a cloning system for *Frankia*.

C. REGENERATION OF *FRANKIA* PROTOPLASTS

The first successful procedure for regeneration of *Frankia* protoplasts was presented in 1984 by Tisa and Ensign at an international symposium on *Frankia* and actinorhizal plants and was published in 1987.[142] Tisa and Ensign described the optimal conditions for protoplast regeneration which, for four isolates, consisted of 0.3 *M* sucrose, 5 m*M* $CaCl_2$, and 5 m*M* $MgCl_2$ at 25°C. The first colonies were observed in 14 days, and maximum efficiency was obtained after 40 days. Further investigations to improve regeneration conditions were restricted to two isolates. Regeneration efficiency was increased by addition to the medium mentioned above of 25 m*M* MOPS buffer, pH 6.0—6.5, for strain CpI1, and pH 7.0 for strain EAN1pec. Finally, regeneration was optimized when *Frankia* protoplasts were sandwiched between this medium containing 1.4% agar and a layer of low-melting point agarose containing 0.3 *M* sucrose, 5.0 m*M* $CaCl_2$ and 5.0 m*M* $MgCl_2$. Under these conditions, efficiency of regeneration was 37% for CpI1, 27% for EAN1pec, 20% for EUI1c, but only 1.3% for ACN1AG.[142]

Normand et al.[146] obtained protoplast regeneration using a medium which contained 0.5 *M* sucrose, 5 m*M* $MgCl_2$, 10 m*M* sodium succinate, 0.1% (w/v) Difco Noble Agar, 5 m*M* $CaCl_2$, and 25 m*M* TES-NaOH pH 7.2. The agar overlay consisted of 0.5 *M* sucrose, 5 m*M* $MgCl_2$, 5 m*M* $CaCl_2$ and 0.8% (w/v) Difco Noble Agar. Scanning electron micrographic observations demonstrated that the protoplasts appeared to fuse and form hyphal-like structures, which later developed into true colonies (Figure 2B). Because of problems in counting protoplasts, results were not expressed as a regeneration percentage but rather according to the number of colonies visualized. Polyethylene glycol (PEG) with a molecular weight of 1000 or 6000, frequently used in the transformation and fusion of *Streptomyces* protoplasts,[147] was found to increase the number of colonies regenerated relative to an untreated control. In an attempt to increase regeneration efficiency further, other parameters were studied such as the concentrations of Mg^{2+} and Ca^{2+} as well as osmolarity, age of cells, or agar quality, but results were similar to those described by Tisa and Ensign in 1984. A temperature range of 20 to 30°C was found to be suitable for regeneration; some delays were found with different temperatures.

D. PLASMID VECTORS AND HOST STRAINS

Frankia plasmids have been discussed in detail in a recent review.[7] They vary in size and so far no phenotype has been ascribed to 26 of 27 described indigenous plasmids. The nitrogenase genes have been localized on plasmid pFQ69, a 190-kb plasmid detected in strain ARgP5[AG].

Small size, presence of unique restriction sites, high copy number, and stability are some of the requirements for a good cloning vector. The 8.3-kb plasmid, pFQ31 detected in strain ArI3, meets some of these requirements and was analyzed physically.[135] Single sites for *Bcl*I and *Eco*RI have been determined. There are no restriction sites for *Bam*HI,

*Bst*EII, *Eco*RV, *Hin*dIII, *Pst*I, *Xba*I and *Xho*I. Moreover, pFQ31 is present at high copy numbers; values of 50 to 200 have been determined, and the plasmid is reasonably stable. Because three different strains, isolated from geographically distinct locations, possess the same plasmid, it can be expected that pFQ31 is a broad-host-range plasmid capable of replicating in a number of *Frankia* strains.

However, there are certain difficulties with using derivatives of pFQ31 as cloning vectors. Introduction of derivatives of pFQ31 into ArI3 and strains which normally possess it as an indigenous plasmid could be hindered by incompatibility between the two plasmids. To overcome this problem, one could clone into other *Frankia* strains or other actinomycetes for which no plasmid or no plasmid incompatible with pFQ31 has been detected. However, when a pFQ31 derivative with an antibiotic marker was transformed into *Streptomyces lividans,* there was no success.[135] Another approach is to cure strain ArI3 of its native plasmids. Plasmid curing has been achieved by a number of methods including heat treatment,[148] treatment with chemical agents[149] such as acridine or ethidium bromide, or by regenerating protoplasts.[147] In the course of regenerating *Frankia* protoplasts, Normand et al.[146] tested 17 ArI3 and 20 AcoN24d colonies derived from regenerated protoplasts from the presence of two genetic markers, plasmids, and various symbiotic properties. All of the regenerated colonies were still infective, forming nitrogen-fixing nodules on *Alnus glutinosa* seedlings. However, one of the ArI3 regenerants, designated LC2, had lost its two plasmids. Evidence for the loss of the plasmids was obtained by both agarose gel electrophoresis of *in-situ* lysed cells[39] and hybridization with radioactive probes. For other regenerants, the plasmid copy number was found to be higher than in the parent ArI3 strain indicating that plasmid content is affected by the isolation and regeneration procedure or that plasmid copy number control was mutated by the treatment.

Several reasons may explain the difficulty in obtaining a cloning system for *Frankia*. First, the conditions for protoplast formation and regeneration are still not optimal and the conditions for transformation are unknown. New procedures for introducing plasmid DNA by electroporation, as has been used for various bacteria, including *Streptomyces lividans,*[150] have not yet been used for *Frankia*. Lack of transformation might also be due to the presence of restriction systems for plasmid DNA as was found in *Streptomyces fradiae.*[151] Highly transformable mutants of *S. fradiae* defective in several restriction systems have been obtained after repeated mutation and selection.[152] Absence of transformation might also be due to the lack of selective markers, such as antibiotic resistance. Recently, *Frankia* strains have been isolated which produce antimicrobial/antifungal metabolites.[43a] Production of and resistance against these antibiotics can become useful tools as specific markers.

E. TESTING GENE FUNCTION IN *FRANKIA*

So far, attempts to generate *Frankia* mutants by classical mutagenesis protocols have met with limited success. In order to assign gene functions to specific DNA fragments, it is necessary to develop methods to mutagenize *Frankia* strains not only by using molecular biological tools but also through classical mutagenesis techniques. In this way, symbiotic genes in *Frankia* can be more readily identified. When reliable mutagenesis techniques are established, an effective means of reintroducing the DNA fragments of interest into the bacteria to confirm their suspected gene function must also be developed. Hence, a successful transformation system also needs to be established.

Some *Frankia* mutants have been described. Antibiotic resistance and pigment synthesis have been used extensively in genetic studies of bacteria. Normand and Lalonde[7] showed background sensitivity of different *Frankia* strains to various antibiotics. Some natural strains showed resistance to streptomycin while most others are sensitive.[7] So far, only few natural antibiotic resistant *Frankia* strains have been reported. Dobritsa[153] reported six strains resistant to nine antibiotics (ampicillin, penicillin G, kanamycin, clindamycin, lincomycin, polymixin B, erythromycin, oleandomycin, and streptomycin). After treatment of various *Frankia* strains with *N*-methyl-*N*-nitro-*N*-nitrosoguanidine (NTG), various kanamycin-resistant AGP and AGN strains have been isolated.[153a] This kanamycin resistance remained stable even in the absence of kanamycin and after reisolation of the strains from nodules, indicating its usefulness as a marker. The nature of the resistance is still unknown. DNA from the kanamycin-resistant strain AgpM2 did not hybridize with the neomycin phosphotransferase gene (NPT) from *Streptomyces fradiae*.[153b] In addition, Lie et al.[154] reported the isolation of NTG-mutants of *Frankia* AvcII that were ineffective on *Alnus glutinosa*. Preliminary results indicate that strain Ar13 is sensitive to erythromycin and thiostrepton. No pigment marker has been detected in strain Ar13 although different pigments are synthesized by various other *Frankia* strains.[155-157]

Some naturally occurring strains of *Frankia* may be useful genetic tools once a transformation system is developed. For example, Normand et al.[57] reported the isolation of strain AiR12 from *Alnus rugosa*, which may be a Nif deletion mutant. When AiR12 was reinoculated onto young *Alnus* roots, nodules were induced (Nod[+]) but the nodules were ineffective (Fix[-]).[198] Using a cloned *nifHDK* probe derived from strain Ar13, these authors were unable to detect any signal under conditions where DNA from other *Frankia* strains hybridized strongly to the probe. Similar results have been obtained with ineffective *Frankia* strains on *Alnus glutinosa*,[158] whose DNA fails to hybridize with a *Frankia nifHDK* probe.[158a] These natural Nif deletion mutants may prove to be a useful candidate for testing the restoration of the Nif[+] phenotype concomitant with the introduction of *nif* DNA via protoplast transformation.

IV. GENETICS OF THE HOST

So far, we have not considered the plant partner in the *Frankia*-actinorhizal symbiosis. In part, this is because little is known regarding the genetics and/or molecular biology of the host. As mentioned previously, there are a number of developmental parallels that are observed between the *Rhizobium*-legume and the *Frankia*-actinorhizal plant symbiosis. The developmental steps leading to the establishment of a nitrogen-fixing, actinorhizal nodule have been considered in a number of recent reviews and, hence, will not be discussed here.[159,160] As discussed earlier, a number of symbiotic genes are conserved between the two prokaryotes, namely, *nif*, *hup*, GS2 and perhaps others. We can now ask the question as to whether various plant genes are also conserved

Plant proteins specifically formed in response to the plant-microbe interaction are termed "nodulins"[161] (see also Chapter 8). In many legumes, approximately 20 to 30 nodulins are detected in effective, nitrogen-fixing nodules. For example, in effective root nodules of alfalfa, 17 nodule-specific translation products have been identified.[213] Several of the nodulins have been identified, including uricase,[162] sucrose synthetase,[163] nodule-specific glutamine synthetase,[164] and leghemoglobin.[165]

Leghemoglobin (Lb) is one of the most characteristic proteins found in legume root nodules. Approximately 20 to 25% of the total protein found in nitrogen-fixing nodules of some species is comprised of Lb. In soybean, Lb represents a small gene family and some Lb genes have been sequenced.[165,166]

Nodules of a number of actinorhizal plants contain hemoglobin on the basis of spectrophotometric measurements but others do not.[83,165,167] Those nodules that contain hemoglobin include those present on roots of *Casuarina* and *Myrica* (found at quite high levels), and also nodules of *Alnus* and *Elaeagnus*, albeit at low levels.[83] The nonlegume *Parasponia* (Ulmaceae),[168] which is nodulated by *Bradyrhizobium* sp., develops nodules which contain Lb-like proteins. Recently, Flemming et al.[169,170] described the purification, characterization, and amino acid sequencing of a soluble hemoglobin from root nodules of *Casuarina glauca*.

This protein resembles other known plant hemoglobins, with similar oxygen-binding properties. Western blot analysis showed close immunological relationships between *Casuarina* and *Parasponia* hemoglobins and a weaker relationship between these two proteins and soybean hemoglobin. The amino acid sequence of *Casuarina* hemoglobin shows extensive homology with other plant hemoglobins: 52% with *Parasponia andersonii* hemoglobin I, 49% with lupin Lb II, and 43% with soybean Lb a.[170]

Several techniques for isolating high molecular weight DNA from actinorhizal plants have been developed. Such procedures facilitate studies of gene expression in actinorhizal plants. Restrictable DNA has been prepared from seeds,[171] leaves,[172] and callus and suspension cultures.[173] In the latter case, protoplasts were first isolated from the callus or cell suspensions by enzymatic treatment. To prevent the formation of a complex between the DNA and an orange pigment found in leaves, 0.1 *M* disodium diethyldithiocarbamate was added to the extraction buffer used by Roberts et al.[172] The presence of various pigments (mainly quinones and polyphenols) which complex with nucleic acids has hindered the isolation of translatable RNA from nodules and roots of actinorhizal plants. Hirsch and Bisseling[172a] have isolated RNA from nodules and roots of *Alnus glutinosa* and *Datisca cannabina* and have used this RNA on Northern blots, but found that the RNA does not translate in an *in vitro* rabbit reticulocyte system.

Leghemoglobin-like DNA sequences in a number of actinorhizal plants have been reported following DNA-DNA hybridization with Lb cDNA clones derived from soybean mRNA. A partial cDNA clone from soybean Lb mRNA hybridized quite strongly to DNA isolated from *Alnus glutinosa*, *Casuarina glauca*, *Ceanothus americanus*, and *Elaeagnus pungens*.[172] Using several soybean Lb cDNA probes, Hattori and Johnson[171] detected hybridization to genomic DNA isolated from *Alnus crispa*, *A. rugosa*, and *Myrica gale*, as well as two from nonnodulating plants, *Ceratonia siliqua*, a legume, and *Betula alleghaniensis*, a nonlegume taxonomically related to *Alnus*.

However, such relatedness between legume and nonlegume hemoglobin is not necessarily maintained at the protein level. Using antibodies made to pea and soybean Lb, no signal was detected on Western blots to nodule protein extracts of *Alnus glutinosa*.[172a] Similar results were obtained by Flemming,[169] using an antibody against soybean Lb and protein extracts of *Casuarina glauca*. Until the DNA sequences exhibiting hybridization to soybean Lb are cloned and a corresponding mRNA is found to be expressed in actinorhizal nodules, there is no evidence to suggest that these Lb-like DNA sequences are indeed functional. Interestingly, soybean leghemoglobin and *Parasponia* hemoglobin exhibit a protein homology of about 40%, but a soybean Lb cDNA did not hybridize to *Parasponia* DNA or nodule RNA, even under conditions of low stringency.[174] Recently, cDNA and genomic clones for hemoglobin have been isolated from two *Parasponia* and sequenced.[174] The *Parasponia* gene showed more than 50% nucleotide sequence similarity with hemoglobin genes of soybean and kidney bean and also hybridized to hemoglobin genes of *Casuarina* and to DNA from *Trema*, a close, but nonnodulating, relative of *Parasponia*.[6,21] Whether this gene is functional in *Trema* and, if it is, what it encodes are unknown. These authors speculate that hemoglobin-like genes may be expressed cryptically in nonsymbiotic tissues in all plants.

Nodules of *Datisca glomerata* have shown to lack hemoglobin as measured by spectrophotometry of nodule segments, but they demonstrate rates of nitrogen fixation equivalent to nodules of *Casuarina*, which do contain hemoglobin.[83] Also, little or no hemoglobin has been detected by spectrophotometric measurements in nodules of *Ceanothus americanus*,[175] yet a relatively strong signal is detected on a Southern blot using a partial soybean Lb cDNA clone as a DNA probe.[172]

Why do some actinorhizal nodules have hemoglobin and others not? Tjepkema et al.[175] have postulated that the absence or low concentration of hemoglobin in certain actinorhizal nodules may be related to the relatively free access of O_2 to the microsymbiont, while in other nodules, O_2 access is restricted. For example, *Casuarina* nodules contain relatively high levels of hemoglobin. The host cell walls surrounding the microsymbiont, which exists exclusively in the hyphal stage, are suberized[85,176] and this suberization may restrict O_2 diffusion to the *Frankia* cells. Likewise, "fixation" threads (threads confining *Bradyrhizobium* cells) in nodules of *Parasponia*, which also contain hemoglobin, have a suberin-like component.[177] The large variation in presence of hemoglobins in actinorhizal plants, in the occurrence of vesicles by *Frankia* (see above), in suberization of cell walls, and in the presence of O_2-protection enzymes, indicate that various strategies might have been developed in plant-*Frankia* symbiosis to overcome O_2 diffusion limitations and damage of nitrogenase by free O_2.

In contrast to Lb, antibodies made to *Phaseolus* glutamine synthetase (GS)[178] cross-reacted with proteins isolated from *Datisca cannabina* and *Alnus glutinosa* on Western blots, and preliminary results suggested that there may be nodule-specific or nodule-enhanced isoenzymes.[172a] The majority of the actinorhizal plants investigated (*Elaeagnus* sp., *Comptonia peregrina, Ceanothus americana,* and others) to date transport asparagine or glutamine (amide exporters) while *Alnus* spp. and others transport citrulline (ureide transporters).[179] Blom et al.[180] have detected GS activity in nodules of *Alnus glutinosa* and this enzyme has been localized in the host cytoplasm.[180,181] In addition to GS, other nitrogen assimilation enzymes have been found in actinorhizal nodules, including ornithine carbamoyl transferase (OCT), which is involved in the conversion of fixed ammonia to citrulline, and glutamate dehydrogenase (GDH).[180,182] Although frequently enhanced in nodules, none of these proteins has been demonstrated unequivocally to be nodule-specific, i.e., to be nodulins. Therefore, so far, we have very little knowledge of how many or what type of nodule-specific proteins are produced by actinorhizal nodules.

V. TAXONOMY OF *FRANKIA*

Since 1978[33] an increasing number of *Frankia* have been isolated,[183] and attempts have been made to classify the strains. So far all strains have been classified within the genus *Frankia*.[184] This taxon is characterized by:

a.   The ability to nodulate plants;
b.   The ability to fix nitrogen;
c.   Its unique morphological properties, vesicles, and sporangia;
d.   The presence of a sugar, 2-*o*-methyl-D-mannose;[185]
e.   The presence of cell wall type III and phospholipid content of type I of actinomycetes;[156] and
f.   A high G + C% in the range of 68 to 72%.[31]

As recalled by Lechevalier,[3] the first classification within the genus *Frankia* was based on host plant relationships using crushed nodules as inocula sources.[184] However, further investigations with pure cultures, reviewed by Normand and Lalonde[7] and Baker,[186] demonstrated that the host specificity groups were quite different. Lechevalier[3] summarized the problems to define species within the genus *Frankia* and proposed that new criteria had to be developed for classification. Particularly, the development of new biochemical and molecular biological techniques are powerful tools to understand the phylogenetic position of *Frankia*. These include methods to measure large phylogenetic distances of species-kingdom ranks (5/16S rRNA oligonucleotide cataloging), of species-order ranks (DNA:16/23S rRNA pairing and sequencing of proteins with conserved primary structures), of species-family ranks (RNA secondary structure and comparative immunology), and methods to measure small phylogenetic distances among closely related species (DNA:DNA pairing and electrophoretic pattern of proteins).[187] So far, the information on the phylogenetic position of actinomycetes in general, and *Frankia* in particular, is limited. Most information is now available on parameters which provide information on discrimination between closely related strains, while more basic information on phylogenetic relationships with other actinomycetes are still limiting.

A. 16S RIBOSOMAL RNA SEQUENCES

16S ribosomal RNA (rRNA) oligonucleotide analyses[188] have been used to determine phylogenetic relationships among bacteria. This method has recently been applied to actinomycetes, including one *Frankia* strain.[189] These observations have been extended to other *Frankia* strains.[189a] (Figure 3), indicating that at least two *Frankia* strains are phylogenetically related to *Geodermatophylus obscurus*. This confirms the previous classification of *Frankia* and *Geodermatophilus* in a taxon "Multilocular sporangia", based on morphological features and cell wall type (III).[187] By measuring homology values of 16S RNA, further information has been obtained on the phylogenetic relationship of *Frankia* with actinomycetes. Recent data by Hahn et al.[189a] show the relationship of *Frankia* Ag45/Mut15 with a number of actinomycetes (Figure 4). Unfortunately, no information is available on the 16S rRNA sequences of the "cell-wall type III" actinomycetes in the upper rank of Figure 3. Although these techniques seem quite promising, more strains have to be analyzed before general conclusions can be drawn.

B. THE NITROGENASE GENES AS A TAXONOMIC TOOL

Since it was demonstrated that the nitrogenase structural genes from *K. pneumoniae* hybridized strongly to DNA of all N_2-fixing bacterial species,[45] it was first thought that *nif* genes were a recent "purchase" of these prokaryotes.[190-192] The very interesting results obtained by Hennecke et al.,[58] showing that nitrogenase genes and 16S rRNA of several bacteria had evolved in parallel would speak against this idea. For instance, they have compared *nif*H sequences (the most conserved gene) and 16S rRNA of slow- and fast-growing rhizobia and found that they possessed a low degree of similarity, confirming differences detected by other taxonomic criteria (see Reference 58). Normand and co-workers[57] have determined the complete nucleotide sequence of *nif*H from *Frankia* strain ArI3 and compared it to other published sequences. They have found that this *Frankia nif*H gene had about 70% positional similarity with *A. chroococcum*, *R. meliloti*, and *K. pneumoniae* DNA *nif* sequences[193] with some areas having a higher similarity degree of 80% and even 90% over short stretches of about 50 bp. If the putative amino-acid sequence of *Frankia* was found to be closest to *Anabaena* (80%), the degree of similarity was found to be lowest with *Methanococcus voltae* (50%),[194] the second lowest degree of similarity being found with the other Gram-positive bacterium *Clostridium pasteurianum* (60%). The latter finding would imply that at least in the case of *Frankia*, lateral gene transfer did occur, a conclusion contrary to that of Hennecke et al.[58] Recently, Ligon and Nakash[195] also sequenced *nif*K *from Frankia* FaC1. However, there are few published *nif*K sequences from other microbes to which comparison can be made. Nevertheless, there is amino-acid sequence similarity with *Anabaena* and *Azotobacter chroococcum* of 50 and 65%, respectively.

The question remains whether nitrogenase genes can provide a taxonomic tool within the genus *Frankia*. Of course, it is unrealistic to determine *nif*H sequences routinely from a large number of strains. Moreover, this technique has mainly been applied for discerning phylogenetic relationships and could be successful in delineating species. However, various experiments would also favor a different interpretation. First of all, using DNA-DNA hybridizations[30,195a] there are indications that the genetic diversity among *Frankia* is large, a very low level of homology being found between some isolates. Using the *Frankia* ArI3

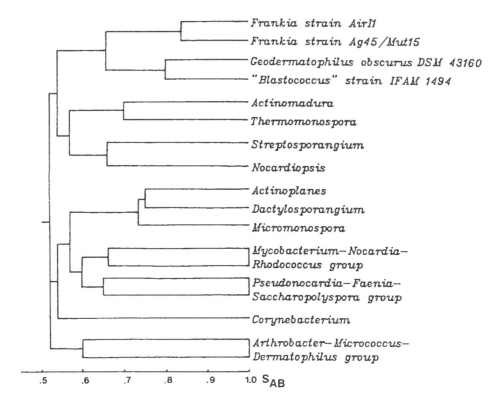

FIGURE 3. Dendrogram showing the phylogenetic position of *Frankia* within the Actinomycetes, determined by comparative 16S rRNA oligonucleotide analysis.[189a]

nifH gene as a probe against DNA from different nitrogen-fixing microorganisms, Simonet et al.[58a] have demonstrated that most *Frankia* strains have lost the *nif*-hybridizing band under strong stringencies. Under similar stringency conditions, *nif*-hybridizing bands also disappeared in *R. meliloti, Azospirillum lipoferum,* or *Pseudomonas paucimobilis*. On the other hand, the other *Alnus*-compatible strain ACoN24d exhibited a more resistant signal, indicating a very high level of similarity.

Sequencing the two genes would confirm this very high level of similarity predicted also by physical analysis of the *nif* region in ArI3 and ACoN24d. The use of seven restriction enzymes to produce a map has shown that ArI3 and ACoN24d are almost identical (except for a small deletion of about 50 bp in ACoN24d) for a region of 7.7 kb carrying *nifHDK* and pectate lyase *(pel)* genes.[58a] On the other hand, hybridization experiments and restriction mapping of two *Elaeagnus*-compatible strains and one *Casuarina*-compatible strain have shown that the respective *nif* regions were more divergent than between the *Alnus*-compatible strains. In all cases, similarity was highest in a zone corresponding to *nifH* gene and much less in the zone of *nifD*.

Normand and co-workers[57] have used the *nifHD* region of *Frankia* ArI3 as an homologous probe to detect *nif* RFLPs. For most *Alnus*-compatible strains tested, hybridization occurred with an 8 kb *Bam*HI fragment and a 10-kb *Bgl*II fragment. With *Sst*I, which has more sites, hybridization was strictly restricted to the *nif* genes and occurred in all *Alnus*-compatible strains with two fragments of 1.3 and 0.4 kb. These data and those obtained with other *Frankia* strains confirm the observations of Hennecke et al.[58] on nitrogenase genes of other nitrogen-fixing microorganisms. Among nitrogenase genes, *nifH* appeared to be the most conserved. The degree of similarity is very high for strains for which close relationships have been determined with other criteria, and less for more divergent strains. Similar ob-

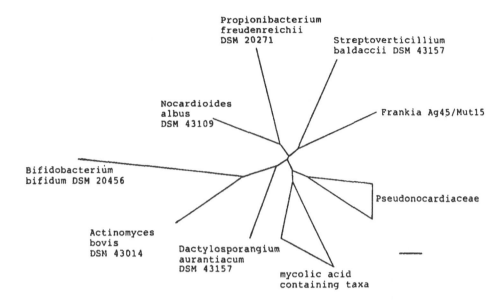

Propionibacterium
freudenreichii
DSM 20271

Streptoverticillium
baldaccii DSM 43157

Nocardioides
albus
DSM 43109

Frankia Ag45/Mut15

Bifidobacterium
bifidum DSM 20456

Pseudonocardiaceae

Actinomyces
bovis
DSM 43014

Dactylosporangium
aurantiacum
DSM 43157

mycolic acid
containing taxa

FIGURE 4. Dendrogram of relationships displaying the phylogenetic position of *Frankia* and some actinomycetes, determined by 16S rRNA sequencing. Additional information on actinomycetes of the cell wall type III other than *Frankia* has not been available.[189a]

servation were found for *nifD* gene with a higher amplitude of the divergences observed for more divergent strains. This means that nitrogenase genes are some of the most highly conserved portions of the genome, as mentioned by An et al.,[30] however, sufficiently divergent to provide information on relationships between strains. Hennecke et al.[58] have demonstrated that the nitrogenase genes and 16S rRNA have evolved in parallel, which provides powerful tools to study nitrogen-fixing bacteria. Such results argue against the idea of lateral transfer of genes between nitrogen-fixing bacteria, and for the notion that *nif* genes have evolved in parallel and independently in the bacteria that carry them. These observations further suggest that measurement of *nif* sequence similarity could be used as a taxonomic criterion in addition to 16S rRNA.

C. DNA HOMOLOGY

Genomic DNA-DNA hybridization has been used extensively in bacterial taxonomy and is considered useful in the determination of species. Comparing these data with those obtained by other criteria, Brenner[196] established the 70% boundary whereby two isolates belong or do not belong to the same species. Analyses using genomic DNA-DNA or Southern hybridization techniques have demonstrated the usefulness of these particular methods for studying genetic diversity in *Frankia*. An et al.,[30] Simonet,[195a] and Dobritsa and co-workers[196a] have found that genetic diversity among *Frankia* strains is large; in fact, a low level of homology is found between some isolates. *Alnus*-type strains tested for DNA sequence similarity against strain ArI4 (isolated from *Alnus rubra* nodules) exhibited hybridization in the range of 67 to 94%.[30] In contrast, hybridization levels of the *Alnus*-compatible strains to an *Elaeagnus*-compatible strain were always lower than 50%. Even within the *Elaeagnus* group, a low level of DNA-DNA hybridization was observed indicating that strains within this group are genetically diverse. However, genomic DNA-DNA hybridization has been applied to only a few *Frankia* strains because this method requires a substantial number of cells as starting material to isolate enough DNA. The production of large amounts of cellular material still remains a stumbling block with *Frankia*.

D. PROTEIN AND ISOZYME PATTERNS

Identification of strains by relatively simple and reliable methods is often needed to discriminate between closely related organisms. In addition to the use of antisera,[197-199] restriction analysis of the genome[32,38,200] and specific DNA or RNA probes (see above), determination of whole cell protein patterns[199,201-203] proved to be promising, particularly one-dimensional SDS-PAGE analysis. This method demonstrates the large diversity in *Frankia* populations, even within the same nodule and isolates from the same place in soil. Based on 1-D SDS-PAGE protein analysis, Gardes and Lalonde[203] characterized and identified 35 strains isolated from various hosts of different geographic origin. They distinguished two groups: group A comprising strains isolated from several *Alnus* spp., *Comptonia peregrina* and *Myrica gale,* and group B, composed mainly of strains of the *Elaeagnus* host specificity group. However, five strains isolated from nodules of *Myrica gale* or obtained in the laboratory from *Alnus crispa* plantlets inoculated with soil possess electrophoretic characteristics typical of the *Elaeagnus* strains. Saint Laurent and Lalonde[204] have found that three of these strains, MGX34a, MGX34b, and MGX34c, isolated from nodules of *Myrica gale,* were able to nodulate both *Alnus* and *Elaeagnus* plants. This would confirm the idea that nodulation tests alone are not sufficient for classification purposes.

Related strains can also be distinguished on the basis of electrophoretic separation of isozymes. Gardes et al.[205] characterized *Frankia* strains by the electrophoretic separation of isozymes of eight different enzymes. They analyzed diaphorase, leucine aminopeptidase, peptidase, phosphoglucose isomerase, esterase, malate dehydrogenase, phosphoglucomatase, and superoxide dismutase. The results obtained from this analysis are beyond the scope of this review, but they allowed the authors to classify strains by numerical analysis based on pair-wise similarity coefficients, and further, to propose the delineation of groups of strains.

Previously, Steele and Stowers[82] observed that *Frankia* possesses relatively high activities of SOD, and catalase. These workers found two superoxide dismutases, one, a MnSOD, that functions when cells are grown with ammonium as the nitrogen source, and the second, a FeSOD that is present along with the MnSOD under nitrogen-fixing conditions. Using PAGE, Puppo et al.[206] determined the patterns of these and several other enzymes in an attempt at *Frankia* strain identification. When *Frankia* cells were grown in complete medium, one SOD isozyme with the same relative mobility was detected for all the *Casuarina*-compatible strains tested. However, the relative mobility of this enzyme was different in the *Alnus*-compatible strains. In contrast, there was significant heterogeneity when other enzymes, namely, oxidase, esterase, and diaphorase, were analyzed in various strains isolated from *Casuarina* nodules, most likely because some strains isolated from these nodules are not *Casuarina*-compatible (see Section E). However, this heterogeneity was significantly reduced when only *Casuarina*-compatible strains were considered. Further details on other strains[205] confirmed the usefulness of this method to strain identification and delineation of groups of strains.

E. HOST-COMPATIBILITY GROUPS

1. General

As recalled by Lechevalier,[3] the first classification within the genus *Frankia* was based on host-plant relationships, using crushed nodules as inocula.[184] However, further investigations with pure cultures demonstrated that the host specificity groups were quite different than originally proposed. This subject has been reviewed recently in more detail.[7] Recently, Baker[186] performed cross-inoculation studies using 50 strains of *Frankia* and 6 species of actinorhizal plants. The most important conclusion drawn from this study is that host specificity groups are found to be more complex than previously thought, and also that they overlap. For example, the actinorhizal plant *Myrica cerifera* is nodulated by all strains which nodulate *Alnus* species, and also by *Frankia* strains which nodulate *Casuarina* species. But according to Baker,[186] this type of host "promiscuity" appears restricted to *Myrica cerifera* because neither *M. rubra* nor *M. californica* possess so broad an endophyte range.[207] Baker[186] has also demonstrated that there are two types of *Frankia* strains in the *Elaeagnus* compatibility group, and they either nodulate or do not nodulate *M. cerifera*. From these observations, Baker[186] has proposed at least four *Frankia* cross-inoculation groups: one group with strains infective on *Alnus, Comptonia,* and *Myrica;* one group with strains able to nodulate *Casuarina* species and *Myrica cerifera;* a third group composed of strains which nodulate Elaeagnaceae and *M. cerifera;* and a fourth group of strains strictly restricted to host plants of the family Elaeagnaceae. However, there still is little information available on the nodulation ability of many *Frankia* strains because so few strains have been isolated and grown in pure culture. The fact that some *Frankia* strain-host plant combinations are delayed in nodule development also causes errors concerning the nodulation ability of a particular strain. Significant delay in nodulation has been observed with strains which induce ineffective nodules on *Alnus*. Hahn et al.[158] described the nodulation pattern of nonnitrogen-fixing *Frankia* on *A. glutinosa*. The ability to nodulate was found to be both dependent on the plant clone and the actinomycete strain. These observations confirm previous observations by van Dijk[208] on host-dependent nodulation of ineffective endophytes. Some ineffective, Nif⁻ strains gave great delay in nodulation or even failed to nodulate.[158] Those organisms which fail to nodulate, which do not form vesicles, and which do not fix nitrogen are difficult to distinguish from other soil actinomycetes. Their identity can be confirmed by using specific probes against unique sequences in the 16S rRNA.[186a] In addition, some strains are capable of crossing barriers and nodulating different hosts. Indeed, host groups alone are insufficient to insure the definition of *Frankia* species. However, the response to certain hosts must be considered as the expression of specific *Frankia* genes, as of yet uncharacterized, and this response must be used as one of the numerous criteria on which taxonomy has to be based.

2. *Alnus*-Compatible Strains

Various analyses which offer potential for studies on *Frankia* taxonomy have been developed in a number of laboratories. Table 1 was constructed to determine whether results obtained with protein analysis and DNA-DNA hybridization could be correlated. We compared the groups generated by the investigation of Gardes et al.,[205] the groups and subgroups defined by gel electrophoresis protein patterns,[203] data generated from *nif* RFLPs,[57] and the groups obtained using DNA-DNA hybridization.[30,195a] The three large isozyme groups delineated by Gardes et al.[205] correlate well to the electrophoresis subgroups defined by SDS-PAGE.[203] Two strains in the A2 protein group, CPI1 and ACNIAG, exhibit a relatively high level of DNA-DNA hybridization and also hybridize to an ArI4 *nifHD* probe. The size of the DNA fragment hybridizing to the probe is almost identical for all the *Frankia* strains in this subgroup. A 8-kb *Bam*HI fragment was detected only in the A2 protein group. These strains were sampled from geographically different areas (U.S., Canada, and France), and represent several host plant species. Thus, a good correlation between isozymes and protein patterns on one hand, and hybridization to a specific probe on the other, can be noted. In addition, DNA-DNA hybridization analysis indicates a strong similarity level (80%) for these strains, but a comparable level of hybridization was detected for the slightly less similar (on the basis of other criteria) AGN1g strain. A 10-kb *Bam*HI DNA fragment from the strain ARgX17c, which falls into a different isozyme subgroup, was found to hybridize to the *nifHD* probe. Other strains, not examined by Gardes,[203,205] can be allied to this group. For example, Benson and Hannah[201] described four strains, Ai20-4, Ai21, Ai22, and Ai23-2, which are quite similar to strain CpI1 (protein subgroup A2) in protein pattern. Strains ArI5 and AvN16a also exhibit hybridization patterns to a *nifHD* probe similar to that observed for the typical A2 subgroup.[205a] Furthermore, An et al.[30] have determined that DNA from

TABLE 1
Relationships among *Frankia* Strains

Strains (compat. group.)[a]	Isozyme group (205)	Protein group (203)	ArI4 (30)	EUIIa (30)	EUNIf (**)	HRNI8a (**)	AGNIIa (**)	BamHI (57)	BglII (57)	SstI (57)
			DNA-DNA hybridization with (%)					Size of the *nifHD* bands (RFLPs)		
ACNIᴬᴳ(A)	1	A2	90	30				10;8	10	1.3;0.4
ANNI(A)	1	A2						8		
ArI3(A)	1	A2						8	10	1.3;0.4
CpII(A)	1	A2	81	14				8	10	1.3;0.4
TNI8bᴬᶜ(A)	1	A2						8	10	1.3;0.4
TX38bᴬᶜ(A)	1	A2								
94/////////ARbN4b (30)										
TX41bᴬᶜ(A)	1	A2								
ACoN24d(A)	1	A2						8	10	1.3;0.4
ACN14a(A)	1	A2								
CPX32b(A)	1									
A2////Ai22; Ai23; Ai20-4; Ai21 (201)										
ACX50h(A)	1	A5								
ARgX15i(A)	1	A5								
ArI5, AVN16a(***) /////////8/////// 10 /// 1.3;0.4										
ARgN22d(A)	1	A5								
ARgX17c(A)	1							10	10	1.3;0.4
CPX31e(A)	1									
AGN1g(A)	1	A6	81							
CPX34g(A)	1	A6								
72//////AiRII (30 and ***)////////////////10 10////1.3;0.4										
MPX1aᴹᴳ(A)	1	A6								
72///////// MPII (30)										
ASN1cᴬᴳ(A)	3	A8								
67///////// AvCII (30)										
TX10eᴬᶜ(A)	3	EI								
TX10rᴬᶜ(A)	3	EI								
ARgP5ᴬᴳ(A)	3	A4						6; 5	10; 12	1.3;0.4
MGX35a(A)	3	A7								
FaCl (56)/////////////////// 27										
MGX35b(A)	3	A7								
MGX39b(A)	3	A7								
MGX39c(A)	3	A7								
MGX34a(AE)	2	E2								
MGX34b(AE)	2	E2								
MGX34c(AE)	2	E2								
MGX34d(AE)	2									
(57 and **) HRN18a/// 51 ///// 100 /////// 45 ////// 3.8 ////12; 5///1.2; 0.5										
HRX40a /////////////////////////////6; 5/////12///1.6; 0.9; 0.7										
SCN10a(E)	2	E3								
TX30mˢᴬ(E)	2	E3								
TX30bˢᴬ(E)	2									
EANIᴱᴬ(E)	2	E5	21	15						
EANIpec(E)	2	E5	39				21			
TX31eHR(E)	2	E5								
EUNIf(E)	2	E4			100	49	53	4.5;1.4	20;12	1.5;0.4
EUNIfs20(E)	2	E4	18	13						
ARgX10b(A)		A3								
AVP3d(A)		A1								

[a] A, *Alnus* group; E, *Elaeagnus* group.
[b] Simont, P., unpublished results.
[c] Normand, P., Simont, P., and Bardin unpublished results.

ARbN4b hybridized at the 94% level with ArI4 DNA. These data strongly suggest that hybridization criteria are sufficient for classification of *Frankia* strains. The fact that DNA from strain AiRI1 hybridized to ArI4 DNA at 72%,[30] and carries *nifHD* genes on a 10-kb *Bam*HI fragment would tend to confirm the validity of the correlation. However, additional investigations using other parameters are needed.

This combination of data clearly demonstrates that *Frankia* strains from different plants of various geographical origins belonging to the *Alnus* host-specificity group are close enough to be classified as a single species. No doubt, the strains classified in the A2 protein subgroup belong to the same species. However, do we include in this species the strains belonging to the A5 and A6 subgroups? Recalling Brenner's[196] proposal that DNA-DNA hybridization at the 70% level is the limit as to whether strains belong to the same or different species, we can consider protein groups A2, A5, and A6, all of which are in isozyme group 1, as a single species.

Other *Alnus*-compatible strains which do not belong to isozyme group 1 have been classified separately. It is interesting to note that some of these have been classified in the *Elaeagnus* protein group (e.g., strains TX10eAC and TX10rAC), and six other strains have been placed in three different *Alnus* protein subgroups (A8, A4, and A7). Gardes et al.[205] have determined that four of these strains (ASN1AG, TX10eAC, TX10rAC, and ARgP5AG) contain large amounts of 2.0 methyl-D-mannose, a sugar generally encountered in the *Elaeagnus* host specificity group. Unfortunately, DNA-DNA hybridization data are not available. Such values are necessary to delineate the relationships of these strains with other *Frankia* isolates. Only one strain, ARgP5AG, has been examined for DNA fragments hybridizing to a *nifHD* probe. The size of the *Bam*HI and *Bgl*II DNA fragments from ARgP5AG differ from that of the first isozyme group; however, the small *Sst*I bands hybridizing to the *nif* probe are the same size (1.3 and 0.4 kb) as found in the other strains. FaC1, another *Alnus*-compatible isolate, is also not classified with the first isozyme group. Benson et al.[202] have shown that proteins isolated from strain FaC1 do not exhibit the same pattern on 2-D PAGE as proteins isolated from the well-characterized strain CpI1 (isozyme group 1). In addition, Ligon and Nakas[56] have observed a different organization of *nif* genes in FaC1; the *nifH* gene is not contiguous with *nifDK*. These differences would suggest that these strains are taxonomically separate from the first group of *Alnus*-compatible isolates.

3. *Elaeagnus*-Compatible Strains

The relationships among the *Elaeagnus*-compatible strains is even less clear than in the *Alnus* situation because there is greater heterogeneity of the samples used in the various studies. Gardes et al.[205] have shown that the four strains isolated from the same *Myrica gale* plant, MGX34a, MGX34b, MGX34c, and MGX34d, belong to isozyme group 3, as do other *Elaeagnus*-compatible strains. This confirms the results obtained with SDS-PAGE[203] and infectivity tests because these strains nodulate both *Alnus* and *Elaeagnus* plants.[204] Furthermore, *nif* hybridization analysis[57] generates results similar to those produced by comparing protein patterns,[203] isozymes,[205] or DNA-DNA hybridizations.[30,195a] Whatever criteria are used, it appears that the *Elaeagnus* host specificity group is very heterogeneous. The low level of DNA-DNA hybridization (13 to 51%) among *Elaeagnus*-compatible strains suggests that a new species would have to be created for each isolate tested. However, results of isozyme tests on strains EAN1pec and TX31eHR indicate that these two strains from different geographical areas (Ohio and Quebec) may be closely related. More information is needed about these strains before species can be delineated.

4. *Casuarina*-Compatible Strains

Because a number of problems were encountered in the isolation of *Frankia* from nodules of *Casuarina*,[209] little information is available concerning this *Frankia* group. Two host specificity groups have been defined according to the ability to nodulate *Casuarina* or *Allocasuarina* species.[210,211] To date, no *Casuarina*-isolated strains have been found to be infective on *Allocasuarina* plants, while two *Allocasuarina* isolates (HFPA11I1 from nodules of *A. lehmaniana* and ORS 022602 from *A. torulosa* nodules) nodulate *Casuarina* species and some species of *Allocasuarina*.[206] Other investigations have suggested that more than one *Frankia* strain may be found in some nodules. This could explain why some of the strains isolated from *Casuarina* nodules were unable to reinfect the *Casuarina* host plant, but effectively nodulate *Hippophae rhamnoides*.[212] An et al.[30] examined strains ORS 020602 (D11) and ORS 020604 (G2) using DNA-DNA hybridization methods. Unfortunately, neither of these isolates nodulates *Casuarina* and although there appears to be a high level of DNA-DNA hybridization between the two strains, it is difficult to apply this information to cross-inoculation groups or to species determinations. Seven *Frankia* strains isolated from *Casuarina* nodules have been analyzed using PAGE.[206] Preliminary results suggest that superoxide dismutase could be used as a marker for identifying "typical" *Casuarina*-type strains.

VI. CONCLUDING REMARKS

Although much has been accomplished using interspecies homology to clone and characterize *Frankia* symbiotic genes, there is still much to do. *R. meliloti* and *K. pneumoniae* *nifHDK* genes, *R. meliloti* *nifAB* genes, *K. pneumoniae* *nifEN* genes, and *E. chrysanthemi* *pel* (pectate lyase) genes have exhibited enough similarity to DNA sequences in *Frankia* to permit detection of these genes from a *Frankia* gene library. Likewise, the use of heterologous probes such as a cDNA clone made from soybean mRNA has suggested that there are Lb-like sequences in actinorhizal plants. Such sequence similarities have allowed the demonstration that the three *nif* structural genes (*HDK*) are contiguous in some *Frankia* strains, but that a different organization is found in other isolates. For example, in strain FaC1, the *nifH* gene is separated from *nifDK* and in strain ARgP5Ag, some nitrogenase genes are reiterated on a second plasmid-borne replicon.

Evidence for symbiotic gene clustering such as is found in *Rhizobium* is also given by evidence derived by interspecies DNA-DNA hybridization. This is the case of *nifE* and *nifN* which are adjacent to the *nifDK* genes in strain FaC1 and for *nifAB* genes which are approximately 4.5 kb away from the *nifK* gene in strain ArI3. In addition, *Frankia* sequences which hybridize to *pel* genes are positioned close to the structural *nif* genes in ArI3, indicating that a number of symbiosis-involved genes are clustered.

Nodulation sequences (*nod* genes) no doubt are present in *Frankia*, but their isolation and characterization remain elusive. The position of *Frankia* *nod* sequences should be determined in order to see if they, like the *nod* genes in *Rhizobium*, are clustered near the *nif* genes. This will be important not only to determine whether gene organization is similar in the two nitrogen-fixing prokaryotes but also to understand gene regulation in *Frankia*. DNA sequence determination of *nif* genes has given a better understanding of the taxonomy of *Frankia* strains and also of the evolution of the two distinct nitrogen-fixing symbioses. Once *Frankia* *nod* sequences have been isolated, it will be important to sequence the genes and compare the DNA sequence to that of the common *nod* (*nodABC*) genes of *Rhizobium* and *Bradyrhizobium*. Preliminary studies suggest that DNA sequence similarity is slight, however.

Various questions come to mind to pursue regarding these symbiotic genes. For example, what kind of promoter sequence characterizes the *Frankia* *nod*-like gene(s)? Are *Frankia* *nod*-like sequences arranged as operons? Do plant factors regulate the expression of the *nod*-like genes as in *Rhizobium*? Is there a constitutively expressed regulatory gene (*nodD*-like) present in the *Frankia* genome? Are the *nod*-like genes present in more than one copy in *Frankia*?

However, interspecies homology and very likely, functional complementation, is restricted to very a few genes and progress will be stymied in *Frankia* genetics until a transformation system is developed. This will be the next challenge with *Frankia*. Transformation with plasmid vectors via protoplast regeneration seems to be the method to develop, but only a part of this challenge has been solved with the studies described herein regarding production and regeneration of *Frankia* protoplasts. With a transformation system, cloned genes can be reintroduced intro *Frankia*. Mutual analysis to deduce the function of cloned genes can then be pursued and new strains of *Frankia* eventually can be constructed. Successful introduction of heterologous DNA by protoplast fusion between *Frankia* and *Streptomyces* has now been reported[144] and future exploitation of such hybrids in the pharmaceutical industry can be expected.

On the plant side, techniques for the isolation of translatable, good quality RNA must be developed. Once this has been accomplished, cDNA libraries can be constructed and investigations to study nodule-specific gene expression in actinorhizal plants can be established. Questions for future study include the following: are nodulin genes conserved between the legumes and nonlegumes? Are there as many nodulins in actinorhizal plants with their more root-like appearing nodules as there are in legumes? What is the function of the DNA regions exhibiting sequence similarity to soybean Lb cDNA probes? What are the early nodulin genes in the *Frankia*-actinorhizal symbiosis?

Methods in molecular biology in combination with protein analysis have enabled us to achieve a better understanding of the relationships among *Frankia* strains and allow us to discriminate between Nif⁻, nonnodulating (noninfective) frankias and other soil actinomycetes. Because a strong correlation exists between proteins and DNA, the delineation of *Frankia* strains as *species* seems more feasible now. A possible *Frankia* species consists of some *Alnus*-compatible strains isolated from various plant species from different geographic areas. Each strain exhibits a close relationship as evaluated with modern taxonomic criteria. However, the genus *Frankia* contains very dissimilar strains as well, indicating that great heterogeneity exists within the family Frankiaceae.

The long-term outcome of research on the *Frankia*-actinorhizal symbiosis will be in the definition of the most efficient host-endophyte association which can be used for wood production and environmental protection. An important goal for laboratory scientists should be the engineering of *Frankia* strains with high nitrogen-fixing potential. However, other properties to be considered are persistence in the soil and ability to compete with indigenous soil microorganisms. Hence, genetic and molecular biological pursuits in the laboratory must not exist in a vacuum which means that a better utilization of the *Frankia*-actinorhizal symbiosis will not occur without thorough ecological studies. To that end, techniques in molecular genetics and biology may provide interesting tools for competition tests, for identification of engineered strains released into the soil, and for detection of these organisms within nodules, eliminating the need for propagation in pure culture.

REFERENCES

1. **Smith, D. C. and Douglas, A. E.**, *The Biology of Symbiosis*, Edward Arnold, London, 1987.
2. **Dixon, R. O. D. and Wheeler, C. T.**, *Nitrogen Fixation in Plants*, Blackie, Glasgow, 1986.
3. **Lechevalier, M. P.**, The taxonomy of the genus *Frankia*, *Plant Soil*, 78, 1, 1984.
4. **Torrey, J.G.**, Endophyte sporulation in root nodules of actinorhizal plants, *Physiol. Plant.*, 70, 279, 1987.
5. **Moiroud, A. and Gianninazzi-Peason, V.**, Symbiotic relationships in actinorhizae, in *Genes Involved in Microbe-Plant Interactions*, Verma, D. P. S. and Hohn, T. H., Eds., Springer Verlag, New York, 1984, 205.
6. **Akkermans, A. D. L. and van Dijk, C.**, Non-leguminous root-nodule symbioses with actinomycetes and *Rhizobium*, in *Nitrogen Fixation*, Vol. 1, Ecology, Broughton, W. J., Ed., Oxford University Press, London, 1981, 57.

7. **Normand, P. and Lalonde, M.,** The genetics of actinorhizal *Frankia,* a review, *Plant Soil,* 90, 429, 1986.
8. **Torrey, J. G. and Tjepkema, J. D., Eds.,** Proc. Workshop Petersham, *Symbiotic Nitrogen Fixation in Actinomycete-Nodulated Plants, Bot. Gaz.,* 140s, 1979.
9. **Gordon, J. C., Wheeler, C. T., and Perry, D. A., Eds.,** Proc. Workshop Corvallis, *Symbiotic Nitrogen Fixation in the Management of Temperate Forests,* Oregon University Press, Corvallis, 1979.
10. **Torrey, J. G. and Tjepkema, J. D., Eds.,** Proc. Int. Conf. Madison, *The Biology of Frankia, Can. J. Bot.,* 61, 2765, 1983.
11. **Akkermans, A. D. L., Baker, D., Huss-Danell, K., and Tjepkema, J. D., Eds.,** Proc. Int. Workshop Wageningen, *Frankia* Symbioses, *Plant Soil,* 78, 1, 1984.
12. **Lalonde, M., Camiré, C., and Dawson, J. O., Eds.,** Proc. Int. Symp. Laval, *Frankia and Actinorhizal Plants, Plant Soil,* 87, 1, 1985.
13. **Huss-Danell, K. and Wheeler, C. T., Eds.,** Proc. Int. Symp. Umeå, *Frankia and Actinorhizal Plants, Physiol. Plant.,* 70, 235, 1987.
14. **Gordon, J. C. and Wheeler, C. T., Eds.,** *Biological Nitrogen Fixation in Forest Ecosystem: Foundations and Applications.* Martinus Nijhoff/Junk, The Hague, 1983.
15. Office of Technology Assessment, Technologies to Sustain Forest Resources, OTA-F-214, U.S. Congress, Washington, D.C., 1984.
16. **Lawrence, D., Schoenike, R., Quispel, A., and Bond, G.,** The role of *Dryas drummondii* in vegetation development following ice recession at Glacier Bay, Alaska, with special reference to its nitrogen fixation by root nodules, *J. Ecology,* 55, 447, 1967.
17. **Silvester, W.,** Ecological and economic significance of the nonlegume symbioses, in *Proc. 1st Int. Symp. Nitrogen Fixation,* Vol. 2, Newton, W. E. and Nyman, C., Eds., Washington State University Press, Corvallis, 1974, 489.
18. **Torrey, J. G.,** Nitrogen fixation by actinomycete-nodulated angiosperms, *Bioscience,* 28, 586, 1978.
19. **Anon.,** *Firewood Crops, Shrubs and Tree Species of Energy Production,* National Academy of Sciences Press, Washington, D.C., Vol. 1, 1980.
20. **Anon.,** *Casuarina: Nitrogen Fixing Trees for Adverse Sites, Innovations in Tropical Reforestation (series),* National Academy of Sciences Press, Washington, D.C., 1984.
21. **Akkermans, A. D. L., Abdulkadir, S., and Trinick, M. J.,** N$_2$-fixing root nodules in Ulmaceae: *Parasponia* or (and) *Trema* spp.?, *Plant Soil,* 47, 711, 1978.
22. **Trinick, M. J.,** Symbiosis between *Rhizobium* and the non-legume *Trema aspera, Nature (London),* 244, 459, 1973.
23. **Cen, Y., Bender, G. L., Trinick, M. J., Morrison, N. A., Scott, K. F., Gresshoff, P. M., Shine, J., and Rolfe, G. B.,** Transposon mutagenesis in rhizobia which can nodulate both legumes and the non-legume *Parasponia, Appl. Environm. Microbiol.,* 43, 233, 1982.
24. **Shine, J., Scott, K. F., Fellows, F., Djordjevic, M. A., Schofield, P., Watson, J. M., and Rolfe, B. G.,** Molecular anatomy of the symbiotic region in *R. trifolii* and *R. parasponia,* in *Molecular Genetics of the Bacteria-Plant Interaction,* Puhler, A., Ed., Springer Verlag, Berlin, 1983, 204.
25. **Akkermans, A. D. L. and Roelofsen, W.,** Symbiotic Nitrogen Fixation by Actinomycetes in *Alnus*-type root nodules, in *Nitrogen Fixation* (Proc. Symp. Phytochemical Society of Europe, Sussex), Stewart, W. D. P. and Gallon, J., Eds., Academic Press, New York, 1980, 279.
26. **Bond, G.,** Taxonomy and distribution of non-legume nitrogen-fixing systems, in *Biological Nitrogen Fixation in Forest Ecosystems: Foundations and Applications,* Gordon, J. C. and Wheeler, C. T., Eds., Martinus Nijhoff/Junk, The Hague, 1983, 55.
27. **Faria, S. M. de, Sutherland, J. M. and Sprent, J. I.,** A new type of infected cell in root nodules of *Andira* spp. (Leguminosae), *Plant Sci.,* 45, 143, 1986.
28. **Trinick, M. J.,** Structure of nitrogen-fixing nodules formed by *Parasponia andersonii* Planch., *Can. J. Microbiol.,* 25, 565, 1979.
29. **An, C. S., Wills, J. W., and Mullin, B. C.,** DNA relatedness of *Frankia* isolates, in *Advances in Nitrogen Fixation Research,* Veeger, C. and Newton, W. E., Eds., Martinus Nijhoff/Junk Pudoc, The Hague, 1984, 369.
30. **An, C. S., Riggsby, W. S., and Mullin, B. C.,** Relationships of *Frankia* isolates based on deoxyribonucleic acid homology studies, *Int. J. Syst. Bacteriol.,* 35, 140, 1985.
31. **An, C. S., Wills, J. H., Riggsby, W. S., and Mullin, B. C.,** Deoxyribonucleic acid base composition of 12 *Frankia* isolates, *Can. J. Bot.,* 61, 2859, 1983.
32. **Dobritsa, S. V.,** Molecular organization of the *Frankia* genome, in *Problems in Biochemistry and Physiology of Microorganisms,* Academy of Sciences SSSR, Institute of Biochemistry and Physiology of Microorganisms, Pushchino, SSSR, 1985, 294 (in Russian).
33. **Callaham, D., Del Tredici, P., and Torrey, J. G.,** Isolation and cultivation in vitro of the actinomycete causing root nodulation in *Comptonia, Science,* 199, 899, 1978.
34. **Quispel, A. and Tak, T.,** Studies on the growth of the endophyte of *Alnus glutinosa* (l.) Vill. in nutrient solutions, *New Phytol.,* 81, 587, 1978.

35. Hopwood, D. A., Bibb, M. J., Chater, K. F., Kieser, T., Burton, C. J., Kieser, M. H., Lydiate, D. J., Smith, C. P., Ward, J. M., and Schrempf, H., *Genetic Manipulation of Streptomyces: a Laboratory Manual.* John Innes Foundation, Norwich, England, 1985.

36. Chater, K. F. and Hopwood, D. A., *Streptomyces* genetics, in *The Biology of the Actinomycetes*, Goodfellow, M., Mordarski, M., Williams, S. T., Eds., Academic Press, London, 1983, 229.

37. Normand, P., Simonet, P., Butour, J. L., Rosenberg, C., Moiroud, A., and Lalonde, M., Plasmids in *Frankia* sp., *J. Bacteriol.*, 155, 32, 1983.

38. Dobritsa, S. V., Restriction analysis of the *Frankia* spp. genome, *FEMS Microbiol. Lett.*, 29, 123, 1985.

39. Simonet, P., Capellano, A., Navarro, E., Bardin, R., and Moiroud, A., An improved method for lysis of *Frankia* with achromopeptidase allows detection of new plasmids, *Can. J. Microbiol.*, 30, 1292, 1984.

40. Roop, D. R., Mundt, J. O., and Riggsby, W. S., Deoxyribonucleic acid hybridization studies among some strains of group K and N streptococci, *Int. J. Syst. Bacteriol.*, 24, 330, 1974.

41. Marmur, J., A procedure for the isolation of deoxyribonucleic acid from its thermal denaturation temperature, *J. Mol. Biol.*, 3, 208, 1961.

42. Chater, K. F., Hopwood, D. A., Kieser, T., and Thompson, C. J., Gene cloning in *Streptomyces*, *Curr. Top. Microbiol. Immunol.*, 96, 69, 1982.

43. Heath, L. S., Sloan, G. L. and Heath, H. E., A simple and generally applicable procedure for releasing DNA from bacterial cells, *Appl. and Environm. Microbiol.*, 51, 1138, 1986.

43a. Akkermans, A. D. L., unpublished results.

44. Reeck, R. H., Haen, de, C., Teller, D. C., Doolittle, R. F., Fitch, W. M., Dickerson, R. E., Chambon, P., McLachlan, A. D., Margoliash, E., Jukes, T. H., and Zuckerkandl, E., "Homology" in proteins and nucleic acids: a terminology muddle and a way out of it, *Cell*, 50, 667, 1987.

45. Ruvkun, G. B. and Ausubel, F. M., Interspecies homology of nitrogenase genes, *Proc. Natl. Acad. Sci. U.S.A.*, 77, 191, 1980.

46. Simonet, P., Normand, P., and Bardin, R., Heterologous hybridization of *Frankia* DNA to *Rhizobium meliloti* and *Klebsiella pneumoniae nif* genes, *FEMS Microbiol. Lett.*, 55, 141, 1988.

47. Ruvkun, G. B., Sundaresan, V. and Ausubel, F. M., Directed transposon Tn5 mutagenesis and complementation analysis of *Rhizobium meliloti* symbiotic nitrogen fixation genes, *Cell*, 29, 551, 1982.

48. Corbin, D., Barran, L., and Ditta, G., Organization and expression of *Rhizobium meliloti* nitrogen fixation, *Proc. Natl. Acad. Sci. U.S.A.*, 80, 3005, 1983.

49. Downie, J. A., Ma, Q. S., Knight, C. D., Hombrecher, G., and Johnston, A. W. B., Cloning of the symbiotic regions of *Rhizobium leguminosarum:* the nodulation genes are between the nitrogenase genes and a *nif*A-like gene, *EMBO J.*, 2, 947, 1983.

50. Schofield, P. R., Djordjevic, M. A., Rolfe, B. G., Shine, J., and Watson, J. M., A molecular linkage map of nitrogenase and nodulation genes in *Rhizobium trifolii*, *Mol. Gen. Genet.*, 192, 459, 1983.

51. Rolfe, B. G. and Shine, J., *Rhizobium*-Leguminosae symbiosis: the bacterial point of view, in *Genes Involved in Microbe-Plant Interactions*, Verma, D. P. S. and Hohn, T. H., Eds., Springer Verlag, New York, 1984, 95.

52. Prakash, R. K. and Atherly, A. G., Plasmids of *Rhizobium* and their role in symbiotic nitrogen fixation, *Int. Rev. of Cytol.*, 184, 1, 1986.

53. Fuhrmann, M. and Hennecke, H., Coding properties of cloned nitrogenase genes from *Rhizobium japonicum*, *Mol. Gen. Genet.*, 187, 419, 1982.

54. Kaluza, K., Furhmann, M., Hahn, M., Regensburger, B., and Hennecke, H., In *Rhizobium japonicum* the nitrogenase genes *nif*H and *nif*KD are separated, *J. Bacteriol.*, 155, 915, 1983.

55. Scott, K. F., Rolfe, B., and Shine, J., Nitrogenase structural genes are unlinked in the non-legume symbiont *Parasponia Rhizobium*, *DNA*, 2, 141, 1983.

56. Ligon, J. M. and Nakas, J. P., Isolation and characterization of *Frankia* sp. strain FaC1 genes involved in nitrogen fixation, *Appl. Environ. Microbiol.*, 53, 2321, 1987.

57. Normand, P., Simonet, P., and Bardin, R., Conservation of *nif* sequences in *Frankia*, *Mol. Gen. Genet.*, 213, 238, 1988.

58. Hennecke, H., Kaluza, K., Thony, B., Fuhrmann, M., Ludwig, W., and Stackebrandt, E., Concurrent evolution of nitrogenase genes and 16S rRNA in *Rhizobium* species and other nitrogen fixing bacteria, *Arch. Microbiol.*, 142, 342, 1985.

58a. Simonet, P., Normand, P., and Bardin, R., Heterologous hybridization of *Frankia* DNA to *Rhizobium meliloti* and *Klebsiella preummiae nif* genes *FEMS Microbial. Lett.*, 55, 141, 1988.

59. Thony, B., Kaluza, K., and Hennecke, H., Structural and functional homology between the a and b subunits of the nitrogenase MoFe protein as revealed by sequencing the *Rhizobium japonicum nif*K gene, *Mol. Gen. Genet.*, 198, 441, 1985.

60. Simonet, P., Haurat, J., Normand, P., Bardin, R., and Moiroud, A., Localization of *nif* genes on a large plasmid in *Frankia* sp. strain ULQ0132105009, *Mol. Gen. Genet.*, 204, 492, 1986.

61. Quinto, C., Vega de la, H., Flores, M., Fernandez, L., Ballado, T., Soberon, G., and Palacios, R., Reiteration of nitrogen fixation gene sequences in *Rhizobium phaseoli*, *Nature (London)*, 299, 724, 1982.

62. **Barbour, W. M., Mathis, J. N., and Elkan, O. H.,** Evidence for plasmid and chromosome-borne multiple *nif* genes in *Rhizobium fredii, Appl. Environ. Microbiol.,* 50, 41, 1985.

63. **Mazur, B. J., Rice, D., and Haselkorn, R.,** Identification of blue green algal nitrogen fixation genes by using heterologous DNA hybridization probes, *Proc. Natl. Acad. Sci. U.S.A.,* 77, 186, 1980.

64. **Rice, D., Mazur, B. J., and Haselkorn, R.,** Isolation and physical mapping of nitrogen fixation genes from the cyanobacterium 7120, *J. Biol. Chem.,* 257, 13157, 1982.

65. **Scolnick, P. A. and Haselkorn, R.,** Activation of extra copies of genes coding for nitrogenase in *Rhodopseudomonas capsulata, Nature (London),* 307, 289, 1984.

66. **Jones, R., Woodley, P., and Robson, R.,** Cloning and organization of some genes for nitrogen fixation from *Azotobacter chroococum* and their expression in *Klebsiella pneumoniae, Mol. Gen. Genet.,* 197, 318, 1984.

67. **Bishop, P. E., Jarlenski, D. M. L., and Hetherington, D. R.,** Evidence for an alternative nitrogen fixation system in *Azotobacter vinelandii, Proc. Natl. Acad. Sci. U.S.A.,* 77, 7342, 1980.

68. **Premakumar, R., Lemos, E. M., and Bishop, P. E.,** Evidence for two dinitrogenase reductases under regulatory control by molybdenum in *Azotobacter vinelandii, Biochem. Biophys. Acta,* 797, 64, 1984.

69. **Derylo, M., Głowacka, M., Skorupska, A., and Lavrkiewicz, Z.,** *Nif* plasmid from *Lignobacter, Arch. Microbiol.,* 130, 322, 1981.

70. **Singh, M., Kleeberger, A., and Klingmuller, W.,** Location of nitrogen fixation *(nif)* genes on indigenous plasmids of *Enterobacter agglomerans, Mol. Gen. Genet.,* 190, 373, 1983.

71. **Dobritsa, S. V.,** Extrachromosomal circular DNAs in endosymbiont vesicles from *Alnus glutinosa* root nodules, *FEMS Microbiol. Lett.,* 15, 87, 1982.

72. **Dobritsa, S. V.,** Large plasmids in an actinomycete, *FEMS Microbiol. Lett.,* 23, 35, 1984.

73. **Tomashevsky, A. Y. and Dobritsa, S. V.,** Hybridization of DNA from actinomycetes of the genus *Frankia* with the nitrogenase structural genes (*nif*HDK) of *Klebsiella pneumoniae* and the *nod*-genes of *Rhizobium meliloti, Molecular Genetics, Microbiology and Virology* (USSR), 3, 27, 1987 (in Russian).

74. **Dobritsa, S. V. and Tomashevsky, A. Yu.,** Homology between structural nitrogenase genes (*nif*HDK) from *Klebsiella pneumoniae* and extrachromosomal DNAs from microsymbiont vesicles of *Alnus glutinosa, Biopolymers and Cell,* 4, 44, 1988 (in Russian).

75. **Meesters, T. M.,** The Function of Vesicles in the Actinomycete *Frankia,* Ph.D. thesis, Wageningen Agricultural University, Wageningen, The Netherlands, 1988.

76. **Meesters, T. M., van Genesen, S. Th., and Akkermans, A. D. L.,** Growth, acetylene reduction activity and localization of nitrogenase in relation to vesicle formation in *Frankia* strains Cc1.17 and Cp1.2, *Arch. Microbiol.,* 143, 137, 1985.

77. **Huss-Danell, K., Sellstedt, A., Flower-Ellis, A., and Sjöström, M.,** Ammonium effects on function and structure of nitrogen-fixing root nodules of *Alnus incana* (L.) Moench, *Planta,* 156, 332, 1982.

78. **Noridge, N. A. and Benson, D. B.,** Isolation and nitrogen-fixing activity of *Frankia* sp. strain Cp11 vesicles, *J. Bacteriol.,* 166, 301, 1986.

79. **Meesters, T. M.,** Localization of nitrogenase in vesicles of *Frankia* sp. Cc1.17 by immunogoldlabelling on ultrathin cryosections, *Arch. Microbiol.,* 146, 327, 1987.

80. **Meesters, T. M., van Vliet, W. M., and Akkermans, A. D. L.,** Nitrogenase is restricted to the vesicles in *Frankia* strain EANlpec, *Physiol. Plant.,* 70, 267, 1987.

81. **Benson, D., Arp, D., and Burris, R.,** Cell-free nitrogenase and hydrogenase from actinorhizal root nodules, *Science,* 205, 688, 1979.

82. **Steele, D. B. and Stowers, M. D.,** Superoxide dismutase and catalase in *Frankia, Can. J. Microbiol.,* 32, 409, 1986.

83. **Tjepkema, J. D.,** Hemoglobins in the nitrogen-fixing root nodules of actinorhizal plants, *Can. J. Bot.,* 61, 2924, 1983.

84. **Tjepkema, J. D. and Asa, D. J.,** Total and CO-reactive heme content of actinorhizal nodules and the roots of some nonnodulated plants, *Plant and Soil,* 100, 225, 1987.

85. **Berg, R. H. and McDowell, L.,** Endophyte differentiation in *Casuarina* actinorhizae, *Protoplasma,* 136, 104, 1987.

86. **Parsons, R., Silvester, W. B., Harris, S., Gruijters, W. T. M., and Bullivant, J.,** *Frankia* vesicles provide inducible and absolute oxygen protection for nitrogenase, *Plant Physiol.,* 83, 728, 1987.

87. **Akkermans, A. D. L., Hafeez, F., Roelofsen, W., Chaudhary, A. H., and Baas, R.,** Ultrastructure and nitrogenase activity of *Frankia* grown in pure culture and in actinorhizas of *Alnus, Colletia* and *Datisca* spp., in *Advances in Nitrogen Fixation Research,* Veeger, C. and Newton, W. E., Eds., Nijhoff/Junk Pudoc, The Hague, 1984, 311.

88. **Torrey, J. G.,** The site of nitrogenase in *Frankia* in free-living culture and in symbiosis, in *Nitrogen Fixation Research Progress,* Evans, H. J., Bottomley, P. J. and Newton, W. E., Eds., Nijhoff, Dordrecht, 1985, 293.

89. **Dixon, R., Kennedy, C., Kondorosi, A., Krishnapillai, V., and Merrick, M.,** Complementation analysis of *Klebsiella pneumoniae* mutants, *Mol. Gen. Genet.,* 157, 189, 1977.

90. **Hodgson, A. L. M. and Stacey, G.,** Potential for *Rhizobium* improvement, *Crit. Rev. Biotechnol.*, 4, 1, 1986.

91. **Szeto, W. W., Zimmerman, J. L., Sundaresan, V., and Ausubel, F. M.,** A *Rhizobium meliloti* symbiotic regulatory gene, *Cell*, 36, 1035, 1984.

92. **Buikema, W. J., Szeto, W. W., Lemley, P. V., Orme-Johnson, W. H., and Ausubel, F. M.,** Nitrogen fixation specific regulatory genes of *Klebsiella pneumoniae* and *Rhizobium meliloti* share homology with the general nitrogen regulatory gene *ntrC* of *K. pneumoniae*, *Nucleic Acids Res.*, 13, 4539, 1985.

93. **Nair, S. K., Jara, P., Quiviger, B., and Elmerich, C.,** Recent developments in the genetics of nitrogen fixation in *Azospirillum*, in *Azospirillum II, Experentia Suppl.*, 48, Klingmüller, W., Ed., Birkauser Verlag, Basel, 1983, 29.

94. **Adams, T. H., McClung, C. R., and Chelm, B. K.,** Physical organization of the *Bradyrhizobium japonicum* nitrogenase region, *J. Bacteriol.*, 159, 857, 1984.

95. **Ahombo, G., Willison, J. C., and Vignais, P.,** The *nif* HDK genes are contiguous with a *nif*A-like regulator gene in *Rhodobacter capsulatus*, *Mol. Gen. Genet.*, 205, 442, 1986.

96. **Ausubel, F. M.,** Regulation of nitrogen fixation genes, *Cell*, 37, 5, 1984.

97. **Rossen, L., Ma, Q.-S., Mudd, E. A., Johnston, A. W. B., and Downie, J. A.,** Identification of DNA sequences of *fix*Z, a *nif*B-like gene from *Rhizobium leguminosarum*, *Nucleic Acids Res.*, 12, 7123, 1984.

98. **Buikema, W. J., Klingensmith, J. A., Gibbons, S. L., and Ausubel, F. M.,** Conservation of structure and location of *Rhizobium meliloti* and *Klebsiella pneumoniae nif*B genes, *J. Bacteriol.*, 169, 1120, 1987.

99. **Earl, C. D., Ronson, C. W., and Ausubel, F. M.,** Genetic and structural analysis of the *Rhizobium meliloti fix*A, *fix*B, *fix*C and *fix*X genes, *J. Bacteriol.*, 169, 1127, 1987.

100. **Long, S. R., Buikema, W. J., and Ausubel, F. M.,** Cloning of *Rhizobium meliloti* nodulation genes by direct complementation of Nod mutants, *Nature (London)*, 298, 485, 1982.

101. **Prakash, R. K. and Atherly, A. G.,** Reinteration of genes involved in symbiotic nitrogen fixation by fast-growing *Rhizobium japonicum*, *J. Bacteriol.*, 160, 785, 1984.

102. **Masterson, R. V., Prakash, R. K., and Atherly, A. G.,** Conservation of symbiotic nitrogen fixation gene sequences in *Rhizobium japonicum* and *Bradyrhizobium japonicum*, *J. Bacteriol.*, 163, 21, 1985.

103. **Broughton, W. J., Heycke, N., Meyer, Z. A. H., and Pankhurst, C. E.,** Plasmid linked *nif* and *nod* genes in fast-growing rhizobia that nodulate *Glycine max, Psophocarpus tetragonolobus* and *Vigna unguiculata*, *Proc. Natl. Acad. Sci. U.S.A.*, 81, 3093, 1984.

104. **Long, S. R.,** Genetics of *Rhizobium* nodulation, in *Plant-Microbe Interactions*, Vol. 1, Kosuge, T. and Nester, E. W., Eds., Macmillan, New York, 1984, 265.

105. **Fogher, C., Dusha, I., Barbot, P., and Elmerich, C.,** Heterologous hybridization of *Azospirillum* DNA to *Rhizobium nod* and *fix* genes, *FEMS Microbiol. Lett.*, 30, 245, 1985.

106. **Drake, D., Leonard, J. T., and Hirsch, A. M.,** Symbiotic genes in *Frankia*, in *Nitrogen Fixation Research Progress*, Evans, H. J., Bottomly, P. J., and Newton, W. E., Eds., Martinus Nijhoff, Dordrecht, 1985, 147.

106a. **Reddy, A., Torrey, J. G., and Hirsch, A. M.,** unpublished.

107. **Staskawicz, B., Dahlbeck, D., Keen, N., and Napoli, C.,** Molecular characterization of cloned avirulence genes from race 0 and race 1 of *Pseudomonas syringae* pv. *glycinea*, *J. Bacteriol.*, 5789, 1987.

108. **Marvel, D. J., Kuldau, G., Hirsch, A., Richards, E., Torrey, J. G., and Ausubel, F. M.,** Conservation of nodulation genes between *Rhizobium meliloti* and a slow-growing *Rhizobium* strain that nodulates a nonlegume host, *Proc. Natl. Acad. Sci. U.S.A.*, 82, 5841, 1985.

109. **Lalonde, M. and Knowles, R.,** Ultrastructure, composition, and biogenesis of the encapsulation material surrounding the endophyte in *Alnus crispa* var. *mollis* root nodules, *Can. J. Bot.*, 53, 1951, 1975.

110. **Callaham, D., Newcomb, W., Torrey, J. G., and Peterson, R. L.,** Root hair infection in an actinomycete-induced root nodule in *Casuarina, Myrica* and *Comptonia*, *Bot. Gaz.*, 140, S1, 1979.

111. **Newcomb, W.,** Fine structure of the root nodules of *Dryas drummondii* Richards (Rosaceae), *Can. J. Bot.*, 59, 2500, 1981.

112. **Hafeez, F., Akkermans, A. D. L., and Chaudhary, A. H.,** Observations on the ultrastructure of *Frankia* sp. in root nodules of *Datisca cannabina* L., *Plant Soil*, 79, 383, 1984.

113. **Miller, I. M. and Baker, D. D.,** The initiation, development and structure of root nodules in *Elaeagnus angustifolia* L. (Elaeagnaceae), *Protoplasma*, 128, 107, 1985.

114. **Garibaldi, A. and Bateman, D. F.,** Pectic enzymes produced by *Erwinia chrysanthemi* and their effects on plant tissue, *Physiol. Plant Pathol.*, 1, 25, 1971.

115. **Starr, M. P. and Chatterjee, A. K.,** The genus *Erwinia*: enterobacteria pathogenic to plants and animals, *Annu. Rev. Microbiol.*, 26, 389, 1972.

116. **Vandersman, C., Andro, T., and Bertheau, Y.,** Extracellular proteases in *Erwinia chrysanthemi*, *J. Gen. Microbiol.*, 132, 899, 1986.

117. **Chatterjee, A. K. and Starr, M. P.,** Genetics of *Erwinia* species, *Annu. Rev. Microbiol.*, 34, 645, 1980.

118. **Kotoujansky, A., Diolez, A., Boccara, N., Bertheau, Y., Andro, T., and Coleno, A.,** Molecular cloning of *Erwinia chrysanthemi* pectinases and cellulase structural genes, *EMBO J.*, 4, 781, 1985.

119. **Van Gijsegem, F., Toussaint, A., and Schoonejans, E.,** *In vivo* cloning of the pectate lyase and cellulase genes of *Erwinia chrysanthemi, EMBO J.,* 4, 787, 1985.

120. **Reverchon, S., Van Gijsegem, F., Rouve, M., Kotoujansky, A., and Robert-Baudouy, J.,** Organization of a pectate lyase gene family in *Erwinia chrysanthemi, Gene,* 49, 215, 1986.

121. **Safo-Sampah, S. and Torrey, J. G.,** Polysaccharide-hydrolyzing enzymes of *Frankia* (Actinomycetales), *Plant Soil,* 112, 89, 1988.

122. **Barras, F., Boyer, M. -H., Chambost, J. -P., and Chippaux, M.,** Construction of a genomic library of *Erwinia chrysanthemi* and molecular cloning of cellulase gene, *Mol. Gen. Genet.,* 197, 513, 1984.

123. **Benson, D. R., Arp, D. J., and Burris, R. H.,** Hydrogenase in actinorhizal root nodules and root nodule homogenates, *J. Bacteriol.,* 142, 138, 1980.

124. **Albrecht, S. L., Maier, R. J., Hanus, F. J., Russell, S. A., Emmerich, D. W., and Evans, H. J.,** Hydrogenase in *Rhizobium japonicum* increases nitrogen fixation by nodulated soybeans, *Science,* 203, 1255, 1979.

125. **DeJong, T. M., Brewin, N. J., Johnston, A. W. B., and Phillips, D. A.,** Improvement of symbiotic properties in *Rhizobium leguminosarum* by plasmid transfer, *J. Gen. Microbiol.,* 128, 1829, 1982.

126. **Tait, R. C., Andersen, K., Cangelosi, G., and Lim, S. T.,** Hydrogen uptake (Hup) plasmids: characterization of mutants and regulation of the expression of hydrogenase, in *Genetic Engineering of Symbiotic Nitrogen Fixation,* Lyons, J. M., Valentine, R. C., Phillips, D. A., Rains, D. W., and Huffaker, R. C., Eds., Plenum Press, New York, 1981, 131.

126a. **Benson, D. R.,** personal communication.

127. **Edmonds, J., Noridge, N. A., and Benson, D. R.,** The actinorhizal root nodule symbiont *Frankia* sp. strain CpI1 has two glutamine synthetases, *Proc. Natl. Acad. Sci., U.S.A.,* 84, 6126, 1987.

128. **Casadesus, J. and Olivares, J.,** General transduction in *Rhizobium meliloti* by a thermosensitive mutant of bacteriophage DF2, *J. Bacteriol.,* 139, 316, 1979.

129. **Sik, T., Hovath, J., and Chatterjee, S.,** Generalized transduction in *Rhizobium meliloti, Mol. Gen. Genet.,* 178, 511, 1980.

130. **Finan, T. M., Hartwieg, E., Lemieux, K., Bergman, K., Walker, G. C., and Signer, E.,** General transduction in *Rhizobium meliloti, J. Bacteriol.,* 159, 120, 1984.

131. **Martin, M. O. and Long, S. R.,** Generalized transduction in *Rhizobium meliloti, J. Bacteriol.,* 159, 125, 1984.

132. **Buchanan-Wollaston, V.,** Generalized transduction in *Rhizobium leguminosarum, J. Gen. Microbiol.,* 112, 135, 1979.

133. **Shah, K., Sousa, S., and Modi, V. V.,** Studies on transducing phage M-1 for *Rhizobium japonicum* D211, *Arch. Microbiol.,* 130, 262, 1981.

134. **Shah, K., Patel, C., and Modi, V. V.,** Linkage mapping of *Rhizobium japonicum* D211 by phage M-1 mediated transduction, *Can. J. Microbiol.,* 29, 33, 1983.

134a. **Tisa, L. S. and Akkermans, A. D. L.,** unpublished results.

135. **Normand, P., Downie, J. A., Johnston, A. W. B., Kieser, T., and Lalonde, M.,** Cloning of a multicopy plasmid from the actinorhizal nitrogen-fixing bacterium *Frankia* sp. and determination of its restriction map, *Gene,* 34, 367, 1985.

136. **Simonet, P., Normand, P., Moiroud, A., and Lalonde, M.,** Restriction enzyme digestion patterns of *Frankia* plasmids, *Plant Soil,* 87, 49, 1985.

137. **Hopwood, D. A. and Chater, K. F.,** Cloning in *Streptomyces:* Systems and strategies, in *Genetic Engineering,* Vol. 4, Setlow, J. K. and Hollaender, A., Eds., Plenum, New York, 1982, 119.

137a. **Akkermans, A. D. L., Jaurin, B., and Tisa, L. S.,** unpublished results.

138. **Baltz, R. H. and Matsushima, P.,** Protoplast fusion in *Streptomyces:* conditions for efficient genetic recombination and cell regeneration, *J. Gen. Microbiol.,* 127, 137, 1981.

139. **Baltz, R. H. and Matsushima, P.,** Advances in protoplast fusion and transformation in *Streptomyces, Experientia* (Suppl.), 46, 143, 1983.

140. **Ochi, K., Hitchcock, M. J. M., and Katz, E.,** High-frequency fusion of *Streptomyces parvulus* or *Streptomyces antibioticus* protoplasts induced by polyethylene glycol, *J. Bacteriol.,* 139, 984, 1979.

141. **Ochi, K.,** Protoplast fusion permits high-frequency transfer of a *Streptomyces* determinant which mediates actinomycin synthesis, *J. Bacteriol.,* 150, 592, 1982.

142. **Tisa, L. S. and Ensign, J. C.,** Formation and regeneration of protoplasts of the actinorhizal nitrogen-fixing actinomycete *Frankia, Appl. Environ. Microbiol.,* 53, 53, 1987.

142a. **Dobritsa, S. V. and Akkermans, A. D. L.,** unpublished results.

143. **Tisa, L. S.,** Formation and Characterization of Vesicles, the Site of Nitrogen Fixation by *Frankia,* Ph.D. Thesis, University of Wisconsin, Madison, 1987.

144. **Prakash, R. K. and Cummings, B.,** Creation of novel nitrogen-fixing actinomycetes by protoplast fusion of *Frankia* with streptomyces, *Plant Mol. Biol.,* 10, 281, 1988.

145. **Faure-Raynaud, M., Bonnefoy, M.-A., Perradin, Y., Simonet, P., and Moiroud, A.,** Protoplast formation from *Frankia* strains, *Microbios,* 41, 159, 1984.

146. **Normand, P., Simonet, P., Prin, Y., and Moiroud, A.,** Formation and regeneration of *Frankia* protoplasts, *Physiol. Plant.,* 70, 259, 1987.
147. **Hopwood, D. A.,** Genetic studies with bacterial protoplasts, *Annu. Rev. Microbiol.,* 35, 237, 1981.
148. **Higashi, S., Uchiumi, T., and Abe, M.,** Analysis of Sym-plasmid-expression by immunoaffinity chromatography, in *Advances in Nitrogen Fixation Research,* Veeger, C. and Newton, W. E., Eds., Nijhoff/Junk, The Hague, 1984, 709.
149. **Singer, J. T. and Finnerty, W. R.,** Genetics of hydrocarbon-utilizing microorganisms, in *Petroleum Microbiology,* Atlas, R. M., Ed., Macmillan, New York, 1984, 299.
150. **MacNeil, D. J.,** Introduction of plasmid DNA into *Streptomyces lividans* by electroporation, *FEMS Microbiol. Lett.,* 42, 239, 1987.
151. **Matsushima, P. and Baltz, R. H.,** Efficient plasmid transformation in *Streptomyces ambofaciens* and *Streptomyces fradiae* protoplasts, *J. Bacteriol.,* 163, 180, 1985.
152. **Matsushima, P., Cox, K. L., and Baltz, R. H.,** Highly transformable mutants of *Streptomyces fradiae* defective in several restriction systems, *Mol. Gen. Genet.,* 206, 393, 1987.
153. **Dobritsa, S. V.,** *Antibiotiki,* 7, 511, 1982 (in Russian).
153a. **Akkermans, A. D. L. and Van Vliet, M.,** unpublished results.
153b. **Akkermans, A. D. L., Jaurin, B., and Van Vliet, M.,** unpublished results.
154. **Lie, T. A., Akkermans, A. D. L., and van Egeraat, A. W. S. M.,** Natural variation in symbiotic nitrogen-fixing *Rhizobium* and *Frankia* spp. *Antonie van Leeuwenhoek,* 50, 489, 1984.
155. **Gerber, N. N. and Lechevalier, M. P.,** Novel benzo (alpha) naphtacene quinones from an actinomycete, *Frankia* G2 (ORS 020604), *Can. J. Chem.,* 62, 2818, 1984.
156. **Lechevalier, M. P., Horrière, F., and Lechevalier, H.,** The biology of *Frankia* and related organisms, *Develop. Indust. Microbiol.,* 23, 51, 1982.
157. **Burggraaf, A. J. P. and Valstar, J.,** Heterogeneity within *Frankia* sp. LDAgpI studied among clones and reisolates, *Plant Soil,* 78, 29, 1984.
158. **Hahn, D., Starrenburg, M. J. and Akkermans, A. D. L.,** Variable compatibility of cloned *Alnus glutinosa* ecotypes against ineffective *Frankia* strains, *Plant Soil,* 197, 233, 1988.
158a. **Hahn, D.,** unpublished results.
159. **Berry, A.,** Cellular aspects of root nodule establishment in *Frankia* symbiosis, in *Plant-Microbe Interactions. Molecular and Genetic Perspectives,* Vol. 2, Kosuge, T. and Nester, E. W., Eds., Macmillan, New York, 1987, 194.
160. **Wood, S. M. and Newcomb, W.,** Morphogenesis and fine structure of *Frankia* (Actinomycetales): a microsymbiont of nitrogen-fixing actinorhizal root nodules, *Int. Rev. Cytol.,* 109, 1, 1988.
161. **Van Kammen, A.,** Suggested nomenclature for plant genes involved in nodulation and symbiosis, *Plant Mol. Biol. Rep.,* 2, 43, 1984.
162. **Nguyen, T., Zlechowska, M., Foster, V., Bergmann, H., and Verma, D. P. S.,** Primary structure of the soybean nodulin-35 gene encoding uricase II localized in peroxizomes of uninfected cells of nodules, *Proc. Natl. Acad. Sci. U.S.A.,* 83, L5040, 1985.
163. **Thummler, F. and Verman, D. P. S.,** Nodulin-100 of soybean is the subunit of sucrose synthase regulated by availability of free heme in nodules, *J. Biol. Chem.,* 262, 14730, 1987.
164. **Cullimore, J. V., Gebhardt, C., Saarelainen, R., Miflin, B. J., Idler, K. B., and Barker, R. F.,** Glutamine synthetase of *Phaseolus vulgaris* L.: organ-specific expression of a multigene family, *J. Mol. Appl. Genet.,* 2, 589, 1984.
165. **Hyldig-Nielson, J. J., Jensen, E. O., Paludan, K., Wiborg, O., Garrett, R., Jorgensen, P., and Marcker, K. A.,** The primary structure of two leghemoglobin genes from soybean, *Nucleic Acids Res.,* 10, 689, 1982.
166. **Brisson, N. and Verma, D. P. S.,** Soybean leghemoglobin gene family: normal, pseudo, and truncated genes, *Proc. Natl. Acad. Sci. U.S.A.,* 79, 4055, 1982.
167. **Davenport, H. E.,** Haemoglobin in the root nodules of *Casuarina cunninghamiana, Nature (London),* 186, 653, 1960.
168. **Appleby, C. A., Tjepkema, J. D., and Trinick, M. J.,** Hemoglobin in a nonleguminous plant, *Parasponia:* possible genetic origin and function in nitrogen fixation, *Science,* 220, 551, 1983.
169. **Flemming, A. I., Wittenberg, J. B., Wittenberg, B. A., Dudman, W. F., and Appleby, C. A.,** The purification, characterization and ligand-binding kinetics of hemoglobins from root nodules of the non-leguminous *Casuarina glauca-Frankia* symbiosis, *Biochem. Biophys. Acta,* 911, 209, 1987.
170. **Flemming, A. I., Appleby, C. A., Dudman, W. F., Kortt, A. A., Wittenberg, B. A. and Wittenberg, J. B.,** Characterization of hemoglobin purified from root nodules of the *Casuarina glauca-Frankia* symbiosis, in *Proc. 8th Australian Nitrogen Fixation Conference,* Australian Institute of Agricultural Science, Parkville, Victoria, 1987, 85.
171. **Hattori, J. and Johnson, D. J.,** The detection of leghemoglobin-like sequences in legumes and non-legumes, *Plant Molec. Biol.,* 4, 285, 1985.

172. **Roberts, M. P., Jafar, S., and Mullin, B. C.**, Leghemoglobin-like sequences in the DNA of four actinorhizal plants, *Plant Molec. Biol.*, 5, 333, 1985.

172a. **Hirsch, A. and Bisseling, T.**, unpublished results.

173. **Giasson, L. and Lalonde, M.**, Restriction pattern analysis of deoxyribonucleic acid isolated from callus and suspension of actinorhizal and non-actinorhizal Betulaceae, *Physiol. Plant.*, 70, 235, 1987.

174. **Landman, J., Dennis, E. S., Higgins, T. J. V., Appleby, C. A., Kortt, A. A., and Peacock, W. J.**, Common evolutionary origin of legume and non-legume plant haemoglobins, *Nature (London)*, 324, 166, 1986.

175. **Tjepkema, J. D., Schwintzer, C. R., and Benson, D. R.**, Physiology of actinorhizal nodules, *Annu. Rev. Plant Physiol.*, 37, 209, 1986.

176. **Berg, R. H.**, Preliminary evidence for the involvement of suberization in infection of *Casuarina, Can. J. Bot.*, 61, 2910, 1983.

177. **Smith, C. A., Skvirsky, R. C., and Hirsch, A. M.**, Histochemical evidence for the presence of a suberin-like compound in *Rhizobium*-induced nodules of the nonlegume *Parasponia rigida, Can. J. Bot.*, 64, 1474, 1986.

178. **Cullimore, J. V. and Miflin, B. J.**, Immunological studies on glutamine synthetase using antisera raised to two plant forms of the enzyme from *Phaseolus* root nodules, *J. Expt. Bot.*, 35, 581, 1984.

179. **Schubert, K. R.**, Products of biological nitrogen fixation in higher plants: synthesis, transport and metabolism, *Annu. Rev. Plant Physiol.*, 37, 539, 1986.

180. **Blom, J., Roelofsen, W., and Akkermans, A. D. L.**, Assimilation of nitrogen in root nodules of alder *(Alnus glutinosa), New Phytol.*, 89, 321, 1981.

181. **Hirel, B., Perrot-Rechenman, C., Maudinas, B., and Gadal, P.**, Glutamine synthesis in alder *(Alnus glutinosa)* root nodules. Purification, properties, cytoimmunochemical localization. *Physiol. Plant.*, 55, 197, 1982.

182. **Scott, A., Gardner, I. C., and McNally, S. F.**, Localization of citrulline synthesis in the alder root nodule and its implication in nitrogen fixation, *Plant Cell Reports*, 1, 21, 1981.

183. **Lechevalier, M. P.**, Catalog of *Frankia* strains, *The Actinomycetes*, 19, 131, 1986.

184. **Becking, J. H.**, Family III, Frankiaceae Becking, 1970, 201. In *Bergey's Manual of Determinative Bacteriology*, Buchanan, R. E. and Gibons, N. E., Eds., Williams, & Wilkins Co., Baltimore, 1974, 701.

185. **Mort, A., Normand, P., and Lalonde, M.**, 2-*o*-methyl-D-mannose, a key sugar in the taxonomy of *Frankia, Can. J. Microbiol.*, 29, 993, 1983.

186. **Baker, D. D.**, Relationships among pure cultured strains of *Frankia* based on host specificity, *Physiol. Plant.*, 70, 245, 1987.

186a. **Hahn, D. Dorsch, M., Stackebrandt, E., and Ahlermans, A. D. L.**, Synthetic oligonucleotide probes for identification of *Frankia* stains, *Plant Soil*, in press.

187. **Goodfellow, M.**, Actinomycete systematics: present state and future prospects, in *Biological, Biochemical and Biomedical Aspects of Actinomycetes*, Szabo, G., Biro, S., and Goodfellow, M., Eds., Symposia Biologica Hungarica, Vol. 32, Akademiai Kiado, Budapest, 1986, 487.

188. **Stackebrandt, E., Ludwig, W., and Fox, G. E.**, 16S ribosomal RNA oligonucleotide cataloguing, in *Methods in Microbiology*, 18, 75, 1985.

189. **Stackebrandt, E.**, The significance of "wall types" in phylogenetically based taxonomic studies on actinomycetes, in *Biological, Biochemical and Biomedical Aspects of Actinomycetes*, Szabo, G., Biro, S., and Goodfellow, M., Eds., Symposia Biologica Hungarica, Vol. 32, Akademiai Kiado, Budapest, 1986, 497.

189a. **Hahn, D., Lechevalier, M. P., Fischer, A., and Stackebrandt, E.**, Evidence for a close phylogenetic relationship between members of the genera *Frankia, Geodermatophilus*, and "*Blastococcus*" and emendation of the family, Frankiaceae, *System. Appl. Microbiol.*, in press.

190. **Postgate, J. R.**, Evolution within nitrogen-fixing systems, in *Evolution of The Microbial World*, Carlisle, M. and Skehel, J., Eds., Symp. Soc. Gen. Microbiol., Cambridge Univ. Press, Cambridge, 1974, 263.

191. **Brill, W. J.**, Nitrogen fixation, *The Bacteria*, Vol. 7, Gunsalus I. C., Ed., Academic Press, New York, 1979.

192. **Pühler, A., Burkhardt, H. J., and Klipp, W.**, Cloning in *Escherichia coli* the genomic region of *Klebsiella pneumoniae* which encodes genes responsible for nitrogen fixation, in *Plasmids of Medical, Environmental and Commerical Importance*, Timmis, K. N. and Pühler, A., Eds., Elsevier-North Holland, Amsterdam, 1979, 435.

193. **Ausubel, F. M., Brown, S. E., de Bruijn, F. J., Ow, D. W., Riedel, G. E., Ruvkun, G. B., and Sundareasan, V.**, Molecular cloning of nitrogen fixation genes from *Klebsiella pneumoniae* and *Rhizobium meliloti*, in *Genetic Engineering: Principles and Methods*, Setlow, J. K. and Hollaender, A., Eds., Plenum Press, New York, 1982, 169.

194. **Souillard, N. and Sibold, L.**, Primary structure and expression of a gene homologous to *nif*H (nitrogenase Fe protein) from the archaebacterium *Methanococcus voltae, Mol. Gen. Genet.*, 203, 21, 1986.

195. **Ligon, J. M. and Nakas, J. P.**, Nucleotide sequence of *nifK* and partial sequence of *nifd* from *Frankia* species strain FaCl, *Nucleic Acids Res.*, 16, 11843, 1988.

195a. **Simonet, P.**, unpublished.

196. **Brenner, D. J.**, Deoxyribonucleic acid reassociation in the taxonomy of enteric bacteria, *Int. J. Syst. Bacteriol.*, 23, 298, 1973.

196a. **Dobritsa, S. V.**, unpublished results.

197. **Baker, D., Pengelly, W., and Torrey, J. G.**, Immunochemical analysis of relationships among isolated frankiae (Actinomycetales), *Int. J. Syst. Bacteriol.*, 31, 148, 1981.

198. **Lechevalier, M. P., Baker, D., and Horrière, F.**, Physiology, chemistry, serology, and infectivity of two *Frankia* isolates *Alnus incana* subsp. *rugosa*, *Can. J. Bot.*, 61, 2826, 1983.

199. **Parson, W. L., Robertson, L. R., and Carpenter, C. V.**, Characterization and infectivity of a spontaneous variant isolated from *Frankia* sp. WEY 0131391, *Plant Soil*, 87, 31, 1985.

200. **An, C. S., Riggsby, W. S., and Mullin, B. C.**, Restriction pattern analysis of genomic DNA of *Frankia* isolates, *Plant Soil*, 87, 43, 1985.

201. **Benson, D. R. and Hanna, D. G.**, *Frankia* diversity in an alder stand as estimated by sodium dodecyl sulfate-polyacrylamide gel electrophoresis of whole-cell proteins, *Can. J. Bot.*, 61, 2919, 1983.

202. **Benson, D. R., Buchholz, S. E., and Hanna, D. G.**, Identification of *Frankia* strains by two-dimensional polyacrylamide gel electrophoresis, *Appl. Environ. Microbiol.*, 47, 489, 1984.

203. **Gardes, M. and Lalonde, M.**, Identification and subgrouping of *Frankia* strains using sodium dodecyl sulfate-polyacrylamide gel electrophoresis, *Physiol. Plant.*, 70, 237, 1987.

204. **St.-Laurent, L. and Lalonde, M.**, Isolation and characterization of *Frankia* strains isolated from *Myrica gale*, *Can. J. Bot.*, 65, 1356, 1987.

205. **Gardes, M., Bousquet, J., and Lalonde, M.**, Isozyme variation among 40 *Frankia* strains, *Appl. Environ. Microbiol.*, 53, 1596, 1987.

205a. **Normand, P. and Simonet, P.**, unpublished results.

206. **Puppo, A., Dimitrijevic, L., Diem, H. G., and Dommergues, Y. R.**, Homogeneity of superoxide dismutase patterns in *Frankia* strains from Casuarinaceae, *FEMS Microbiol. Lett.*, 30, 43, 1985.

207. **Huang, J. B., Zhao, Z.-Y., Chen, G.-X., and Liu, H.-C.**, Host range of *Frankia* endophytes, *Plant Soil*, 87, 61, 1985.

208. **Van Dijk, C. and Sluimer-Stolk, A.**, An ineffective strain of *Frankia* AG in the dune area of Voorne, *Prog. Rep. Inst. Ecol. Res. Verh. Kon. Ned. Akad. Wetensch. 2ᵉ Reeks*, 82, 56, 1984.

209. **Gauthier, D. L., Diem, H. G., and Dommergues, Y. R.**, Tropical and subtropical actinorhizal plants, *Pesq. Agropec. Bras.*, 19 s/n, 119, 1984.

210. **Zhang, Z., Lopez, M. F., and Torrey, J. G.**, A comparison of cultural characteristics and infectivity of *Frankia* isolates from root nodules of *Casuarina* species, *Plant Soil*, 78, 79, 1984.

211. **Zhang, Z. and Torrey, J. G.**, Studies of an effective strain of *Frankia* from *Allocasuarina lehmanniana* of the Casuarinaceae, *Plant Soil*, 87, 1, 1985.

212. **Diem, H. G., Gauthier, D., and Dommergues, Y. R.**, Isolation of *Frankia* from nodules of *Casuarina equisetifolia*, *Can. J. Microbiol.*, 28, 526, 1982.

213. **Norris, J. H., Macol, L. A., and Hirsch, A. M.**, Nodulin gene expression in effective alfalfa nodules and in nodules arrested at three different stages of development, *Plant Physiol.*, 88, 321, 1988.

Chapter 5

LEGUME NODULE BIOCHEMISTRY AND FUNCTION

Robert B. Mellor and Dietrich Werner

TABLE OF CONTENTS

I. METABOLISM

A. INTRODUCTION

Nodules on legumes and also the nonlegume *Parasponia* exhibit different morphologies. In addition to being either annular or irregular, the two major types are either cylindrical or spherical (for review, see Reference 1). Cross-infection experiments prove these differences to be host-plant determined. Insofar as they do not affect the basic biochemistry of N_2 fixation in nodules, they will not be considered further here. Readers interested in this subject are referred to excellent articles by Corby[2] and Kidby and Goodchild.[13] The interplay between the partners involves an intensive mutual question-and-answer dialogue. The wrong answer from either partner leads to some kind of failure, ranging from immediate abortion to less effective nodules. Thus, ineffectiveness can be either plant or bacteria determined. Due to the relative ease of genetic manipulations in bacteria, plant-determined ineffectiveness is not as biochemically well characterized as is bacterial ineffectiveness. Those interested in host-determined ineffectiveness are referred to Vance.[4] In this chaper, the biochemistry of nodule function (metabolism) and the structural biochemistry will be treated separately.

B. CARBON METABOLISM

1. Vascular Sap to Acetyl CoA

Nodule cells can be considered to be bathed in a medium containing C sources provided by plant photosynthesis. Sucrose is considered to be the main energy source transported into the nodules for bacterial nitrogen fixation. N_2 fixation is reduced in low light intensities for prolonged periods or increased after the addition of extra leaves by grafting or CO_2 enrichment studies (see Reference 5 for review). That photosynthate supply is not the limiting factor in N_2 fixation and that sucrose itself is not the form of carbon used by the bacteroids is spoken for by experiments of Werner and Krotzky,[6] who showed increases in N_2 fixation with raised partial pressures of O_2 (also see Chapter 7). Sucrose and other carbon sources enter the plant cell into the cytoplasm. Sucrose is then broken into a catabolically usable form by sucrose synthase (catabolic). This protein in nodules is a nodulin, nodulin 100. It is a four-subunit enzyme controlling C metabolism and carbon supply to bacteroids.[7]

After the conversion of sucrose to glucose-6-phosphate, some of the carbon is drawn to a separate pool to deposit starch in amyloplasts. This pool presumably acts as a buffer to cover energy needs during diurnal rhythm or stress periods, since N_2 fixation continues for some days after photosynthate cutoff to mature nodules.[8]

A second pool of carbon is siphoned off to form several compounds, seemingly inert and slowly turning over (for review see Reference 9). These are cyclic alcohols, malonate, the disaccharide trehalose, and poly-β-hydroxybutyrate (PHB). The last two are assembled by the bacteria, trehalose apparently being released back into the host cell, whereas the PHB is stored in the microsymbiont. The cyclic alcohols (cyclitols) are major carbohydrates in legumes, myoinositol in *Glycine*, oronitol in *Pisum* and pinitol in *Trifolium*, *Lupinus*, and *Medicago*. However, *Phaseolus* nodules contain only traces of cyclitols. Malonate concentration in nodules reaches up to 5 mM and appears to have no influence on the utilization of dicarboxylic acids by bacteroids, nor to be significantly utilized itself, although free-living bacteria may take it up in large quantities.[10] PHB is the storage carbohydrate for bacteria and presumably helps released bacteroids survive after nodule senescence. Although the roles of cyclitols, trehalose, and malonate are not known, it is also possible that they help bacteroid survival after release by influencing the rhizosphere. Cytoplasmic trehalose is also thought to help eukaryotic cells withstand stress conditions.[11]

A third pool of organic carbon is carried from glucose-6-P down the Emden-Meyerhoff pathway to acetyl-CoA (see Figure 1). Nodule cytoplasm contains PEP carboxylase, the onset of activity being concomitant with that of N_2 fixation, reaching a maximum activity in nodules 50 times greater than that in roots.[12] The Emden-Meyerhoff pathway is tapped at various points, not only to provide amino acids necessary for cell growth, but also it has been postulated that NH_4^+ excreted from the bacteroid inhibits pyruvate kinase, allowing greater production of oxaloacetate by PEP carboxylase, the excess oxaloacetate being used for the assimilation of ammonia (see Section I.D).[13] The end product of the Emden-Meyerhoff pathway, acetyl-CoA, enters the mitochondria (Figure 2). Whether NH_4^+ levels ever get sufficiently high, in view of high nodule glutamine synthetase levels, to affect such regulation *in vivo*, has not been determined.

2. The TCA Cycle

Due to lack of experimental evidence, we are in almost complete ignorance about biochemical differences between mitochondria from infected and uninfected nodule cells, root cells, ineffectively infected cells, etc. Mitochondria of infected cells appear morphologically modified (see Section II.B.5). In nodules, free O_2 concentration is around 10 nM, a concentration at which oxidative phosphorylation is no longer possible, although mitochondria are clustered around the cell periphery, where O_2 concentrations of 20 to 26 mM are postulated to exist,[14] allowing some oxidative phosphorylation. Rawsthorne and La Rue[15] found ATP/O ratios and respiratory control ratios typical for aerobic mitochondria rather than those from anaerobic tissue. This may be due to their taking whole nodules as starting material, rather than separating the aerobic and anaerobic cells first. Further uninvestigated factors include malonate, present in high concentration in nodules (Section I.B.1). Malonate is a classic inhibitor of succinate dehydrogenase, and if concentration of malonate in the mitochondria is sufficient, this could lead to accumulation of succinate needed for N_2 fixation (Section I.C.1). In a similar study, Day et al.[15a] compared mitochondria from infected (nodule) and uninfected (root) soybean cells isolated over Percoll gradients. Nodule mitochondria were found to oxidize NADH and oxoglutarate rapidly. Nodule mitochondria had rotenone and KCN-sensitive respiration. Rooted cotyledon mitochondria, in contrast, were highly cyanide- and rotenone-insensitive. These findings were corroborated by two further published papers from Rawsthorne and LaRue.[15b,15c]

C. ENERGY FOR N_2-FIXATION
1. Sugars

The supply of carbon for the bacteroid is assumed to be completely plant determined and is complicated by the existence of a double barrier — the bacteroid membrane and the peribacteroid membrane — between it and the plant host. Transport across the peribacteroid membrane is an area relatively untouched, due partially to the difficulty of working with isolated membranes, representing variable proportions of inside-out and right-side-out vesicles.[16]

The active uptake of disaccharides into bacteroids has been demonstrated for fast-growing *Rhizobium* species, whereas slow-growers (i.e., *Bradyrhizobium*) accumulate disaccharides only by passive diffusion.[17] Glucose can enter bacteroids of both classes, although only slowly; however, new results indicate that at low oxygen tensions glucose uptake is significant. Fast-growing species (i.e., *Rhizobium*) possess the complete set of enzymes for sugar catabolism by the Entner-Doudoroff pathway in contrast to the slow-growing *Bradyrhizobium*, which are unable to use this pathway, preferring the formation of pentose phosphates.[18] While these transformations are probably used to provide amino acids for bacterial growth, for reserve compounds for the free-living part of the bacterial life cycle, and perhaps for other compounds important to specific stages of development, they are not to any great extent involved in nitrogen fixation. First, observed uptake rates are far too low to account for the energy expenditure of N_2 fixation, and second, mutants unable to metabolize sugars, as also pyruvate dehydrogenase mutants, almost always form nodules which are *fix*+ (able to fix nitrogen).[19]

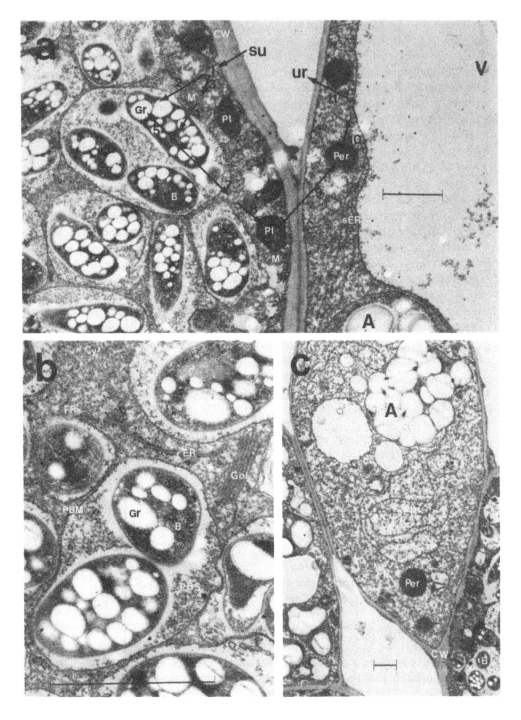

FIGURE 1

2. Dicarboxylic Acids

Although some strains of *Bradyrhizobium* are able to fix nitrogen under free-living conditions,[20] it is unlikely that this is of great importance in the soil. There they might prefer to take up organic nitrogen. This led to the question, why do bacteria fix nitrogen in nodules, a nitrogen-rich environment? This consideration led to the view that bacteria are protected from external nitrogen by their membrane systems and that nitrogen-free carbon would be used to support N_2 fixation. More precise details were forthcoming with the discovery that dicarboxylate transport *(dct)* mutants were *Fix,* $^-$ as were succinate dehydrogenase mutants.[21] The high levels of TCA cycle enzymes in bacteroids and the level of uptake of succinate into bacteroids (up to 50 times higher than that of carbohydrates) has led to the generally held belief that N_2 fixation is powered using dicarboxylic acids (mainly succinate but perhaps also malate and fumarate) as the electron source. It is worth noting here that *dct* mutant bacteroids fill host cells and appear morphologically normal, indicating that other carbon sources (probably sugars, see Section I.C.1) are used for growth, and C_4-dicarboxylic acids specifically for N_2 fixation, although here additional roles, e.g., precursors for heme synthesis, cannot yet be ruled out. Ronson and Astwood[22] have found a *nifA*-regulated promoter sequence upstream of the chromosomal *dct* structural gene, which would provide the coupling between the energy-demanding N_2 fixation and the import of necessary substrate. Udvardi et al.[22a] isolated peribacteroid units from soybean nodules and characterized the dicarboxylate transporter in the PBM. It was found to have a higher affinity for the monovalent malate^{-1} anion that for the succinate^{-1} anion (K_m = 2 and 15 μM, respectively), although the V_{max} for malate^{-1} was lower than for succinate^{-1} (V_{max} = 11 and 30 nmol $\cdot$min$^{-1}\cdot$ mg^{-1} protein, respectively). This suggests that the PBM strongly regulates the movement of dicarboxylate into the peribacteroid space and to the bacteroid itself. Whether this transporter is auodulin is one of the key questions for further investigation.

3. Amino Acids

Rhizobia fix nitrogen under conditions of nitrogen starvation. They must therefore be nitrogen starved in nodules. Under conditions of nitrogen limitation, many other changes in *Rhizobium* metabolism take place, e.g., glutamine synthase and GOGAT (see Section I.D) are stimulated and a high affinity $NH_4{}^+$ uptake system is induced, so that all environmental nitrogen in whatever quantities can also be scavenged. Whereas these changes appear simultaneously in free-living bacteria, in N_2-fixing bacteroids GS/GOGAT levels are very low and the $NH_4{}^+$ uptake system is not switched on, suggesting N_2 starvation in the presence of glutamate and $NH_4{}^+$. An elegant explanation has been put forward by Kahn et al.,[23] who postulate that the transport form of carbon is bound to nitrogen which is then deaminated

FIGURE 1.   Thin sections through an effective soybean (*Glycine max*) nodule showing: (a) Infected and uninfected cells. Extracellular sucrose (SU) crosses the cell wall (CW) into the host cell where breakdown is catalyzed by cytoplasmic sucrose synthetase (1) (See Reference 7). Although bacterial cell growth may be accomplished using monosaccharides (2) (see Reference 9), electron sources destined to feed nitrogen fixation are provided by glycolysis (3) and the TCA cycle in mitochondria (M). Either dicarboxylic acids or amino acids enter the bacteroid (B) (4) where they may also contribute to metabolism (5) (represented here by the formation of PHB granules [Gr]), or fuel nitrogenase (6) (see Reference 18). Fixed nitrogen as $NH_4{}^+$ is incorporated first into glutamine by cytoplasmic glutamine synthetase (7) (see Reference 27), which is then used for purine biosynthesis (8) in the plastid (PL) (see Reference 82). Upon leaving the plastid, purine nucleotide is broken down until eventually urate enters peroxisomes (Per) of uninfected cells (9). Allantoin formed may be further catabolized in the smooth endoplasmic reticulum (sER) of uninfected cells (10). The net effect (11) is the export of ureides (UR) (see Reference 87). V = vacuole, A = amyloplast, bar = 1 μm. (b) Bacteroid (B) with PHB granules (Gr), contained in peribacteroid membrane (PBM). PBS = peribacteroid space. Host cell cytoplasm (Cyt) contains free ribosomes (FR), rough endoplasmic reticulum (rER), and Golgi apparatus (Gol). Bar = 1 μm. (c) Uninfected cell showing large peroxisome (Per) and extensive smooth endoplasmic reticulum (sER). The cytoplasm is dense with free ribosomes (FR). The prominent amyloplast (A) stores starch reserves. Bacteroids (B) are shown in neighboring infected cell. CW = cell wall, bar = 1 μm.

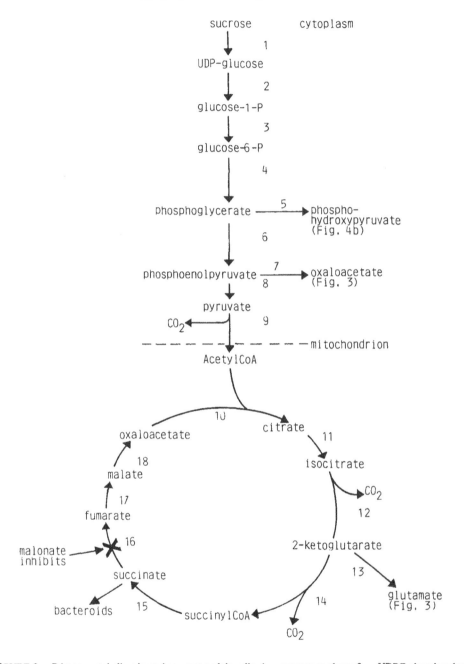

FIGURE 2.   Primary metabolism in soybean root nodule cells. 1 = sucrose synthase; 2 = UDPG phosphorylase; 3 = phosphoglucomutase; 4 = glycolysis; 5 = PGA dehydrogenase; 6 = phosphopyruvate hydratase; 7 = phosphopyruvate carboxylase; 8 = pyruvate kinase; 9 = pyruvate dehydrogenase; 10 = citrate synthase; 11 = aconitate hydratase; 12 = isocitrate dehydrogenase; 13 = glutamate dehydrogeanse; 14 = ketoglutarate dehydrogenase; 15 = succinyl CoA synthetase; 16 = succinate dehydrogenase; 17 = fumarase; 18 = malate dehydrogenase.

and the waste product, ammonium, returned to the plant in addition to the ammonia fixed by nitrogenase.

The theory of Kahn et al.[23] focuses on glutamate because of the effect of glutamate on cell morphology and its association with nitrogen fixation, and because its catabolism is normally accomplished accompanied by NH_4^+ excretion; furthermore, glutamate catabolism mutants are Fix$^-$. This theory fits almost all known facts dealing with the supply of energy for nitrogen fixation, but as yet, given the relative newness of the theory, few have had time to test it properly. Recently, Humbeck and Werner[24] determined pools of organic acids in the peribacteroid space of soybean nodules. Malate and 2-oxoglutarate were several-fold enriched in this compartment compared to the host cytoplasm, while succinate and malonate pools were hardly detectable. This is in agreement with the Kahn model, but it does not exclude the uptake of succinate by the high affinity uptake system,[25] leaving only very small concentrations of succinate in the peribacteroid space. There are also no data to suggest that all bacteroids in an infected cell are fixing equally and that all have the same carbon source. Indeed, in an effectively infected plant cell with mitochondria close to the plasma membrane, one expects a decreasing gradient of succinate concentration from the outside, over the bacteroids (taking up succinate) to the cell center. Assuming all bacteroids to be exporting NH_4^+ and ammonia assimilation to be uniformly spread in the cytoplasm, an opposing gradient of amino acid would be built, high concentrations in the cell center, low at the cell periphery where excess is continuously exported (Section I.D). Thus, there is no reason why bacteroids in the cell interior should not use more amino acid as carbon source relative to C_4-dicarboxylic acid than do their counterparts on the outer areas of the plant cell.

D. ASSIMILATION OF FIXED N$_2$
1. GS/GOGAT vs. GDH

Ammonia, fixed by nitrogenase (E.C.1.18.6.1.), is excreted, moving freely across membranes. Thus, it is assumed to flow out of the bacteroid by diffusion down the concentration gradient caused by its continual removal in the host cytoplasm. Ammonia assimilation in plants and yeast can be over either the glutamate dehydrogenase (GDH) or the glutamine synthetase (GS)/glutamate synthase (GOGAT = glutamine: 2-oxoglutarate aminotransferase) pathways (for review see Reference 26) (Figure 3). Research over the past 10 years has revealed that only in yeast is the GDH pathway of importance in NH_4^+ assimilation and in almost all green plants the GS/GOGAT is the preferred way, although ATP is needed.[27] In root nodules, the enzymes for both pathways are present, but the high K_m of GDH for NH_4^+ and its low specific activity rule it out for assimilatory purposes. Glutamine thus provided may be used for either asparagine synthesis or purine synthesis, depending on whether the plant is an amide-transporting legume or a ureide-transporting legume (see Section I.D.2).

2. Ureides vs. Amides

Legumes entering into N$_2$-fixing symbioses can be broken down into two groups; those whose nodules export amides (e.g., pea) or ureides (e.g., soybean) (see Reference 28). The basis for this differentiation is unclear, as are the possible benefits of one N-transport form over the other. One possible explanation has been proposed by Kahn et al.[23] in connection with the model of carbon supply to bacteroids via amino acids (see Section I.C.3) where *Bradyrhizobium* catabolism could supply the plant with glycine, a purine synthesis precursor, thus giving an advantage in soybeans to ureide synthesis.

The basic differences between the two pathways are as follows. In amide plants, cytoplasmic glutamate synthase and asparagine synthetase are presumed to be used, although forms of both enzymes are also found in plastids (Figure 4a).

The synthesis of purines and their subsequent breakdown for transport are more complicated. Basically, N from glutamine is channeled into purine biosynthesis in the plastid of the infected cell (Figure 4b). Since many processes are energy requiring, this may explain the many mitochondria-plastid associations seen in infected cells. Plastids, which are also normally found on the periphery of infected cells, export, then, purine nucleotide to interstitial, uninfected cells where the large prominent peroxisomes there play a role. These

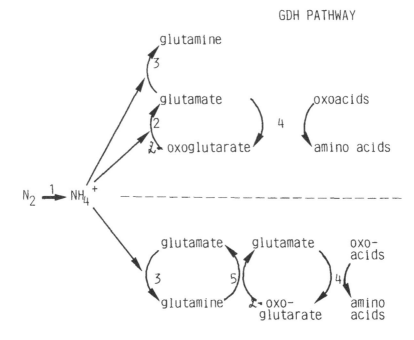

FIGURE 3.   Primary nitrogen assimilation. The GDH pathway is believed to be relatively unimportant in green plants. 1 = nitrogenase; 2 = glutamate dehydrogenase (GDH); 3 = glutamine synthetase (GS); 4 = aminotransferase; 5 = glutamate synthase (GOGAT).

peroxisomes, in contrast to those of infected cells, contain the nodulin uricase II,[29,30] which converts urate to allantoin. This allantoin is then converted by allantoinase (located in the endoplasmic reticulum, or ER) into forms suitable for xylem transport (Figure 5). Van den Bosch and Newcomb[30] have noted that uninfected cells contain long strands of smooth ER.

The products of biological nitrogen fixation have recently been reviewed by Schubert.[30a]

E. HYDROGEN METABOLISM

Hydrogen is a product of all nitrogenase reactions according to

$$N_2 + 8H^+ + 8e^- + 16MgATP \rightarrow 2NH_3 + H_2 + 16\ MgADP + 16Pi$$

It is formed by replacement of enzyme-bound hydrogen by N_2 according to Mortensen[31] by the equation

$$MoFe{:}2H + N_2 \rightarrow MoFe{:}N_2 + H_2$$

This explains a molar ratio of N_2 fixation to hydrogen evolution of 1:1, observed in symbiotic and nonsymbiotic nitrogen-fixing organisms. An explanation of a higher ratio of H_2 evolution to N_2 fixation is given by the proposal of Loew et al.[32] according to

$$MoFe{:}2H \quad \text{and} \quad MoFe{:}3H \xleftrightarrow{\text{reverse reactions}} MoFe + H_2$$

Reduced ratios of hydrogen evolution to N_2 fixation are a consequence of hydrogen uptake and utilization in legume nodules, according to the scheme of Evans et al.[33]

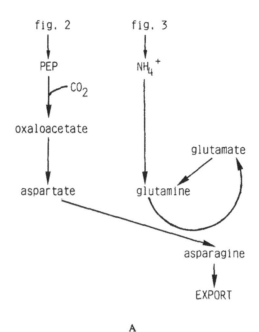

A

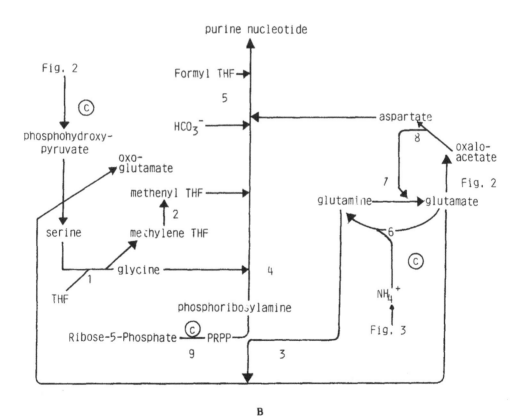

B

FIGURE 4.   Nitrogen export pathways in root nodules of (A) amide and (B) ureide legumes. While in the first case, the processes are thought to be cytoplasmic, the plastid plays a central role in the latter, where enzymes (numbered) are as follows: (1) serine hydroxymethylase, (2) methylene THF dehydrogenase, (3) phosphoribosylamidotransferase, (4) glycinamide ribonucleotide synthetase, (5) aminoimidazole ribonucleotide carboxylase, (6) glutamine synthetase, (7) glutamate synthase, (8) asparagine aminotransferase, and (9) phosphoribosylpyrophosphate synthetase. THF = tetrahydrofolate, Ⓒ = cytoplasmic location.

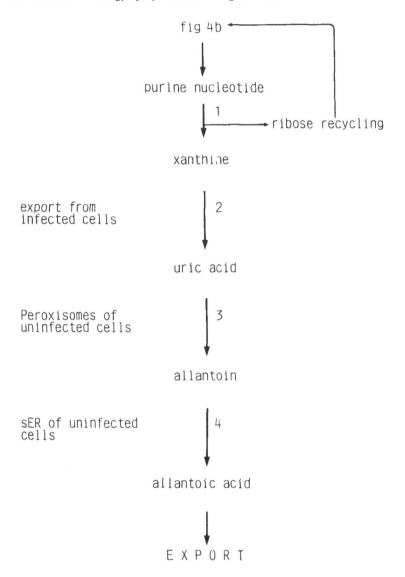

FIGURE 5.   1 = The exact catabolism of purine nucleotide to xanthine is not yet known;
2 = xanthine dehydrogenase; 3 = uricase; 4 = allantoinase

The inoculation of legumes with efficient hydrogen uptake (Hup⁺) strains of *Rhizobium* and *Bradyrhizobium* reduces the loss of electron flux through nitrogenase as H_2 from over 30 to about 4%. Hup⁺ strains of *Bradyrhizobium japonicum* can grow also by chemolithotrophy, using H_2 as energy source and assimilating CO_2 by ribulose-bisphosphate carboxylase,[34] but under these conditions, no nitrogen fixation was observed.

The uptake hydrogenase in *B. japonicum* is a membrane-bound protein, similar to other membrane-bound hydrogenases. It has a molecular weight of about 96 kDa, with one subunit of 63 kDa and one smaller subunit with 33 kDa. It is an iron-sulfur protein with probably one nickel atom per enzyme molecule. The K_m for H_2 is relatively low and on the order of 1 μM.

The hydrogen oxidation pathway starts with a still unidentified physiological electron acceptor via a number of electron carriers to O_2 as the final acceptor. As electron carrier from H_2 to O_2, ubiquinone is involved[35] and probably a cytochrome b_{559} "upstream" of ubiquinone, since only H_2, but not succinate or NADH, reduce the membrane-bound cytochrome b_{559} to a maximum extent.

ATP formation by H_2 oxidation has been successfully demonstrated with bacteroids from *Rhizobium*[36] and *Bradyrhizobium*.[37] Utilization of oxygen in this hydrogen-recycling system also reduces the oxygen tension in nodules and is another oxygen protection mechanism in addition to oxygen transport by leghemoglobin, respiration of the plant cell and bacteroids, and the diffusion barriers of the cork layer and the water barriers in the intracellular spaces of the nodules. Alternatively, it may be argued that, if nodules actually are O_2 limited, hydrogenase activity removes scarce O_2 from oxidative phosphorylation, resulting in no significant alteration in the net ATP status of the cell.

II. COMPARTMENTATION

A. THE MICROSYMBIOTIC ORGANELLE

1. Bacteroids

Bacteroids are defined as the specialized stage of the life cycle of symbiotic bacteria, occurring when they inhabit host cells. Bacteroids may be effective (N_2 fixing) or ineffective (not N_2 fixing). The host cells are normally root derived but may be from other parts of the plant.[38]

Bacteroids can exhibit a very irregular morphology or appear more or less normal. This and the number of bacteroids contained inside each peribacteroid membrane have been proved by cross-inoculation experiments to be under host control.[39] As a generality, however, one may note the *Rhizobium* produces single, enlarged bacteroids, while *Bradyrhizobium* in symbiosis with tropical legumes has multiple bacteroids per peribacteroid vesicle and that they are of the size of vegetative bacteria. After infection, the bacteroid DNA content per cell can increase to four to eight times the level of the free-living form[40] in *Rhizobium*. This is, however, due to the multicellular form of bacteroids, where cells increase their size without division. In *B. japonicum* it has been shown that per unit cell volume DNA content of bacteroids and free-living cells is the same.[41,42]

As bacteroids multiply and fill the host cell, there is a rapid rise in bacterial protein synthesis. It appears that all bacteroid proteins are synthesized on bacteroid ribosomes.[43] Due to obvious technical difficulties it still cannot be absolutely ruled out that plant-produced proteins sometimes get into the bacteroids or, vice versa, that bacteroid-produced proteins or other components are exported to the plant (as bacteriodins).

The cell wall of free-living *Bradyrhizobium* and *Rhizobium* consists of two bilayered membranes in the typical Gram-negative bacterial form. The inner (cyctoplasmic) membrane is separated from the outer by a periplasmic space. the outer cell wall consists of structural peptidoglycon (murein) elements delimited by a thin outer layer. Upon infection and the attainment of the bacteroid state[44] cell walls become thinner and more flexible. Van Brussel and co-workers[45] attribute this to a loss of cross-linking in the murein layer. Lipopolysaccharide (LPS) components remain, however, constant[46] (for review see Reference 47). In mature bacteroids the outer layers are sloughed off[48] with the practical consequence that, for culturing microsymbionts from nodules, high osmo-protecting media may be needed.[49] It is tempting to hypothesize that saccharidases present in the bacteroid's immediate environment are responsible for this change, but Mellor et al.[50] reported no detectable interaction between these enzymes and bacterial or bacteroidal cell wall or exopolysaccharide components. Concomitant with the differentiation of bacteria into bacteroids reserve polysaccharide (glycogen) disappears and reserve granules filled with poly (beta)-hydroxybutyric acid (PHB), which is present only in small amounts in the free-living stage, become dominant in the bacteroid morphology of *B. japonicum*. However, other bacteroids (as from *R. meliloti* in alfalfa or *R. leguminosarum* in broad bean) show hardly any PHB accumulation.[51] On the other hand, inside infection threads, very obvious PHB granules were observed in *R. leguminosarum*.[52] In *Bradyrhizobium*, polyphosphate bodies associate with the nucleoid body in bacteroids.[53] The function of these is unknown. This forms one of the differences which have led to the classification of the slow-growing *Rhizobium* as the distinct taxonomic group *Bradyrhizobium*. Numerous reports suggest that bacteroids retain the ability to form colonies after reisolation from nodules. Significant variation exists between reports as to the frequency of viability. *Bradyrhizobium* bacteroids generally retain higher regeneration frequencies.

2. The Peribacteroid Space

The peribacteroid space (PBS) is defined as the space between the symbiotic partners. Morphologically it is that space which lies between the outside of the bacteroid and the inside of the peribacteroid membrane. The space is electron-transparent in electron microscopy (EM) but may be isolated and biochemically characterized.[16] The PBS contains several proteins which can be separated by SDS-PAGE or urea IEF (isoelectric focusing). Some lytic activities, including a large proportion of proteases, have been reported in the PBS.[54] The PBS also contains protease inhibitors[55] and it may be on these that effective bacteroids rest in order to escape digestion by the host. The best-characterized protein occurring in the PBS is alpha-mannosidase isoenzyme II. This is found in high concentrations in the PBS and is one of the two vacuolar forms of alpha-mannosidase. The third, isoenzyme III, is found sequestered in the extracellular space.[56] Thus, the PBS is in contact with the host cell vacuole and the bacteroid, at least from the point of view of alpha-mannosidase, sits in a lysosome-like environment and not an extracellular environment as was previously thought.

Due to the action of ATPases in the peribacteroid membrane (PBM) (see next section) H^+ ions are pumped into the PBS. The acid environment protonates NH_3 from the bacteroid. The physiological barrier of the PBS and the PBM must therefore be crossed as NH_4^+, which is notable to cross a membrane without a carrier. The effect of mutations in the bacteria on the PBS composition are not understood, but proteins from PBS from Fix$^-$ symbioses show a different profile after urea IEF than do those from Fix$^+$ symbioses.

3. The Peribacteroid Membrane

The peribacterioid membrane (PBM) forms a continuous envelope around the prokaryote, and various numbers of bacteroids are contained within each membrane, between 1 and 20, depending on species and the age of the nodule. The membrane is provided by the plant partner mostly by membrane flow over the Golgi, but more direct routes for some components from the ER cannot yet be absolutely ruled out.[57] Mature PBM does not arise as previously thought[58] from the plasma membrane (PM), but does exhibit some biochemical similarities to the PM. One such similarity is the ability to form coated vesicles, a property normally associated only with the plasma membrane. Thus, some plasma membrane molecules must be resident in the PBM (see Reference 57).

The PBM can best be imagined as the outer envelope of a N_2-fixing organelle. Bradley et al.[59] have reported it contains patches of bacteroidal LPS, and Bassarab and Werner[60] have characterized a protein kinase activity in the membrane. Details of specific transport processes are not known, but two Mg^{2+}-dependent H^+-pumping ATPases are localized in the PBM.[61,62] Two-dimensional maps of PBM proteins are given in Werner et al.[63] The stability and some other properties of the PBM are determined in some unknown way by the prokaryote. Regensburger et al.[64] found bacteroidal replication not to be a prerequisite for PBM formation from the plant and postulated that, early in the colonization process, the bacteria send a signal to the plant cell nucleus which preprograms the production of PBM. If this is so, then PBM from all mutants must be the same. Werner et al..[63] found differences in PBM from nodules infected with wild-type and mutant bacteria. Thus, they postulated that at least two signals from the bacteria must exist. Fortin et al.[65] reported that the early membrane nodulin, nod 24, a membrane-associated protein, was induced in the precolonization phase, whereas the transmembrane nodulin, nod 26, was only induced later. The situation appears to be that on or before entry into the host cell the bacteria sends a signal to the host cell nucleus for PBM production. This signal may contain leucine since leu⁻ mutants are not released from the infection thread.[66] We speculate that the signal is the product of still unknown symbiotic genes, the "early pbm" gene(s). At this point the PBM consists of about 14 "common proteins". Once inside the cell and replicating, further signals are sent from the bacteria (products of the "late pbm" gene[s] which prompt subsequent changes [e.g., glycosylation] and the appearance of up to eight more proteins. This scheme is represented in Reference 67).

B. HOST CELL CYTOPLASM AND OTHER ORGANELLES
1. Introduction
Nodule cells can be relatively easily fractionated. Standard methods usually include sucrose density gradient centrifugation (see Reference 16). Two points are worth noting here:

1. Damaged plastids and free membrane from broken amyloplasts may cause confusion in the interpretation of results among particulate fractions.
2. Cytoplasmic fractions obtained contain also some soluble matrix components from organelles inevitably opened during homogenization.

Concrete results can therefore only be obtained after correlating cell fractionation experiments closely with ultrastructural (EM) studies.

2. Nucleus
At a certain stage of nodule development cell division in the cortex slows, then stops. DNA synthesis continues, however, and the host cell nucleus becomes polyploid, DNA levels in the bacteroid zone up to 32C being reached (for review, see Reference 68). Grossly enlarged nuclei are to be seen in infected host cells, often with several nucleoli. Nuclei with crenated borders have often been reported, and, thus, probably seem to increase the surface area of the nuclear membrane to allow the passage of more mRNA, etc. into the cytoplasm.

3. Cytoplasm
The host cell cytoplasm after infection becomes dense and granular in EM pictures. The number of free ribosomes increases drastically in all but a few ineffective symbioses. The cytoplasm becomes red due to the appearance of a myoglobin-like protein, leghemoglobin.[69,70] The cytoplasmic nodulin choline kinase II appears in those symbioses where peribacteriod membranes are made,[71,72] as do other early and late nodulins.[73] A host-cytoplasm-located galactosidase is stimulated at this time. Its function, however, remains unknown.

4. Endoplasmic Reticulum
In the early stages of cell colonization, ER vesicles become visibly more numerous throughout the host cell. During this time the ER-located enzymes choline phosphotransferase and GDP-DMP-mannosyltransferase are stimulated by 200 and 300%, respectively.[71] Direct connections between the ER and PBM as reported by Kijne and Planqué[74] are probably not significant, but a direct transfer of components by vesicle shuttle from the ER matrix to the PBS still cannot be ruled out. The case of alpha-mannosidase is an example. The PBS-located isoenzyme II is made on bound ribosomes of the rough ER,[75] remains in the plant vacuome,[56] but has not been found in the Golgi.

5. Golgi Apparatus

The enzyme UDP-galactose-asialoagalactofetuin galactosyltransferase is found in both ER and Golgi[76] and provides subterminal galactose for glycoconjugates. This activity is stimulated 800% in effective symbioses, which is less than the 1600% stimulation recorded for the *N*-acetylgalactosaminyltransferase, which is totally Golgi-located[76] and provides terminal amino-sugar residues for glycoconjugates. Since PBM contains products of both the above enzymes, this provides biochemical support for the observations of Robertson and co-workers[77,78] that the major pathway providing material for the PBM is over the Golgi. Of the putative Golgi marker enzyme, glucan synthetase I, is not measurable in infected cells[79] although IDPase is easily detectable. In transmission electron micrographs dictyosome bundles are often seen adjacent end-on to peribacteroid membrane,[80] and freeze-fracture studies reveal continuities between Golgi and PBM over vesicles.[77]

6. Mitochondria

Mitochondria in infected cells are cristae-rich through intensive folding of the inner mitochondrial membrane and move to the cell periphery.[81] This movement is probably associated with the need for O_2 for oxidative respiration in the mitochondria (see Section I.B.2). pO_2 in the infected cells has been measured with microelectrodes to be 10 n*M*; thus, mitochondrial respiration is O_2 limited[15] and therefore their position at the plasma membrane probably reflects their "wandering" up the cellular O_2 gradient to the highest partial pressure. The biochemical modification in mitochondria in infected cells is more or less unresearched, which is astonishing, considering their central role in N_2 fixation. The limitation of O_2 in the infected cells means mitochondrial metabolism is limited. This, in turn, restricts the supply of C_4 skeletons to the bacteroid, leading to energy limitation and lower N_2 fixation. N_2 fixation can be increased 40% by doubling the atmospheric pO_2.[6] Thus, manipulations in the host cell mitochondria may well lead to higher agricultural yields.

7. Plastids

Plastids have been isolated (e.g., Reference 82), but, as with mitochondria, a thorough biochemical analysis has not yet been undertaken. In EM pictures plastids can often be seen closely associated with mitochondria. This probably reflects the provision of carbon skeletons and energy by the mitochondria for the plastid-located enzymes of NH_4^+ assimilation (Section I.D.2).

C. INFECTED AND UNINFECTED NODULE CELLS
1. Whole Cells

For many years, uninfected cells were ignored as being unimportant — simply cells where infection for one reason or another had been unsuccessful. Protoplasts made from nodule cells were separated in infected and uninfected cells on a Ficoll gradient. In uninfected cells smooth ER proliferates, and the amount of sER-located allantoinase is raised. Soluble aspartate aminotransferase is found almost solely in the cytoplasm of uninfected cells.[83]

2. Peroxisomes

Peroxisomes in infected cells remain small and scattered. In uninfected cells they are larger and are the exclusive site of the nodulin nod 35[29,30] which is uricase II.[84,85] Uricase II appears to be translated on free ribosomes and posttranslationally imported into the organelle as other peroxisomal proteins.[86] Thus, peroxisomes in uninfected cells play an important role in N_2 metabolism.[87]

3. Amyloplasts

Amyloplasts in root nodules have not been well characterized biochemically. They are often seen near mitochondria. They probably function as a carbohydrate reservoir, ensuring an even supply of carbon skeletons for the mitochondria, independent of possible fluctuations in the phloem sugar concentration.[88] The exhaustion of the starch in the amyloplasts correlates with a first drop of nitrogenase in synchronized nodules of *Glycine max*.[89] In uninfected cells, where energy demand for N_2 fixation is understandably lower, more and larger amyloplasts are found.

4. Plasma Membrane

The plasma membrane (PM) is traditionally the hardest membrane to isolate, a situation complicated by its being differentiated into patches (see Reference 90). Thus, although SDS-PAGE profiles of plasma membrane proteins have been published[16] (for a PM/Golgi mixture, see Reference 61), further biochemical characterization has not been carried out. It must also be added that the samples for the gels referred to above were isolated from whole nodules, that is, a mixture of infected and uninfected cells. Although in electron micrographs PM from both types of cells appear morphologically similar, biochemical differences may be expected when we consider the roles of the two cell types. Figure 1 shows that infected cells import sugar residues and export fixed nitrogen. Since energy demand in uninfected cells may be lower than in infected cells, it seems redundant for these cells to import so much carbohydrate. Additionally, while infected cells must continually export fixed nitrogen, uninfected cannot. On the contrary, they must import as much as possible. Thus, here, too, differences in the molecular mechanisms of plasma membrane function must be expected. Uninfected cells export ureides or allantoic acid, and it would be worthless if this was taken up again by any other cells except those leading to the xylem. Thus, we can postulate that differences in the PM exist between different types of specialized nodule cells. The differences between these cells and other root or differentiated cells are at present unknown.

D. CONCLUDING REMARKS

This chapter summarizes the interaction of rhizobial nodule biochemistry, structure, and function. It is recognized that the biochemical analyses require more cellular fractionation studies in conjunction with the use and perhaps the application of both plant and bacterial mutants to correlate molecular changes with function. The review did not encompass the energetics and biochemistry of nitrogenase, which is sufficiently reviewed in conference proceedings.

REFERENCES

1. **Corby, H. D. L., Polhill, R. M., and Sprent, J. I.,** Taxonomy, in *Nitrogen Fixation 3: Legumes,* Broughton, W. J., Ed., Oxford University Press, New York, 1983.
2. **Corby, D. T. L.,** Biological nitrogen fixation in natural and agricultural habitats, *Plant Soil Spec.,* 305, 1971.
3. **Kidby, D. K. and Goodchild, D. J.,** Host influence on the ultrastructure of root nodules of *Lupinus luteus* and *Ornithopus sativus, J. Gen. Microbiol.,* 45, 147, 1966.
4. **Vance, C. P.,** *Rhizobium* infection and nodulation: a beneficial plant disease?, *Annu. Rev. Microbiol.,* 37, 399, 1983.
5. **Emerich, D. W., Lepo, J. E., and Evans, H. J.,** Nodule metabolism, in *Nitrogen Fixation 3: Legumes,* Broughton, W. J., Ed., Oxford University Press, New York, 1983, 213.
6. **Werner, D. and Krotzky, A.,** Die symbiontische Stickstoff-Fixierung der Leguminosen, *Funkt. Biol. Med.,* 2, 31, 1983.
7. **Thummler, F. and Verma, D. P. S.,** Nodulin 100 of soybean is the subunit of sucrose synthase regulated by the availability of free haem in nodules, *J. Biol. Chem.,* 262, 14730, 1987.
8. **Lie, T. A.,** Environmental physiology of the legume-*Rhizobium* symbiosis, in *Nitrogen Fixation 2: Ecology,* Broughton, W. J., Ed., Oxford University Press, New York, 1982, 104.

9. **Streeter, J. G. and Salminen, S. O.**, Carbon metabolism in legume nodules, in *Nitrogen Fixation Research Progress*, Evans, H. J., Bottomley, P. J., and Newton, W. E., Eds., Martinus Nijhoff, Dordrecht, 1985, 277.

10. **Werner, D., Dittrich, W., and Thierfelder, H.**, Malonate and Krebs cycle intermediates utilization in the presence of other carbon sources by *Rhizobium japonicum* and soybean bacteroids, *Z. Naturforsch.*, 37c, 921, 1982.

11. **Keller, F., Schellenberg, M., and Wiemken, A.**, Localization of trehalase in vacuoles, and of trehalose in the cytosol of yeast, *Arch. Microbiol.*, 131, 298, 1982.

12. **Lawrie, A. C. and Wheeler, C. T.**, Nitrogen fixation in the root nodules of *Vicia faba* L in relation to the assimilation of carbon, *New Phytol.*, 74, 437, 1975.

13. **Peterson, J. B. and Evans, H. J.**, Properties of pyruvate kinase from soybean nodule cytosol, *Plant Physiol.*, 61, 909, 1978.

14. **Sheehy, J. E., Minchin, F. R., and Witty, J. F.**, Control of nitrogen fixation in a legume nodule: an analysis of the role of oxygen diffusion in relation to nodule structure, *Ann. Bot.*, 55, 549, 1985.

15. **Rawsthorne, S. and LaRue, T. A.**, Respiration and oxidative phosphorylation of mitochondria from nodules of cowpea, in *Nitrogen Fixation Research Progress*, Evans, H. J., Bottomley, P. J., and Newton, W. E., Eds., Martinus Nijhoff, Dordrecht, 1985, 351.

15a. **Day, D. A., Price, G. D., and Gresshoff, P. M.**, Oxidative properties of mitochondria and bacteroids from soybean root nodules, *Protoplasma*, 134, 121, 1986.

15b. **Rawsthorne, S. and LaRue, T. A.**, Preparation and properties of mitochondria from cowpea nodules, *Plant Physiol.*, 81, 1092, 1986.

15c. **Rawthorne, S. and LaRue, T. A.**, Metabolism under microaerobic conditions of mitochondria from cowpea nodules, *Plant Physiol.*, 81, 1097, 1986.

16. **Mellor, R. B. and Werner, D.**, The fractionation of *Glycine max* root nodule cells: a methodological overview, *Endocyt. Cell Res.*, 3, 317, 1986.

17. **Salminen, S. O. and Streeter, J. G.**, Uptake and metabolism of carbohydrates by *Bradyrhizobium japonicum* bacteroids, *Plant Physiol.*, 83, 535, 1987.

18. **Dilworth, M. and Glenn, A.**, How does a legume nodule work?, *TIBS*, 519, 1984.

19. **Glenn, A. R., McKay, I. A., Arwas, R., and Dilworth, M. J.**, Sugar metabolism and the symbiotic properties of carbohydrate mutants of *Rhizobium leguminosarum*, *J. Gen. Microbiol.*, 130, 259, 1984.

20. **Roberts, G. P. and Brill, W. J.**, Genetics and regulation of nitrogen fixation, *Annu. Rev. Microbiol.*, 35, 207, 1981.

21. **Finan, T. M., Wood, J. M., and Jordan, D. C.**, Succinate transport in *Rhizobium leguminosarum*, *J. Bacteriol.*, 148, 193, 1981.

22. **Ronson, C. W. and Astwood, P. M.**, Genes involved in the carbon metabolism of bacteroids, in *Nitrogen Fixation Research Progress*, Evans, H. J., Bottomley, P. J., and Newton, W. E., Eds., Martinus Nijhoff, Dordrecht, 1985, 201.

22a. **Udvardi, M. K., Price, G. D., Gresshoff, P. M., and Day, D. A.**, A dicarboxylate transporter on the peribacteroid membrane of soybean nodules, *FEBS Lett.*, 231, 36, 1988.

23. **Kahn, M. L., Kraus, J., and Somerville, J. E.**, A model of nutrient exchange in the *Rhizobium*-legume symbiosis, in *Nitrogen Fixation Research Progress*, Evans, H. J., Bottomley, P. J., and Newton, W. E., Eds., Martinus Nijhoff, Dordrecht, 1985, 193.

24. **Humbeck, C. and Werner, D.**, Separation of malate and malonate pools by the peribacteroid membrane in soybean nodules, *Endocyt. C. Res.*, 4, 185, 1987.

25. **Humbeck, C. and Werner, D.**, Two succinate uptake systems in *Bradyrhizobium japonicum*, *Curr. Microbiol.*, 14, 259, 1987.

26. **Müntz, K.**, Stickstoffmetabolismus der Pflanzen, Fischer Verlag, Jena, 1984.

27. **Miflin, B. J. and Lea, P. J.**, The pathway of nitrogen assimilation in plants, *Phytochemistry*, 15, 873, 1976.

28. **Pate, J. S. and Atkins, C. A.**, Nitrogen uptake, transport and utilization, in *Nitrogen Fixation 3: Legumes*, Broughton, W. J., Ed., Oxford University Press, New York, 1983, 245.

29. **Nguyen, T., Zelechowska, M., Forster, V., Bergmann, H., and Verma, D. P. S.**, Primary structure of the soybean nodulin -35 gene encoding uricase II localized in the peroxisomes of uninfected cells of nodules, *Proc. Natl. Acad. Sci. U.S.A.*, 82, 5040, 1985.

30. **Van den Bosch, K. A. and Newcomb, E. H.**, Immunogold localization of nodule-specific uricase in developing soybean root nodules, *Planta*, 167, 425, 1986.

30a. **Schubert, K.**, Products of biological nitrogen fixation in higher plants: synthesis, transport and metabolism, *Annu. Rev. Plant Physiol.*, 37, 539, 1986.

31. **Mortensen, L. E.**, The role of dihydrogen and hydrogenase in nitrogen fixation, *Biochemie*, 60, 219, 1978.

32. **Lowe, D. J., Thornley, R. N. F., and Postgate, J. R.**, The mechanism of substrate reduction by nitrogenase, in *Advances in Nitrogen Fixation Research*, Veeger, C. and Newton, W. E., Eds., Martinus Nijhoff, Dordrecht, 1984, 133.

33. **Evans, H. J., Harker, A. R., Papen, H., Russel, S. A., Hanus, F. J., and Zuber, M.,** Physiology, biochemistry and genetics of the uptake hydrogenase in *Rhizobium, Annu. Rev. Microbiol.,* 41, 335, 1987.

34. **Hanus, F. J., Maier, R. J., and Evans, H. J.,** Autotrophic growth of H_2-uptake positive strains of *Rhizobium japonicum* in an atmosphere supplied with hydrogen gas, *Proc. Natl. Acad. Sci. U.S.A.,* 76, 1788, 1979.

35. **Eisbrenner, G. and Evans, H. J.,** Aspects of hydrogen metabolism in nitrogen fixing legumes and other plant-microbe associations, *Annu. Rev. Plant Physiol.,* 34, 105, 1983.

36. **Dixon, R. O. D.,** Hydrogenase in pea root nodule bacteroids, *Arch. Microbiol.,* 62, 272, 1968.

37. **Emerich, D. W., Ruiz-Argueso, T., Ching, T. M., and Evans, H. J.,** Hydrogen-dependent nitrogenase activity and ATP formation in *Rhizobium japonicum* bacteroids, *J. Bacteriol.,* 137, 153, 1979.

38. **Fyson, A. and Sprent, J. I.,** A light and scanning electron microscope study of stem nodules in *Vicia faba* L., *J. Exp. Bot.,* 31, 1101, 1980.

39. **Dart, P. J.,** Infection and development of leguminous nodules, in *A Treatise on Dinitrogen Fixation, Section III: Biology,* Hardy, R. W. F., Ed., John Wiley & Sons, New York, 1977, 367.

40. **Bisseling, T., van den Bos, R. C., van Kammen, A., van den Ploeg, M., van Duijn, P., and Houwers, A.,** Cytofluorometrical determination of the DNA contents of bacteroids and corresponding broth-cultured *Rhizobium* bacteria, *J. Gen. Microbiol.,* 101, 79, 1977.

41. **Wilcockson, J. and Werner, D.,** On the DNA content of bacteroids of *Rhizobium japonicum, Z. Naturforsch.,* 34, 793, 1979.

42. **Paau, A. S., Oro, J., and Cowles, J. R.,** DNA content of free-living rhizobia and bacteroids of various *Rhizobium*-legume associations, *Plant Physiol.,* 63, 402, 1979.

43. **Sutton, W. D.,** Nodule development and senescence, in *Nitrogen Fixation 3: Legumes,* Broughton, W. J., Ed., Oxford University Press, New York, 1983, 144.

44. **Sutton, W. D., Pankhurst, C. E., and Craig, A. S.,** The *Rhizobium* bacteroid state, *Int. Rev. Cytol. Suppl.,* 13, 149, 1981.

45. **Van Brussel, A. A. N., Planque, K., and Quispel, A.,** The wall of *Rhizobium leguminosarum* in bacteroid and free-living forms, *J. Gen. Microbiol.,* 101, 51, 1977.

46. **Planque, K., van Nierop, J. J., Burgers, A., and Wilkinson, S. G.,** The lipopolysaccharide of free-living and bacteroid forms of *Rhizobium leguminosarum, J. Gen. Microbiol.,* 110, 151, 1979.

47. **Carlson, R. W.,** Surface chemistry, in *Nitrogen Fixation 2: Rhizobium,* Broughton, W. J., Ed., Oxford University Press, New York, 1983, 199.

48. **Bal, A. K., Shantharam, S., and Verma, D. P. S.,** Outer membrane of *Rhizobium* is sloughed off during development of root nodule symbiosis in soybean, *Can. J. Microbiol.,* 26, 1096, 1980.

49. **Gresshoff, P. M. and Rolfe, B. G.,** Viability of *Rhizobium* bacteroids isolated from soybean nodule protoplasts, *Planta,* 142, 329, 1978.

50. **Mellor, R. B., Bassarab, S., Dittrich, W., and Werner, D.,** EPS and LPS from nod^+ nif^+ and nod^+ fix^- *Rhizobium japonicum* and bacteroids, *FEMS Microbiol. Lett.,* 19, 239, 1983.

51. **Wolff, A. and Werner, D.,** Nodule compartmentation and efficiency in *Vicia faba* and *Glycine max, Vortr. Pflanzenzuechtung,* 11, 174, 1986.

52. **Wolff, A., Mörschel, E., Zimmermann, C., Parniske, M., Bassarab, S., Mellor, R. B., and Werner, D.,** Peribacteroid membrane stability and phytoalexin production in legume nodules, in *Physiological Limitations and the Genetic Improvement of Symbiotic Nitrogen Fixation,* O'Gara, F., Manian, S., and Devron, J. J., Eds., Kluwer, Dordrecht, 1988, 65.

53. **Craig, A. S., Greenwood, R. M., and Williamson, K.,** Ultrastructural inclusions of rhizobial bacteroids of *Lotus* nodules and their taxonomic significance, *Arch. Microbiol.,* 89, 23, 1973.

54. **Mellor, R. B., Mörschel, E., and Werner, D.,** Legume root response to symbiotic infection. Enzymes of the peribacteroid space, *Z. Naturforsch.,* 39, 123, 1984.

55. **Garbers, C., Menkbach, R., Mellor, R. B., and Werner, D.,** Protease (thermolysin) inhibition activity in the peribacteroid space of *Glycine max* root nodules, *J. Plant Physiol.,* 132, 442, 1988.

56. **Kinnback, A., Mellor, R. B., and Werner, D.,** Alpha-mannosidase isoenzyme II in the peribacteroid space of *Glycine max* root nodules, *J. Exp. Bot.,* 38, 1373, 1987.

57. **Mellor, R. B. and Werner, D.,** Peribacteroid membrane biogenesis in mature legume root nodules, *Symbiosis,* 3, 75, 1987.

58. **Verma, D. P. S., Kazazian, V., Zogbi, V., and Bal, A. K.,** Isolation and characterization of the membrane envelope enclosing the bacteroids in soybean root nodules, *J. Cell Biol.,* 78, 919, 1978.

59. **Bradley, D. J., Butcher, G. W., Galfre, G., Wood, E. A., and Brewin, N. J.,** Physical association between the PBM and LPS from the bacteroid outer membrane in *Rhizobium*-infected pea root nodule cells, *J. Cell Sci.,* 85, 47, 1986.

60. **Bassarab, S. and Werner, D.,** Calcium dependent proteinkinase in the peribacteroid membrane from nodules of *Glycine max, J. Plant Physiol.,* 130, 233, 1987.

61. **Blumwald, E., Fortic, M. C., Rea, P. A., Verma, D. P. S., and Poole, R. J.,** Presence of host-plasma membrane type H^+-ATPase in the membrane envelope enclosing the bacteroids in soybean root nodules, *Plant Physiol.,* 78, 665, 1985.

62. **Bassarab, S., Mellor, R. B., and Werner, D.,** Evidence for two types of Mg^{++} ATPase in the peribacteroid membrane from *Glycine max* root nodules, *Endocyt. Cell Res.,* 3, 189, 1986.

63. **Werner, D., Mörschel, E., Garbers, C., Bassarab, S., and Mellor, R. B.,** Particle density and protein composition of the peribacteroid membrane from soybean root nodules is affected by mutations in the microsymbiont *Bradyrhizobium japonicum, Planta,* 174, 263, 1988.

64. **Regensburger, B., Meyer, L., Filser, M., Weber, J., Studer, D., Lamb, J. W., Fischer, H. -M., Hahn, M., and Hennecke, H.,** *B. japonicum* mutants defective in root-nodule bacteroid development and nitrogen fixation, *Arch. Microbiol.,* 144, 355, 1986.

65. **Fortin, M. G., Morrison, N. A., and Verman, D. P. S.,** Nodulin 26, a PBM nodulin, is expressed independently of the development of the peribacteroid compartment, *Nucleic Acids Res.,* 15, 813, 1987.

66. **Truchet, G., Michel, M., and Denarie, J.,** Sequential analysis of the organogenesis of lucerne root nodules using symbiotically defective mutants of *R. meliloti, Differentiation,* 16, 163, 1980.

67. **Mellor, R. B., Garbers, C., and Werner, D.,** Peribacteroid membrane nodulin gene induction by *Bradyrhizobium japonicum,* mutants, *Plant Mol. Biol.,* 12, 307, 1989.

68. **Meijer, E. G. M.,** Development of leguminous root nodules, in *Nitrogen Fixation 2: Rhizobium,* Broughton E. J., Ed., Oxford University Press, New York, 1982, 312.

69. **Brisson, N. and Verma, D. P. S.,** Soybean leghemoglobin gene family: normal, pseudo and truncated genes, *Proc. Natl. Acad. Sci. U.S.A.,* 79, 4055, 1982.

70. **Lee, J. S. and Verma, D. P. S.,** Structure and chromosomal arrangement of leghemoglobin genes in kidney bean suggest divergence in soybean leghemoglobin gene loci following tetraploidization, *EMBO J.,* 3, 2745, 1984.

71. **Mellor, R. B., Christensen, T. M. I. E., and Werner, D.,** Choline kinase II is present only in nodules that synthesize stable peribacteroid membranes, *Proc. Natl. Acad. Sci. U.S.A.,* 83, 659, 1986.

72. **Mellor, R. B., Thierfelder, H., Pausch, G., and Werner, D.,** The occurrence of choline kinase II in the cytoplasm of soybean root nodules infected with various strains of *Bradyrhizobium japonicum, J. Plant Physiol.,* 128, 169, 1987.

73. **Fuller, F. and Verma, D. P. S.,** Appearance and accumulation of nodulin mRNA's and their relationship to the effectiveness of root nocules, *Plant Mol. Biol.,* 3, 21, 1984.

74. **Kijne, J. W. and Planqué, K.,** Ultrastructural study of the endomembrane system in infected cells of pea and soybean root nodules, *Physiol. Plant Pathol.,* 14, 339, 1979.

75. **Van der Wilden, W. and Chrispeels, M. J.,** Characterization of the isoenzymes of α-mannosidase located in the cell wall, protein bodies and ER of *Phaseolus vulgaris* cotyledons, *Plant Physiol.,* 71, 82, 1983.

76. **Mellor, R. B. and Werner, D.,** Glycoconjugate interaction in soybean root nodules, *Lectins,* 4, 267, 1985.

77. **Robertson, J. G., Lyttleton, P., Bullivant, S., and Grayston, G. F.,** Membranes of lupin root nodules. I. The role of Golgi bodies in the biogenesis of infection threads and peribacteroid membranes, *J. Cell Sci.,* 30, 129, 1978.

78. **Robertson, J. G. and Lyttleton, P.,** Coated and smooth vesicles in the biogenesis of cell walls, plasmamembranes, infection thread and peribacteroid membranes in root hairs and nodules of white clover, *J. Cell Sci.,* 58, 63, 1982.

79. **Ostrowski, E., Mellor, R. B., and Werner, D.,** The use of colloid gold labelling in the detection of plasma membrane from symbiotic and non-symbiotic *Glycine max* root cells, *Physiol. Plant.,* 66, 270, 1986.

80. **Robertson, J. G. and Lyttleton, P.,** Division of peribacteroid membranes in root nodules of white clover, *J. Cell Sci.,* 69, 147, 1984.

81. **Werner, D. and Mörschel, E.,** Differentiation of nodules of *Glycine max.* Ultrastructural studies of plant cells and bacteroids, *Planta,* 141, 169, 1978.

82. **Boland, M. J., Hanks, J. F., Reynolds, P. H. S., Blevins, D. G., Tolbert, N. E., and Schubert, K. R.,** Subcellular organization of ureide biogenesis from glycolytic intermediates and ammonium in nitrogen-fixing soybean nodules, *Planta,* 155, 45, 1982.

83. **Hanks, J. F., Schubert, K., and Tolbert, N. E.,** Isolation and characterization of infected and uninfected cells from soybean nodules, *Plant Physiol.,* 71, 869, 1983.

84. **Bergmann, H., Preddie, E., and Verma, D. P. S.,** Nodulin 35: a subunit of specific uricase (uricase II) induced and localized in the uninfected cells of soybean nodules, *EMBO J.,* 2, 2333, 1983.

85. **Legocki, R. P. and Verma, D. P. S.,** A nodule-specific plant protein (nodulin 35) from soybean, *Science,* 205, 190, 1979.

86. **Trelease, R. N.,** Biogenesis of glyoxysomes, *Annu. Rev. Plant Physiol.,* 35, 321, 1984.

87. **Reynolds, P. H. S., Boland, M. J., Blevins, D. G., Randall, D. D., and Schubert, K.,** Ureide biogenesis in leguminous plants, *TIBS,* 7, 366, 1982.

88. **Vance, C. P., Heichel, G. H., Barnes, D. K., Bryan, J. W., and Johnson, L. E.,** Nitrogen fixation, nodule development and vegetative regrowth of alfalfa (*Medicago sativa* L.) following harvest, *Plant Physiol.,* 64, 1, 1979.

89. **Werner, D., Mörschel, E., Stripf, R., and Winchenbach, B.,** Development of nodules of *Glycine max* infected with an ineffective strain of *Rhizobium japonicum, Planta,* 147, 320, 1980.
90. **Galbraith, D. W. and Northcote, D. H.,** The isolation of plasma membrane from protoplasts of soybean suspension cultures, *J. Cell Sci.,* 24, 295, 1977.

Chapter 6

THE *RHIZOBIUM/BRADYRHIZOBIUM*-LEGUME SYMBIOSIS

Edward Appelbaum

TABLE OF CONTENTS

I. INTRODUCTION

Rhizobium and *Bradyrhizobium* bacteria are unique among microorganisms in their ability to induce the formation of nitrogen-fixing nodules on leguminous plants. Nodule formation involves a specific recognition between the prokaryotic and eukaryotic partners, invasion of plant cells by bacteria, and many changes in the structure and biochemistry of both organisms as the nodule develops. Not surprisingly, this process is associated with changes in the expression of many genes in both the bacteria and the host plant. Genetic analysis of the bacterial partner has led to the identification of about three dozen "symbiotic genes", that is, genes whose functions are required for the development of a nitrogen-fixing nodule, but not for vegetative growth of the bacteria. This review will consider a number of recent advances in the identification and characterization of symbiotic genes, with particular emphasis on genes controlling early stages of infection and nodule development. It should be kept in mind that many other bacterial genes, not discussed here, may be involved in symbiosis and may also function in other cellular processes. For example, auxotrophic mutants frequently form nodules that do not fix nitrogen or fail to form nodules at all. Still other genes are involved in ancillary functions such as hydrogen recycling, but are not required for symbiosis per se. Several recent papers review various aspects of the genetic control of nodule formation and function.[1-14]

The *Rhizobium/Bradyrhizobium*-legume symbiosis has attracted attention because of its importance in agriculture, and its suitability as a research system for studying basic biological mechanisms in plant development and organogenesis, recognition and exchange of "signals" between different organisms, bacterial ecology, and regulation of prokaryotic and eukaryotic gene expression. It is an attractive area for study since the powerful tools of bacterial genetics can be used to dissect eukaryotic as well as prokaryotic development.

A. *RHIZOBIUM* AND *BRADYRHIZOBIUM* STRAINS

Root nodule bacteria are currently classified into two genera, *Rhizobium* ("fast growers") and *Bradyrhizobium* ("slow growers") (Table 1). These genera are quite distinct in their genetic and physiological characteristics but are both considered members of the family Rhizobiaceae, which also includes *Agrobacterium*.[15,16] DNA homology studies and ribosomal RNA analysis show that *Rhizobium* and *Bradyrhizobium* are not closely related; in fact, *Rhizobium* and *Agrobacterium* are much more closely related to each other than either is to *Bradyrhizobium*.[17-19A] The major characteristic which links *Rhizobium* and *Bradyrhizobium* is the ability to form nodules on legumes. Thus, it is of interest to determine whether these two genera share similar symbiotic genes. An unusual stem-nodulating strain, *Azorhizobium sesbaniae* ORS571, represents a proposed new genus[20] and will not be discussed in this review.

Within each genus, the traditional species names were based on plant host range for the symbiosis. *R. leguminosarum* nodulates peas, *R. trifolii* nodulates clover, etc. (Table 1). This classification has proved inadequate since strains that are genetically unrelated can often form nodules on the same host (e.g., *B. japonicum* and *R. fredii*), and strains that have nearly identical genomes can have different host ranges as a result of differences in one or a few genes (e.g., *R. trifolii* and *R. leguminosarum*).[21] Also, different species sometimes have overlapping host ranges. Thus, many *R. trifolii* and *R. leguminosarum* strains share a common host, subterranean clover, in addition to their specific hosts, white clover and peas, respectively.[22] Consequently, new species designations have recently been proposed on the basis of genetic relatedness.[15] The traditional species names listed in Table 1 will be used in this review in order to be consistent with most of the literature under discussion.

The most rapid advances in the genetic analysis of symbiosis have occurred in *Rhizobium*,

133

TABLE 1
Rhizobium and *Bradyrhizobium* Strains and Hosts

Genus, species	Examples of host plants (genus)[a]
Rhizobium	
R. leguminosarum	Pea (*Pisum*)
R. trifolii	Clover (*Trifolium*)
R. phaseoli	Bean (*Phaseolus*)
R. meliloti	Alfalfa (*Medicago*)
R. loti	Lotus (*Lotus*)
R. fredii (R. japonicum)	Soybean (*Glycine*)
R. sp. (strain NGR234)	Siratro (*Macroptilium*)
Bradyrhizobium	
B. japonicum	Soybean (*Glycine*)
B. sp. (*Lupinus*)	Lupine (*Lupinus*)
B. sp. (*Vigna*)	Cowpea (*Vigna*)
B. sp. (*Parasponia*)	Parasponia (*Parasponia*)
B. sp. (*Lotus*)	*Lotus* (*Lotus*)
B. sp. (*Arachis*)	Peanut (*Arachis*)

[a]   Some host plants can be nodulated by more than one *Rhizobium* or *Brad-ryhizobium* species. Siratro and cowpea can be nodulated by *R. fredii*, *R. sp.* (NGR234), *B. japonicum*, *B. sp.* (*Vigna*), *B. sp.* (*Parasponia*), and *B. sp.* (*Arachis*). Some *Rhizobium* strains can form nodules on multiple hosts. *R. sp.* (strain NGR234) is especially promiscuous, forming nodules on siratro, *Parasponia*, and a variety of tropical legumes.

especially in strains that form nodules on small-seeded temperate legumes, for several reasons:

Growth rate — Colony formation on agar takes about 3 d for rhizobia and 7 d for bradyrhizobia.

Frequency of genetic exchange — Genetic manipulations such as conjugation, transposon mutagenesis, and marker exchange[23] take place at roughly 100-fold lower frequencies in *Bradyrhizobium*.

Intrinsic antibiotic resistance — Many *Bradyrhizobium* strains show high levels of resistance to the antibiotic markers (tetracycline, kanamycin) present on many cloning vectors and transposons.[23]

Plasmids — Many *Rhizobium* symbiotic genes are present on Sym plasmids which can be mobilized from one strain to another or removed altogether (cured) from a strain. In *Bradyrhizobium* there is no evidence for plasmids carrying symbiotic genes.

Gene clusters — Symbiotic genes tend to be more closely clustered in *Rhizobium* (at least in the temperate *Rhizobium* strains, *R. leguminosarum*, *R. trifolii*, and *R. meliloti*) than in *Bradyrhizobium*. For example, the *nifH nifD nifK* genes form a single operon in *Rhizobium* but are present in two widely separated operons (*nifH* and *nifD nifK*) in *Bradyrhizobium*.

Convenience of plant assays — Small-seeded, temperate legumes, such as alfalfa and clover, are routinely grown on agar in test tubes or petri dishes, allowing large numbers of strains and plants to be tested for symbiosis in a small space. Many of the most important tropical legumes that are nodulated by *Bradyrhizobium* and some *Rhizobium* strains (e.g., soybeans) are less convenient to grow. However, there is some compensation for this inconvenience, in that the larger nodules formed by the larger legumes are easier to study biochemically. Also, this inconvenience has been overcome in part by use of somewhat smaller seeded tropical legumes such as siratro (*Macroptilium atropurpureum*) that are nodulated by many *Bradyrhizobium* and tropical *Rhizobium* strains (Table 1).

TABLE 2
Stages of Symbiosis

Stage[a]	Phenotypic code[a]	Genotypic code
1. Root (rhizoplane) colonization/chemotaxis?	Roc	?
2. Root adhesion	Roa	*nod*?
3. Marked root hair curling	Hac	*nod, exo*
4. Infection thread formation	Inf	*nod, exo, ndv*
5. Nodule initiation (meristem formation, differentiation)	Noi	*nod*
6. Bacterial release into plant cells	Bar	*fix*
7. Bacteroid development, multiplication	Bad	*fix*
8. Nitrogen fixation (nitrogenase activity)	Nif	*nif, fix*
9. Complementary functions	Cop	*fix*
10. Nodule persistence	Nop	*fix*

[a]   Abbreviated from Vincent.[24]

In spite of these difficulties, the genetic manipulation of *Bradyrhizobium* has recently proceeded quite rapidly with the advent of recombinant DNA techniques that enable many of the manipulations to be carried out in *Escherichia coli*. There is considerable interest in the genetics of *Bradyrhizobium* strains, because of their great agricultural importance and the large existing literature on the physiology, biochemistry, and ecology of these organisms.

B. STAGES IN SYMBIOSIS

The development of nitrogen-fixing nodules has been characterized as occurring in a series of stages,[24] each of which may be influenced by one or more genes in each symbiotic partner. These stages are listed in Table 2. Recent evidence suggests that these stages actually involve two somewhat independent series of events: infection (invasion) and nodule development involving plant cell division and differentiation. In most *Rhizobium/Bradyrhizobium* interactions with legumes, infection occurs through root hairs. The infection process involves root hair attachment (Roa), various root hair deformations (Had) including a marked root hair curling (formation of "shepherd's crooks") (Hac), infection thread initiation and elongation (Inf), and bacterial release into plant cells (Bar).[24-26] In some symbioses, infection occurs via direct invasion of cortical cells[25,26] or through production of infection threads in the cortex rather than in root hairs.[27-29] In parallel with the infection process, nodule initiation (Noi) occurs. This involves the onset of cortical mitoses, formation of a nodule meristem, and development of nodule structure. Both infection and nodule development begin shortly (within a few hours) after the bacteria and roots come in contact. In soybeans, cell division can be observed within 12 h and does not require physical contact between the bacterium and host cells. Root hair curling can be observed within 6 h, and penetration and infection thread formation 18 h later.[30] Certain strains (mutants of *R. meliloti*[31,32] and *R. phaseoli*,[33] and *Agrobacterium* carrying certain *R. meliloti* genes[34-36]) can induce nodules that undergo development without any evidence of infection. Conversely, some infection thread development can occur in the absence of ongoing nodule development.[22,30,37] Later stages in symbiosis include differentiation of bacteria into bacteroids (bacteroid development, Bad), synthesis of nitrogenase and other gene products required for nitrogen fixation (Nif, Cop), and nodule persistence (Nop).

The characterization of symbiotic genes involves determination of the specific stages that are influenced by that gene (Table 2). Symbiotic genes may also be classified as either common to all strain/host combinations or host-plant specific (see Section II below).

C. IDENTIFICATION OF SYMBIOTIC GENES

Many specialized vectors and techniques have been developed for cloning, mobilization, and mutagenesis of genes in the Rhizobiaceae, as discussed elsewhere.[23] To identify symbiotic genes, these techniques have been used in three general approaches, each of which has particular advantages and disadvantages.

The first approach is to mutagenize a strain and then screen individual colonies on plants for symbiotic defects such as loss of ability to form nodules (Nod⁻ mutants) or formation of nodules which do not fix nitrogen (Fix⁻). In some cases the screening has initially been for other phenotypes (for example, alterations in exopolysaccharide synthesis), followed by testing of specific mutants for symbiotic properties. The mutation frequency can be increased with chemicals, UV light, or transposons, or by stressful culture conditions which lead to loss of plasmids or to genomic rearrangements. Transposon mutagenesis with Tn5 has been especially useful, since the presence of genetic markers (kanamycin/neomycin and streptomycin resistance) on the transposon facilitates localization of the mutation in the genome, cloning of the mutagenized region, and interstrain mobilization (transduction, conjugation, R-prime formation) of the mutated region (see Chapter 2). Complementation of mutations can be used to select clones of a particular gene from a recombinant DNA library prepared from a wild-type strain. Tn5 mutagenesis can be either general (directed randomly at the entire genome) or localized (Tn5 is inserted into specific cloned segments and then recombined into the wild-type genome to create a mutation). A potential disadvantage to the use of Tn5 is that many Tn5-induced mutations are polar. Thus, symbiotic genes that are upstream of a gene essential for vegetative growth might not be detected. Also, phenotypes that might result from alterations in protein structure (as in missense mutations) would be overlooked in a Tn5 mutagenesis program, since Tn5 insertion lends to elimination of the gene product. Finally, it is difficult to carry out plant assays on the number of random mutants that would be needed to ensure a saturation of the entire genome with Tn5 insertions.

The second general approach to locating symbiotic genes has been to mobilize indigenous plasmids or cloned genomic segments from one strain to another, and then to identify transconjugants whose symbiotic phenotypes are altered by the incoming DNA. For example, this has been used to define the particular Sym plasmid regions of *R. trifolii* that are sufficient to extend the *R. leguminosarum* host range to white clover.[21] Potential limitations to this approach are that the genes for a trait may be unlinked, may not be expressed in the recipient, or may not be dominant over genes in the recipient.

The third approach has been to use a cloned gene from one organism to identify homologous DNA sequences in another. For example, genes controlling nitrogen fixation in *Rhizobium* were initially identified using DNA probes containing *nif* genes from the free-living nitrogen-fixing organism *Klebsiella pneumoniae*.[38] Similarly, *Agrobacterium* genes required for tumorigenesis have been used as probes to identify homologous *Rhizobium* genes involved in nodule formation.[39] Once a potentially interesting *Rhizobium* region has been identified in this way, its role in symbiosis can be further characterized by directed mutagenesis or interstrain transfer.

II. BACTERIAL GENES CONTROLLING NODULATION

Individual *Rhizobium* and *Bradyrhizobium* strains are able to form nodules on only a limited number of host legumes. Conversely, leguminous plants are each nodulated by a limited number of strains. Thus, there is a specific recognition between strains and hosts. This recognition can occur at several different stages of symbiosis. In some heterologous strain/plant combinations, there is no interaction, even at the earliest stages of nodulation: root hairs are not markedly curled, and no nodule-like structures are formed. In other combinations, nodules form but are not infected, or are infected but do not fix nitrogen.[24,40] The phenomenon of host specificity has led to the notion that there may be two general classes of bacterial symbiotic genes: those that have a universal function in all strain/host

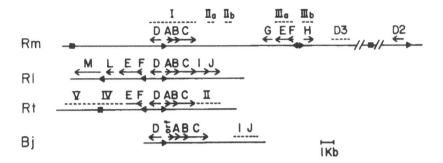

FIGURE 1.   Maps of *nod* regions. The regions shown are for *R. meliloti* (Rm), *R. leguminosarum* (Rl), *R. trifolii* (Rt), and *B. japonicum* (Bj). Genes that have been sequenced are indicated by arrows and letters A through M. Regions identified only by mutagenesis and/or homology with *nod* probes, but not by sequencing, are indicated by dotted lines. The Roman numerals designate regions referred to in some studies.[21,41] Solid boxes and arrows indicate *nod* boxes (promoter regions). Arrows are used when the orientations of *nod* boxes are known. The maps were constructed from published information.[21,41,43,46,48,58-64,71,79,101] The *R. meliloti* *nodFEGH* genes have also been referred to as *hsnABCD*, respectively.[71]

combinations, and those that are specific to particular combinations. In the case of the *nod* genes, which control infection and nodule initiation (Table 2), recent genetic evidence has distinguished two such classes of genes: the "common" and the "host specificity" *nod* genes.

A. *NODABCIJ:* THE COMMON *NOD* GENES

The common *nod* region is a cluster of *nod* genes that are structurally and functionally conserved in many *Rhizobium* and *Bradyrhizobium* strains. This region has been studied in detail in three "temperate strains", *R. meliloti*, *R. trifolii*, and *R. leguminosarum*. DNA sequencing, Tn5 mutagenesis, genetic complementation analysis, and gene product analysis have led to the identification of four genes in two transcriptional units, *nodD* and *nod ABC*.[21,41-45] Two additional genes, *nodI* and *nodJ*, have been identified downstream of *nodC* (and apparently in the same transcriptional unit) in *R. leguminosarum*,[46] and are probably also present in *R. trifolii* (region II) and *R. meliloti* (region IIa). The arrangement of these genes is illustrated in Figure 1. Insertion mutations in *nod*ABC completely block nodulation, while mutations in *nodIJ* cause a delay in the appearance of nodules. Mutations in *nodD* block nodulation in strains such as *R. trifolii* and *R. leguminosarum* which contain a single *nodD*, but cause delayed or reduced nodulation in strains containing multiple *nodD* genes (discussed in detail in Section II.D below).

The common *nod* genes of these temperate rhizobia have been used as probes to identify and clone the corresponding region of several other *Rhizobium* and *Bradyrhizobium* strains. In three *Bradyrhizobium* strains, the organization of the genes differs somewhat from that in the temperate rhizobia. Specifically, *nodD* is separated from *nodABC* by a 700- to 800-base-pair gap which contains a potential open reading frame (ORF) in *B. sp. (Parasponia)*[47] (*nodK*) and in *B. japonicum* strains USDA110[48] (*nodL*-now termed *nodY*) and USDA123.[49] The function, if any, of these open reading frames (ORFs) is unknown, but their presence in a *nod* gene transcriptional unit (see Section II.D below) makes them candidates to be *nod* genes. Deletion of this ORF in *B. japonicum* does not interfere with vegetative cell growth, so the gene does not have essential housekeeping functions.[49] A further difference between *Bradyrhizobium* and the temperate rhizobia is the presence of a roughly 2000-base-pair gap between *nodC* and *nodIJ*.[48,49] The function of any genes in this gap is also unknown, but again, deletion of this region does not affect vegetative cell growth.[49] In the tropical strains, *R. fredii* and NGR234 (derivatives MPIK3030 and ANU240), the *nodABC*-homologous sequences are found many kilobases away from *nodD*.[50-53] The common *nod* genes are carried on Sym plasmids in the "temperate strains", in *R. phaseoli*, and in most *R. fredii* strains.[12] However, in two *R. loti* strains,[54] two *R. fredii* strains,[55] and the bradyrhizobia,[55,57] these genes appear to be chromosomally located.

DNA sequencing[43,47,49a,50,58-64] has revealed a high degree of homology (about 70%) in the amino acid sequences of corresponding common *nod* genes from different *Rhizobium* strains. In many cases, there is very little overall DNA homology between the genomes of the strains; thus, the *nod* genes have been much more strongly conserved than most other genes. It is unknown whether this fact represents recent horizontal transfer of these genes or selection over a long period of time. Common *nod*-hybridizable sequences have not been detected in any genera but *Rhizobium, Bradyrhizobium*,[65] and *Azorhizobium*.[20]

Strains containing mutations in any of the common *nod* genes can be complemented for nodulation of their normal host plant by Sym plasmids or cloned genes from heterologous strains.[47,66-71] This has been taken as an indication that *nodDABCIJ* are not involved in host recognition, but instead serve functions that are common in all strain/host combinations. However, recent studies have suggested that there is a host-specific component to *nodD* function (see Sections II.B and II.D below).

Mutations in *nodABC* prevent marked root hair curling (Hac⁻) and formation of infection threads (Inf⁻).[21,41,42,45,48,63] Mutations in *nodD* also result in a Hac⁻ phenotype[22,42] as expected from the role that *nodD* has in regulation of *nodABC* transcription (see Section II.D below). Cloned DNA segments carrying *R. leguminosarum nodABC* in a broad host-range vector are sufficient to induce root hair curling when carried in Sym plasmid-cured strains.[42] This last result suggests that *nodABC* are the only Sym plasmid genes required for root hair curling in *R. leguminosarum*. However, in *R. meliloti*, a host-specific *nod* gene, *nodH*, also appears to be required for root hair curling (see Section II.B below).[41,71] Also, strains carrying mutations in *exo* genes that are not present on the *R. meliloti* Sym plasmid are Hac⁻.[31,32] In *R. trifolii*, there is also evidence that one or more host-specific *nod* genes are involved in root hair curling.[21] Thus, *nodABC* are clearly necessary, but not sufficient, for normal root hair curling in *R. meliloti* and *R. trifolii*.

The *nodABC* genes also are involved in nodule morphogenesis. This was demonstrated by the observation that *Agrobacterium tumefaciens* containing the cloned *nodDABC* region of *R. meliloti* is able to elicit early stages in nodulation (mitoses in the inner root cortex, establishment of meristem and vascular development). These transconjugant strains do not curl root hairs, and the nodules that develop have no infection threads or intracellular bacteria. Thus, the *nodABC* genes encode functions that stimulate cell division without direct contact between the bacteria and the stimulated plant cells.[34-36]

Another phenotype that is associated with *nodABC* is the production of diffusible factors which affect roots and root hairs. In *R. leguminosarum*, the *nodABC* genes are required for formation of a water-soluble, low molecular weight compound(s) that causes a shortening and thickening of pea roots (Tsr phenotype), root hair induction (Hai), and root hair deformation (Had).[72] An "aromatic compound" (BF-1) produced by *R. trifolii* causes root hair deformations and is not produced by mutants lacking the Sym plasmid that contains *nod* genes.[73] A low-molecular-weight diffusible factor produced by *B. japonicum* has been shown to induce cortical cell divisions in soybeans, but dependence of this factor on *nodABC* has not been examined.[74] Mutations in *R. trifolii* common *nod* genes also reduce root hair attachment, reduce binding of clover lectin, and alter surface polysaccharide composition.[73,75]

nodABC are also required for nodule formation in two systems that do not involve the root hair mode of infection. In the *B.* sp. (*Parasponia*) symbiosis with the nonlegume tree *Parasponia rigida*, the bacteria do not curl root hairs, but do induce cell divisions and initiate infection threads in the cortex.[27-29] Mutations in the *nodABC* genes of *B.* sp. (*Parasponia*) result in a Nod⁻ phenotype on *P. rigida*.[76-78] In the *B.* sp. (*Arachis*) symbiosis with peanut, infection occurs by "crack entry" involving penetration of bacteria between the epidermal cells and passive spread of bacteria in invaded host cells via host cell division.[25,26] A mutation in the *nodABC* region prevents nodulation of peanut. The mutation also blocks nodulation of siratro and pigeon pea, two hosts that are infected by wild-type *B. sp.* (*Arachis*) via root hair infection.[70] These results provide further evidence that *nodABC* are involved in other symbiotic processes besides root hair curling and infection thread formation.

The *nodIJ* region is not absolutely required for symbiosis, since mutations cause only a marginally defective nodulation phenotype (i.e., a slight delay in nodule appearance).[22,41,46] Mutations in the *nodIJ* region do not affect root hair curling in *R. leguminosarum*. In *R. trifolii*, mutations in region II (presumably *nodIJ*) produce an exaggerated root hair curling response (Hac^{++}) and result in infection threads that are often short and aborted. In *R. meliloti*, mutations in regions IIa (presumably *nodIJ*) and IIb cause an unusually high proportion of curled root hairs and infected root hairs, but do not affect nodule number and, in fact, actually delay nodule formation. Thus, *nodIJ* appear to control the efficiency of infection, a role that could be important in competition between strains in natural and agricultural environments.

In summary, the common *nod* region is involved in both the infection pathway and the nodule development pathway. It is not yet known whether these genes control a single biochemical function that somehow affects both pathways, or multiple biochemical functions. In either case, the involvement of *nodABC* in both pathways may enable the bacterial partner to coordinate these pathways, which must occur together temporally and spatially, in order for a successful symbiosis to occur.

B. GENES CONTROLLING HOST SPECIFICITY OF NODULATION

Host-specificity *nod* genes (also referred to as *hsn* genes) are those that influence the specificity of strains for specific host plants. Two criteria have been used to determine which *nod* genes have a role in host specificity: interspecies genetic complementation tests and host-range alteration experiments. *Hsn* genes and regions that have been characterized in detail are *R. meliloti nodFEGH*, *R. leguminosarum nodFE*, and *R. trifolii nodFE*, region IV, region V (Figure 1).

In interspecies complementation tests, transfer of a plasmid carrying a common *nod* gene to a heterologous strain carrying a mutation in the corresponding gene restores the ability of the mutant to nodulate its normal host plant, as discussed previously. In contrast, mutations in host-specific *nod* genes cannot be complemented by Sym plasmids or cloned DNA segments from heterologous strains.[21,41,71] *R. meliloti nodFEGH* mutants, for example, could not be complemented for alfalfa nodulation by a plasmid carrying the entire *R. leguminosarum nod* region.[71] Thus, the function of such genes appears to be specific to particular strain/host combinations. Negative complementation results are generally interpreted as suggesting a role for the mutated gene in recognition between the bacterium and host plant.

However, complementation experiments must be interpreted with caution. Negative results might occur because the incoming gene is not properly transcribed or translated or because its product is unstable. Thus, the lack of complementation might reflect lack of accumulation of the gene product in the heterologous strain rather than inability of the gene product to mediate recognition of a new host plant.

Positive complementation tests can also potentially be misleading when the recipient strain is one that contains a Tn5 insertion or point mutation. These experiments are usually carried out in recombination-proficient strains, in which recombination between the incoming fragment and the mutated gene can occur, thereby restoring the integrity of the mutated gene. Since plant nodulation tests are often carried out with large doses of bacteria (10^6 to

10^9 cells per plant at the time of inoculation, and even more after proliferation in the rhizosphere), it is possible that many such wild-type recombinants will be present in the inoculant and will form nodules. Thus, the positive result would reflect the restoration of the original gene structure and not replacement of the original gene product with the heterologous gene product. Such misleading results could be minimized by the use of recombination-deficient strains that have recently become available but are not yet in widespread use.[78A]

A further complication is that certain genes give positive complementation results with the corresponding gene from some, but not all, heterologous strains. For example, the *R. meliloti nodD1* gene can complement an *R. trifolii nodD*::Tn5 mutant for nodulation of clover, but cannot complement an NGR234 *nodD1* mutant for siratro nodulation.[52,66,80] The *R. fredii nodD1* gene complements a *R. trifolii nodD*::Tn5 mutant, but the *R. fredii nodD2* gene does not, even though both *nodD1* and *nodD2* are required for *R. fredii* symbiosis with soybeans.[49,49a,50] Thus, *nodD* was originally identified as a common *nod* gene, but now appears to have some degree of host specificity (see also Section II.E below).

A more definitive demonstration that a gene is involved in recognition is an observation of host-range alteration. This refers to a situation in which the spectrum of host plants nodulated by a particular strain can be changed, either by mutation in a host-specificity gene or by acquisition of a host-specificity gene (or group of genes) from a heterlogous strain. For example, transfer of a fragment containing the NGR234 *nodD1* gene (and flanking sequences) to *R. meliloti* enabled the transconjugants to nodulate an illegitimate host, siratro.[80] Transfer of *R. trifolii nod* genes (*nodFE* and regions IV, V) to *R. leguminosarum* enabled the transconjugants to form nodules on white clover.[21] Transfer of Sym plasmids or DNA fragments from *R. leguminosarum*,[81] *R. meliloti*,[82] NGR234,[51,52] *R. fredii*,[83,86] and *Bradyrhizobium* sp. (*Parasponia*)[77] to suitable recipients, also extended the host range for nodulation.

Either expansion or contraction of host range can be accomplished by mutagenesis. For example, mutations in *nodFE* (region III) of *R. trifolii* have altered nodulation abilities on several legumes. One mutant, M2, acquired the ability to nodulate a new host, peas, and retained the ability to nodulate white and subterranean clovers. This mutant contained a Tn5 insertion just upstream of *nodF*. Several other *nodFE* mutants lost the ability to form nodules on white clover, but were still able to nodulate subterranean clover and peas.[22] In *R. meliloti* strain 41, Tn5 insertions in *nodH* (*hsnD*) eliminated nodulation of *Medicago sativa* (alfalfa), but not of *Melilotus albus* (white sweet clover), and also extended the host range to *Vicia sativa* and *V. villosa*.[71] Mutations in *nodFE* or *nodH* enabled *R. meliloti* strain RCR2011 to curl clover root hairs.[41] These experiments provide direct evidence that *nodFE* and *nodH* influence host specificity. Mutations which block nodulation of some host plants, but not of others, have also been reported in *B. japonicum*,[48] NGR234,[84] *B.* sp. (*Arachis*),[70] and *B.* sp. (*Parasponia*).[69]

The ability of defined DNA segments to extend the host range of a recipient strain suggests that some host-range genes act positively and are dominant over the host-range genes of the recipient. Narrowing of host range by Tn5 insertions also suggests that some genes act positively. On the other hand, the observation that some Tn5 insertion mutations can extend host range suggests that some *nod* genes are negatively acting.

The *nodFE* genes of *R. meliloti*, *R. trifolii*, and *R. leguminosarum* show strong conservation of their DNA and amino acid sequences, indicating that these host-specific *nod* genes are allelic variants.[58,59,71,85] *nodE* (*hsnB*)- and *nodH* (*hsnD*)- hybridizable sequences could also be easily detected in *R. fredii* and NGR234, but only very weak homology was seen in *R. phaseoli*, *R. lupini*, or *B. japonicum*. Common and host-specificity *nod* genes show no detectable homology to *Agrobacterium* and *E. coli* DNA.[65]

Host-specificity *nod* genes influence root hair curling, infection thread formation, and nodule initiation in various ways. Wild-type *R. meliloti* is Hac$^+$ Inf$^+$ Nod$^+$ on alfalfa and Hac$^-$ Inf$^-$ Nod$^-$ on clover. *NodFE* mutants (*hsnAB* in strain Rm41, region IIIa in strain RCR2011) are Hac$^+$Inf$^-$Nod^d (delayed nodulation, involving rare, delayed infection threads) on alfalfa and Hac$^+$Inf$^-$Nod$^-$ on clover. Wild-type *R. trifolii* is Hac$^+$Inf$^+$Nod$^+$ on white and subterranean clovers and Hac$^{+/-}$Inf$^-$Nod$^-$ on peas. Most *R. trifolii nodFE* mutants are Hac^{++}Inf$^+$Nod$^{+/-}$ on white clover, Hac^{++}Inf$^+$Nod^d on subterranean clover, and Hac$^+$Inf$^+$Nod^d on peas. An *R. leguminosarum nodFE* (region IV) mutant is Hac$^+$Nod^d on peas.[22,41,42,71]

R. meliloti nodG (*hsnC*) mutants are Hac^{d}Inf$^-$Nod^d on alfalfa. *R. meliloti nodH* mutants (*hsnD* in Rm41, region IIIb in RCR2011) are Inf$^-$Nod$^-$ and either Hac$^-$ or "Hac^s" (curled root hairs lack "refractile spot") on alfalfa, and are Hac^{s}Inf$^-$Nod$^-$ on clover. *R. trifolii* region IV appears to have a positive role in root hair curling. Specifically, an *R. trifolii* strain that is cured of its Sym plasmid, but contains the *R. meliloti* common *nod* genes, cannot curl clover root hairs unless *R. trifolii* region IV is introduced. *R. trifolii* regions III, IV, and V must all be introduced into this strain in order for it to infect and nodulate clover normally.[21,41,71]

In summary, *R. meliloti nodH* and *R. trifolii* region IV control the host specificity of root hair curling. *nodFE*, *R. meliloti nodG*, and *R. trifolii* region V are involved in infection thread formation and are required for prompt nodule initiation. It therefore appears that these genes control host specificity at early stages of the infection pathway. This does not rule out the possibility that host specificity is also controlled at other stages of nodule development.

Several recent studies have identified genes or mutations that affect host specificity in broad host-range strains. Tropical *Rhizobium* strain NGR234, and its derivatives ANU240 and MPIK3030, are able to nodulate siratro, cowpea, soybeans, various tropical legumes, and a nonlegume, *Parasponia*. Several different regions of the NGR234 Sym plasmid contain host-specificity loci, some of which are within a few kilobases of common *nod, nifHDK*, or *nod* box sequences. It appears that different host-specificity loci are involved in nodulation of different hosts by NGR234.[51-53] Host-specificity loci have also been identified in *B. japonicum*,[48] *B.* sp. (*Parasponia*),[69] *B.* sp. (*Arachis*),[70] and *R. fredii*,[86] all of which are able to form nodules on cowpea, siratro, and other legumes.

Host specificity can occur with respect to different cultivars, as well as different host genera. An Asian cultivar of peas, cv. Afghanistan, can be nodulated by an Asian *R. leguminosarum* strain, but not by European strains. A strain that is competent for cv. Afghanistan nodulation has a region downstream of *nodABC* which can confer the ability to nodulate cv. Afghanistan when it is transferred into a European strain. This region appears to be an allelic variant of a pea nodulation gene in the European strain.[87]

Another example of cultivar specificity has been observed with *R. fredii*. Most *R. fredii* strains form Fix$^+$ nodules on an Asian soybean cultivar, Peking, and uninfected nodule-like structures (empty nodules) on U.S. commercial cultivars.[88,89] One exceptional strain, USDA191, is Fix$^+$ on some commercial cultivars. Mobilization of the USDA191 Sym plasmid to another *R. fredii* resulted in transconjugants that are Fix$^+$ on these commercial cultivars. The specific gene(s) responsible for cultivar specificity has not been identified.[83]

C. PROTEIN PRODUCTS AND POSSIBLE BIOCHEMICAL FUNCTIONS OF *nod* GENES

As described above, the *nod* genes are involved in several observable events, including root hair curling, infection thread formation, plant cell division, and production of factors that stimulate changes in root morphology. Also, *nod* genes may be involved in production of surface polysaccharides and binding of plant lectins. How do the *nod* gene products mediate these events?

Identification and purification of the *nod* proteins is now getting underway, but *in vitro* assays for their biochemical functions are not yet available. Another approach to this problem has been to predict the amino acid sequences of *nod* polypeptides from the known DNA sequences, and then screen published data bases of polypeptide sequences for homologous proteins. A structural relationship between a *nod* protein and a protein of known function can suggest possible *nod* protein functions that can be tested biochemically. For example, the homology between *nodD* and bacterial regulatory proteins such as *ara*C[59] and *lys*R[49a,50] suggested that *nodD* is a transcriptional regulator. This was supported by the observation that *nodD*⁻ mutants do not properly express *nodABC*, as discussed below (Section II.D). Some of the other predicted *nod* proteins also show interesting homologies. The predicted *nodI* protein shows homology with bacterial inner-membrane ATP-binding proteins involved in active transport. However, the *nodI* protein is not highly hydrophobic, suggesting that it may be membrane associated rather than an integral membrane protein. In contrast, *nodJ* is very hydrophobic and may be an integral membrane protein.[46] The *nodC* protein is hydrophobic at each end. Moreover, the *nodC* protein has been detected immunologically and localized in a membrane fraction in *E. coli* and *R. meliloti*.[90,91] Nodulation of alfalfa by *R. meliloti* and of clover by *R. trifolii* can be inhibited somewhat by *nodC*-specific antibodies, further suggesting a cell surface location.[90] *nodA* and *nodB* are rather hydrophilic. The *nodA* protein has been detected immunologically and appears to be located in the cytosol in *E. coli* and *R. meliloti*. Moreover, *nodA*-specific antibodies do not inhibit nodulation in contrast to the *nodC* results discussed above.[91] The only other known candidates to be common *nod* region genes are the ORFs upstream of *nodABC* in *Bradyrhizobium*. These ORFs do not contain any convincing homologies with known proteins.[48-50]

DNA sequence analysis of host-specificity *nod* genes has also revealed suggestive homologies. *nodF* (*hsnA*) encodes a small polypeptide similar to acyl carrier proteins involved in fatty acid biosynthesis. This suggests a role in lipopolysaccharide (LPS) or exopolysaccharide (EPS) synthesis.[58,59,71,85] *nodG* (*hsnC*) encodes a product with some homology to dehydrogenases (e.g., ribitol dehydrogenase). It has ben hypothesized that *nodG* is required for catabolism of carbon compounds provided to the invading bacteria by the plant.[85] *nodH* encodes an unusually proline-rich protein with a very hydrophobic N terminus, perhaps suggesting a cell surface location.[71,85] *nodM* of *R. leguminosarum* is homologous to amido phosphoribosyl transferases and conceivably could be involved in LPS synthesis.[79] Overall, these homologies have led to the hypothesis that the host-specificity *nod* gene products are involved in the construction of an altered bacterial cell surface that is recognized by the host plant at the stages of root hair curling and infection thread initiation.[7]

D. REGULATION OF *NOD* GENE EXPRESSION

The regulation of *nod* gene expression has been examined by constructing strains in which the *E. coli lacZ* gene, encoding β-galactosidase, is under the control of the *nod* promoter. This allows *nod* expression to be followed, using the sensitive and convenient assays available for β-galactosidase.

The *nodD* gene is expressed constitutively in all of the strains examined (*R. meliloti*, *R. trifolii*, and *R. leguminosarum*).[92-94] Small changes in *nodD* expression may occur during the growth cycle,[95] but treatment of the cells with root or seed exudate or direct exposure to roots have no great effect.

In contrast, all of the other common *nod* and host-specific *nod* genes do appear to be inducible by substances exuded or extracted from legume roots or seeds. This has been demonstrated for *nodABCIJ*, *nodFE*, and *R. trifolii* region IV. Individual strains respond to exudates from their own host and from some other legumes, but not to exudates from nonlegumes. The increases in *nod* gene expression upon treatment are quite large (50-fold or greater in some experiments) and begin quite soon after exposure to inducer (within 3 h

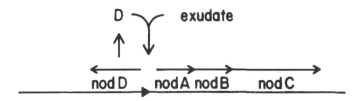

FIGURE 2.   Regulation of *nod* genes. *nodD* is expressed constitutively. In the presence of both *nodD* protein and root exudate, the *nodABC* genes (and host-specific *nod* genes) are induced.[80,92-98]

or less in various strains having doubling times of 3 to 12 h). It has not been established whether high-level *nod* gene expression continues throughout infection and nodule development.[80,92-98]

In most reports the *nod-lac* fusions are carried on multicopy plasmid vectors, which raises the possibility of nonphysiological effects on induction. However, in *R. meliloti* it has been shown that induction takes place when the fusion is inserted into the *Rhizobium* Sym plasmid in its normal location.[93] Also, direct immunoassays for *nod*-encoded proteins have confirmed the increase in gene expression.[99] The induction takes place at the level of transcription, since transcriptional as well as translational fusions are induced. This has also been confirmed by direct measurement of *nod* RNA levels.[100]

The induction of *nodABC* and *nodFE* genes does not occur in strains carrying *nodD* mutations. This indicates that *nodD* has a positive regulatory role (Figure 2).[93,94] These experiments do not establish whether the *nodD* protein acts directly at the level of transcription of *nodABC*, as opposed to a less direct role such as transport or metabolism of exudate components or signal transduction. The homology observed between *nodD* and the well-characterized transcription activator *araC* supports a direct effect for the *nodD* protein on promoter activity of the level of DNA binding.[59] Attempts are underway in several laboratories to purify *nodD* protein and measure its ability to bind *nod* promoters, stimulate transcription *in vitro*, and interact directly with specific exudate components.

R. meliloti has a higher induced level of *nodABC* activity when extra copies of *nodD* are provided on a multicopy vector. Thus, the level of *nodD* protein product may be a limiting factor for *nodABC* expression under these conditions.[93] In *R. leguminosarum*, *nodD* may be autoregulatory, since overproduction of *nodD* inhibits the expression of a *nodD-lacZ* translational fusion.[94] However, this autoregulatory effect has not been observed in *R. meliloti*.[93]

The regions upstream of inducible *nod* genes do not have a consensus *E. coli* promoter sequence, but do have a conserved DNA sequence, referred to as to the "*nod* box".[58,101] This sequence has been found upstream of *nodABC*, *nodFEG*, and *nodH* (Figure 1). A region of about 47 bp contains the most strongly conserved sequences. Analysis of strains mutated in the *nod* box indicates that it is necessary for expression of downstream *nod-lac* fusions, and for nodulation.[101,102] The *in vivo* start site for *nodABC* RNA is approximately 25 base pairs downstream of the *nod* box.[100] Thus, the *nod* box has the characteristics expected for a *nod* promoter.

In temperate rhizobia the translation start sites for the divergent *nodD* and *nodA* genes are about 200 bp apart. The divergent *nodD* and *nodABC* RNA start sites are about 25 bp apart. Although the RNAs themselves do not overlap, their promoters do.[100] The significance of this overlapping promoter arrangement for regulation of gene expression is not yet clear. In contrast, in *R. fredii* and NGR234, the two *nodD* genes are many kilobases away from the *nodABC* genes,[50-53] and only one of the *nodD* genes is located near a *nod* box.[49,49a] In *B. japonicum* and *B.* sp. (*Parasponia*) there is an additional ORF between the *nod* box and *nodA* gene that appears to be part of the *nodABC* operon.[47-49]

143

INDUCER	HYDROXYLATIONS	R. MEL.	R. LEG.	R. TRI.	B. JAP.
FLAVONES					
DHF	4',7'	ND	ND	+	+
APIGENIN	4',5,7	(+)	+	+	+
LUTEOLIN	3',4',5,7	+	+	(+)	−
CHRYSIN	5,7	−	+/−	(+)	−
FLAVANONES					
NARINGENIN	4',5,7	−	+	(+)	−
ERIODICTYOL	3',4',5,7	ND	+	ND	ND
ISOFLAVONES					
DAIDZEIN	4',7	ND	−	−	+
GENISTEIN	4',5,7	ND	−	ND	+
NATURAL INDUCERS:		LUTEOLIN (ALFALFA)	ND	DHF (CLOVER)	DAIDZEIN GENISTEIN (SOYBEAN)

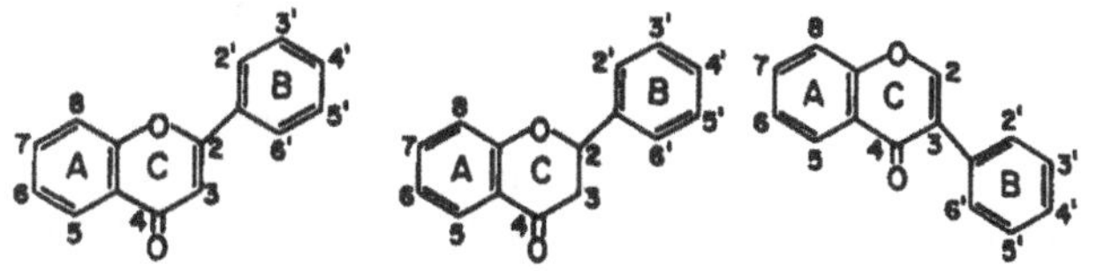

FLAVONES FLAVANONES ISOFLAVONES

FIGURE 3. Inducers of *nod* genes. The response of *nodABC-lacZ* fusions to specific compounds is indicated for *R. meliloti* (*R. mel.*), *R. leguminosarum* (*R. leg*), *R. trofolii* (*R. tri.*), and *B. japonicum* (*B. jap.*). +, strong induction; −, no induction; (+), weak induction; +/−, induction in some studies but not in others; ND, not determined.[96,98,103-105]

The above experiments do not rule out the possibility that other genes also have a role in regulation of *nod* gene expression. In *R. meliloti*, there is evidence that other Sym plasmid genes outside of the *nodDABC* region are needed for full expression of both *nodD* and *nodABC*, but the nature of those genes is unknown.[93] Also, *nodD* and *nodABC* are not highly expressed when carried by *E. coli* or other bacteria outside of the Rhizobiaceae.[98]

The *nod* box is present in other locations. For example, there are six copies of *nod* box sequences on the *R. meliloti* Sym plasmid,three of which are upstream of known *nod* genes (Figure 1).[101] This raises the possibility that additional undiscovered exudate-inducible genes may be present at those locations.

E. PLANT SIGNALS THAT INDUCE *NOD* GENES

Root and seed exudate components that induce *nod* genes have recently been identified. In temperate legume symbioses (alfalfa/*R. meliloti*, pea/*R. leguminosarum*, and clover/*R. trifolii*) the inducers are substituted flavones[96,103-105] and flavanones[65,105] (Figure 3). These phenolic compounds are found in many plant species. In the soybean/*B. japonicum* symbiosis the natural inducers are substituted isoflavones.[97] It is of interest that isoflavonoids as a class include substances (phytoalexins) associated with plant defense against microbial pathogens[106,107] and substances recognized as signals for induction of symbiotic genes. In *R. fredii*, which has a broad host range including soybeans and tropical legumes, *nod* genes are inducible by a broad spectrum of flavones and isoflavones.[97]

The inducing compounds can be isolated from seeds, seedlings, or roots that have not been exposed to rhizobia. Thus, there is no absolute requirement for a bacterial signal to induce flavone or isoflavone production. One natural isolate of *B. japonicum*, Fix⁻ strain 61A24, does trigger the accumulation of the phytoalexin glyceollin I in soybean nodules, but this effect is not seen with Fix⁺ *B. japonicum* strains.[108]

The inducing compounds are extremely potent, showing activities at concentrations of 0.1 μM or less and full induction at 1 to 5 μM. In comparison, the inducers of the metabolic lactose and arabinose operons act at levels that are higher by three to five orders of magnitude.[103]

The structural features of flavonoids and isoflavonoids that are required for induction have been inferred by comparing the inducing abilities of synthetic and natural compounds carrying various substitutions in the flavone and isoflavone backbones. The relative strengths of various inducers are somewhat strain dependent (Figure 3). *R. meliloti nodABC* responds most strongly to luteolin.[103] *R. trifolii nod* genes respond best to 4′,7-dihydroxyflavone, and less strongly to a few other flavones and flavanones.[95,104] *R. leguminosarum* responds well to several flavones and flavanones including naringenin, apigenin, eriodictyol, and luteolin.[72,105] The strongest inducers for *B. japonicum* are the isoflavones daidzein, genistein, 7-hydroxyisoflavone, and 5,7-dihydroxyisoflavone.[97] Thus, hydroxylation at the 7 position is a common feature of all of the inducers. A 4′ hydroxylation is required in some systems but not in others. Substitutions of hydroxyl or other groups at several positions reduce activity.

Flavonoids are commonly found in nature as *O*-glucosides which are less reactive and more water soluble than the aglycones, allowing for storage in the plant cell vacuole.[108A] Apigenin 7-*O*-glucoside is an active inducer of *R. leguminosarum*,[105] but indirect evidence suggests that glycosylated derivatives of isoflavones are not strong inducers of *B. japonicum*.[97] The abilities of *Rhizobium* and *Bradyrhizobium* strains to deglycosylate or otherwise metabolize flavonoids has not been determined. Isoflavonoids are toxic to some *Rhizobium* and *Bradyrhizobium* strains, in accordance with their role in plant defense, but the levels needed to observe toxicity are somewhat higher than the levels needed for induction of *nod* genes.[97,108B]

In *R. leguminosarum*, induction of *nod* genes by pea exudate can be antagonized by the isoflavones daidzein and genistein, by the flavonol kaempferol, and by the acetophenone analogues acetoranillone and 4-hydroxyacetophenone.[105] Inhibition of induction by compounds structurally similar to inducers has also been observed in *R. trifolii*.[95] Thus, inhibition may operate by competition between the inducer and inhibitor for a receptor, transport, or regulatory molecule such as *nodD* protein. It is of interest that compounds which antagonize induction in *R. leguminosarum* are actually inducers in *B. japonicum* and *R. fredii*.

The specificity of different strains for different inducers could conceivably occur at any of several levels, including binding, transport, interaction with regulatory proteins, degradation of potential inducers or inhibitors, or metabolic conversion of noninducers to inducers. It was recently reported that different *nodD* genes show different inducer specificities and influence host specificities of nodulation.[80,108C] Specifically, the *nodD1* gene from the broad host-range strain MPIK3030 mediates induction of *nodABC* by either alfalfa or siratro exudate, whereas the *nodD1* gene of *R. meliloti* mediates induction only by alfalfa exudate and not by siratro exudate. Examination of a hybrid *nodD* gene suggested that the inducer specificity resides in the carboxy terminal half of the polypeptide, and not in the more highly conserved amino terminal end.[80] The *nodD* genes of *R. trifolii*, *R. leguminosarum*, and *R. meliloti* also show differences in their reactions with different flavones.[108C] However, transfer to *R. trifolii* of a *B. japonicum nodD* gene did not enable *R. trifolii* to respond to isoflavones.[97]

A complication in determining the role of *nodD* in inducer specificity is that some individual strains contain multiple *nodD* genes. *R. fredii*[49a,50] and MPIK3030[80] each have two, *R. meliloti* has three[61] and *B. japonicum* has two.[49,97] In contrast, *R. trifolii* and *R. leguminosarum* each have only one *nodD*. These various genes have all been designated *nodD* because of their conserved DNA and predicted amino acid sequences. The two *R. fredii nodD* genes are no more similar to each other in amino acid sequence (70% homology) than either is to *nodD* genes in other strains.[49a,50] The same is true for two *nodD* genes from *B. japonicum*.[49] In *R. meliloti*, two of the three *nodD* genes have been sequenced and are highly homologous (90%) but not identical.[61] The two *nodD* genes of *R. fredii* are functionally different from each other; both are needed for a fully efficient symbiosis, but the phenotypes of *nodD1* and *nodD2* insertion mutants are different, and *nodD2* has an additional role in control of EPS synthesis.[49,49a,50] The different *nodD* genes in a single strain might respond to different inducers or inhibitors, or regulate different promoters, or regulate the same promoters in different ways (i.e., induction or repression). The roles of these different *nodD* genes are under investigation.

The discovery that *nod* genes other than *nodD* are not detectably expressed in laboratory culture implies that many earlier studies on lectin binding, polysaccharide composition, secretion of phytohormones and degradative enzymes, etc., should perhaps be repeated after growth of the strains in the presence of inducers.

Based on the results discussed in Section II.A,B, and C, it appears that the common *nod* proteins might be involved in transport of substances in and/or out of the cell, and perhaps are also involved with synthesis of substances that affect plant roots. The host-specificity *nod* genes appear to be involved in changes in the *Rhizobium* cell surface. Thus, a plausible sequence of events is

1. Synthesis of inducers (flavones or isoflavones) by host plants
2. Induction of the bacterial *nodABCIJ* and host-specificity *nod* genes by the host signals and *nodD* protein
3. Modification of the *Rhizobium* outer surface, and synthesis by the bacteria of plant growth regulators (hormones and/or "oligosaccharins"[109]) that are transported to the bacterial cell surface and root surface (synthesis of these substances could also involve *nod*-mediated transport of plant compounds into the bacterium; synthesis of the new outer surface may involve the *exo* and *ndv* genes as well as *nod* gene discussed below
4. Induction of root hair curling and infection thread formation, and concomitant induction of cell division and plant protein (early nodulin)[110] synthesis.

An interesting parallel to this scenario is the ability of another member of the Rhizobiaceae, *Agrobacterium tumefaciens*, to respond to plant signals and induce new plant cell divisions which result in crown gall tumors. Specifically, plant phenolic compounds such as acetosyringone induce the *vir* genes on the bacterial Ti plasmid, enabling the bacteria to transfer a segment of plasmid DNA (the T-DNA) into the plant genome, leading to synthesis of plant hormones (controlled by genes on the T-DNA) and tumor formation.[111,112] Crown gall tumors and, in some cases, nodules, produce compounds (opines and rhizopines, respectively) that can be catabolized by the associated *Agrobacterium* or *Rhizobium* strains.[112a,112b] However, many of the specific biochemical steps in tumor formation and nodule formation are obviously different, for the following reasons:

1. *nod* Genes show no homology to *vir* or *onc* genes.
2. There is no evidence for transfer of *Rhizobium* DNA into the plant genome.
3. Nodule-specific plant proteins (nodulins) are not found in tumors.
4. The structures of *Agrobacterium*-induced tumors and *Rhizobium*-induced nodules are quite distinct.
5. Nodule formation is repressed by nitrate in the plant growth medium, while tumor formation is not.

The only genetic evidence of common functions are the homologies between *Rhizobium* and *Agrobacterium* genes controlling surface polysaccharides (see Section III below) and observations that *nod* and Ti plasmid genes are able to function in many members of the Rhizobiaceae, but do not appear to function efficiently in other bacteria such as *E. coli*.

III. GENES CONTROLLING SURFACE POLYSACCHARIDES AND NODULE DEVELOPMENT

Rhizobium and *Bradyrhizobium* strains typically produce large amounts of EPS and form mucoid colonies on solid medium. Mutants that are nonmucoid (Muc⁻) are generally defective in EPS production (Exo⁻) and are usually defective in some stage of nodule development.[84,113-116] Symbiotic mutants defective in LPS have also been described.[117,118] Recent genetic studies provide strong evidence that surface polysaccharides are involved in nodulation.

A. *exo* AND *ndv* GENES

R. meliloti mutants that fail to secrete the major acidic EPS, succinoglycan, were recently identified based on their reactions with phages, monoclonal antibodies, and a polysaccharide dye, Calcofluor. On alfalfa the mutants are Hac⁻, but penetrate the epidermis directly, forming nodules that lack infection threads and intracellular bacteria. The mutants fall into six complementation groups, four (*exoA, exoB, exoE, exoF*) on the "second megaplasmid" (i.e., not the Sym plasmid containing *nod* and *fix* genes) and two (*exoC* and *exoD*) on the chromosome.[31,32] This diverse collection of mutants provides convincing evidence that normal EPS is required for infection of alfalfa, but not for induction of cortical cell division and nodule differentiation.

Calcofluor has also been used to identify *Agrobacterium tumefaciens* Exo⁻ mutants, many of which can be complemented by *R. meliloti exo* genes in five complementation groups. All of the *A. tumefaciens* mutants form normal crown galls, except the *exoC* mutants, which are avirulent and deficient in attachment to plant cells, motility, and growth on solid media. *R. meliloti* and *A. tumefaciens exoC* mutants are both deficient in production of the major acidic EPS. In addition, *A. tumefaciens exoC* mutants are deficient in a major neutral EPS, β-1,2-glucan.[119]

Chromosomal genes controlling *A. tumefaciens* virulence (*chvA* and *chvB*) are required for attachment to plant cells as well as tumorigenesis.[120,121] *chvB* mutants lack a 2-linked-β-glucan.[122] *R. meliloti* genes that are structurally and functionally homologous to these genes have been isolated and designated *ndvA* and *ndvB* (for nodule development). *R. meliloti* strains carrying mutations in the *ndv* loci resemble Exo⁻ mutants in that they form "empty" (uninfected) Fix⁻ nodules. They differ from Exo⁻ mutants in that they react normally with Calcofluor and curl root hairs normally.[39] The involvement of *exoC, chvB*, and presumably *ndvB* in β-glucan formation, and of *R. meliloti exoC* and *ndvB* in infection is consistent with a possible role of β-glucan in the infection process.

B. *psi, psr, pss*

R. phaseoli contains a gene, *psi* (polysaccharide inhibition), which affects EPS synthesis. The *psi* gene has been mapped close to *nif* on the Sym plasmid. Strains containing a cloned *psi* gene on a plasmid vector form nonmucoid EPS⁻ colonies and are Nod⁻ on *Phaseolus* beans. Mutants containing *psi*∷Tn5 mutations in the Sym plasmid are Fix⁻. These results suggest that EPS is required for nodulation, but is deleterious for nodule function at later stages if present in excess amounts due to a *psi* mutation. Thus, the role of *psi* may be repression of EPS synthesis in bacteroids. DNA sequence analysis of *psi* shows an open reading frame that encodes a 10 kDa protein that is hydrophobic at the amino terminal end, perhaps indicating an association with membranes. Another Sym plasmid gene, *psr* (poly-saccharide restoration), is able to repress transcription of *psi*. Multicopy *psr* can restore the EPS and nodulation defects caused by multicopy *psi*. Strains with multicopy *psr* induce Fix⁻ nodules, again supporting the notion that *psi* normally serves to prevent production of excess EPS, which inhibits nitrogen fixation. Another gene, *pss* (polysaccharide synthesis) also has a symbiotic role. A *pss*⁻ strain of *R. phaseoli* was found to be EPS⁻ and Nod⁺ Fix⁺ on *Phaseolus* beans, but the same *pss* mutation in *R. leguminosarum* produced an EPS⁻ Nod⁻ phenotype on peas. This suggests that there are different bacterial cell surface requirements in the *R. phaseolus*/bean and *R. leguminosarum*/pea symbioses. Cloned DNA from a plant pathogen, *Xanthomonas campestris*, complements the EPS⁻ and Nod⁻ defects in *pss*⁻ *R. leguminosarum*. Thus, the *pss* gene may be quite widespread among plant-associated bacteria. The *pss* gene is not located on the Sym plasmid in *R. phaseoli*.[123-125]

C. *R. FREDII* nodD2

In *R. fredii*, a *nodD*-related gene appears to be involved in control of EPS production. Specifically, cloned *nodD2* on a plasmid vector changes the USDA191 colony morphology from Muc⁺ to Muc⁻ and inhibits EPS synthesis.[49a,50] Other Sym plasmid genes are also involved, since the colony morphology change does not occur in strains carrying cloned *nodD2* plus certain Sym plasmid deletions. The effect is strain-specific, since *R. trifolii* remains Muc⁺ after introduction of the *R. fredii nodD2* gene.[49a] *nodD2* cannot complement an *R. trifolii nodD*⁻ mutant for nodulation of clover (in contrast to the *R. fredii nodD1* gene which does complement the *R. trifolii nodD*⁻ mutation). Thus, *nodD2* may not be involved in activation of *nodABC* genes. However, *nodD2* does have homology to *lysR*, an *E. coli* transcriptional regulator. *nodD2* mutants are delayed in nodule development, partially defective in nitrogen fixation, and have decreased numbers of bacteroids per infected plant cell.[49a,50] A plausible model for *nodD2* action is that it regulates EPS synthetic genes in response to specific signals during nodule development.

D. OTHER MUTANTS

LPS may be another surface polysaccharide with a role in nodulation. A class of *R. phaseoli* mutations causes a defect in LPS and premature cessation of infection thread development within root hairs. Four such mutations in at least three genes are clustered in the chromosome.[117,126] *B. japonicum* Nod⁻ mutants defective in 0 antigen have been described but not well characterized genetically.[118] Bacteriophage-resistant mutants of *B. japonicum* that are Nod⁻ are also presumed to have alterations in the cell surface, possibly in LPS.[127]

Transformation of Nif⁻ *Azotobacter vinelandii* (a free-living nitrogen-fixing bacterium) with DNA from *R. trifolii* or *B. japonicum* yielded Nif⁺ transformants, some of which were specifically recognized by lectins from clover or soybeans, respectively. Since lectin binding involves surface polysaccharides, the genes controlling the *Rhizobium* polysaccharides are presumably present in the transformants. The *Rhizobium* genes present in the transformant have not yet been identified.[128,129] Recent studies with *R. trifolii nod* mutants suggest that *nod* genes have some influence on the structure of the capsular polysaccharide and binding of clover lectin.[73]

Nonmucoid, EPS-deficient mutants have also been isolated from *R. trifolii* and from the broad host-range tropical strain NGR234 following Tn5 mutagenesis.[84,113,114] The mutants generally form Fix⁻ nodules. The stages at which nodule development is impaired varies with different plant hosts in the case of NGR234. The Tn5 insertion is chromosomally located in all cases. One class of mutants, group 2, forms Fix⁻ calli that lack infection threads on the tropical legume *Leucaena*, and also forms Fix⁻ nodules on siratro. Addition of EPS from wild-type NGR234 allows these Muc⁻ mutants to form normal Fix⁺ nodules on *Leucaena* and siratro. EPS from *R. trifolii* enables an *R. trifolii* Muc⁻ Fix⁻ mutant to form Fix⁺ nodules on clover, but does not restore the symbiotic capacity of the NGR234 Muc⁻ mutants. The Sym plasmid of NGR234 does not appear to have a major role in EPS synthesis, since a Sym-plasmid cured mutant has EPS that is qualitatively and quantitatively indistinguishable from wild-type EPS. These and other studies[33,115,116,130-132] of other EPS-deficient mutants provide further evidence for a role for EPSs in nodule development.

IV. GENES CONTROLLING NITROGEN FIXATION

Fix⁻ mutants, by definition, form nodules that do not fix nitrogen. *Rhizobium* genes that are required for symbiotic nitrogen fixation and are also structurally or functionally homologous to nitrogen fixation (*nif*) genes in *K. pneumoniae* are, by convention, designated *nif* genes. Other *Rhizobium* genes required for symbiotic nitrogen fixation are referred to as *fix* genes. Some Fix⁻ mutants that are blocked in very early stages of the developmental pathway have been reclassified as *ndv* and *exo* mutants, as discussed earlier.

K. pneumoniae is a free-living bacterium that is a member of the Enterobacteriaceae, a family that includes *E. coli*. The genetic basis for nitrogen fixation has been studied in detail in *K. pneumoniae*, using many of the techniques developed for *E. coli*. In *K. pneumoniae*, about 20 contiguous *nif* genes are required for nitrogenase activity and growth on nitrogen-free medium. *nifA* and *nifL* are involved in regulation of all of the other *nif* genes. *nifH*, *nifD*, and *nifK* encode the nitrogenase subunits. *nifQ*, *nifB*, *nifE*, *nifN*, and *nifV* are involved in biosynthesis of the iron-molybdenum cofactor (FeMoco) of nitrogenase. *nifF* and *nifJ* are involved in electron transport to nitrogenase. The roles of several other *nif* genes are less well defined.[133-140]

All *Rhizobium* and *Bradyrhizobium* strains tested to date have one or more copies of genes showing DNA sequence homology with *K. pneumoniae* *nifH*, *nifD*, and *nifK*.[38] In *Rhizobium*, *nifHDK* form a single operon, just as they do in *K. pneumoniae*.[50,141-145] In some cases, complete *nif* genes or the amino terminal end of *nifH* (including the promoter region) are reiterated on the Sym plasmid.[145,146] In *Bradyrhizobium*, *nifH* and *nifDK* are in separate transcriptional units located several kilobases apart.[147-150]

B. japonicum genes homologous to *nifE*, *nifN*, *nifS*, and *nifB* have also been identified in the same genomic region as *nifH* and *nifDK*. A *nifA* gene has been located many kilobases away, near the common *nod* genes.[150-153] In *R. meliloti*, *nifB* and *nifA* (formerly *fixD*) have been identified.[154,155] In *R. leguminosarum*, *nifA*, *nifB* (formerly *fixZ*), and sequences homologous to *nifE* and *nifQ* have been detected.[156,157]

Several *fix* genes have also been characterized by DNA sequence analysis and mutagenesis. In *R. meliloti*, a cluster of genes, *fixA fixB fixC fixX*, form an operon that is upstream of, and divergently transcribed from, the *nifHDK* operon.[158] *fixABC* are also clustered in *R. leguminosarum*,[157] but *B. japonicum fixA* is widely separated from *fixBC*.[150] Genes designated as *fixF*, *fixG*, *fixH*, and *fixI* have been identified.[159] Many other Fix⁻ mutations in Sym plasmids or in chromosomes have been identified but not characterized in detail at the molecular level.[160-162]

The biochemical functions of *Rhizobium nif* genes are presumed to be similar to those of *K. pneumoniae nif* genes. The functions of the *fix* genes are less clear. *fixX* shows some similarity to ferredoxins and could be involved in electron transport to nitrogenase.[158] Mutations in *nifHKD*, *nifB*, and several *fix* genes allow infection thread formation and fairly normal differentiation of bacteroids, although the bacteroids quickly senesce in several mutants. In contrast, *nifA* mutations result in more severe defects in bacterial differentiation.[162-164] Since the *nifA* gene product in known to be involved in regulation of several *nif* and *fix* operons, it is possible that it also regulates other genes, not yet identified, which are involved in bacteroid differentiation. Early stages of nodule development are fairly normal in most *nif* and *fix* mutants, but differences in nodule morphology from wild type have been noted with some mutants.[164]

The regulation of *nif* and *fix* genes has recently been reviewed[165] and will not be discussed in detail here. Briefly, *nif* and *fix* genes generally are expressed only in nodules, or during microaerobic induction conditions in the case of certain *Bradyrhizobium* strains that are capable of *ex planta* induction of nitrogenase activity. The *nifA* gene product is a regulatory protein that must be present in order for several other *nif* and *fix* operons to be transcribed. The promoter regions of *nifA*-dependent operons lack the consensus *E. coli* common promoter sequence at $-35/-10$, but do contain other conserved sequences which are recognized by the *nifA* protein and by an alternative sigma factor of RNA polymerase. During nodule development, the *nifA* gene is expressed, and the *nifA* protein then activates the other *nif* and *fix* promoters, resulting in nitrogenase synthesis and nitrogen fixation. In *R. meliloti*, transcription of *nifA* is induced by low oxygen tension.[166] In *B. japonicum*, the *nifA* protein itself may be oxygen sensitive.[167] Thus, low oxygen tension appears to be a signal for expression of many *nifA*-dependent genes. Since nodules become microaerobic as bacteroids proliferate, and nitrogenase is inactivated by oxygen, the coupling of *nif* gene expression to microaerobiosis makes physiological sense. This does not rule out the possibility that other types of regulation also affect *nifA* transcription or *nifA* protein activity. Low nitrogen levels do not appear to be required for activation of *nifA* or *nifA*-dependent operons during symbiosis. The *ntrC* gene plays a key role in activation of many genes involved in nitrogen assimilation in enteric bacteria and is involved in *nifA* activation in *Rhizobium* during free-living growth at low nitrogen concentrations, but it does not appear to play an important role in induction of *nifA* or *nifA*-dependent genes during symbiosis.

V. SOME POTENTIAL PRACTICAL APPLICATIONS OF RECENT ADVANCES IN *RHIZOBIUM/BRADYRHIZOBIUM* GENETICS

Legume inoculants have been commercially available for more than 80 years and are effective in increasing yields of legumes in low-nitrogen soils which lack indigenous *Rhizobium* populations. However, legume inoculants do not demonstrably affect yields in many soils where indigenous strains are present. One limitation is the problem of competition. Superior nitrogen-fixing inoculum strains are often unable to form a significant proportion of the nodules in a field situation because of competition from indigenous strains. In one recent survey, analysis of 543 nodule isolates from Wisconsin soybean farms showed that none of the nodules were formed by inoculant strains.[168] Thus, improvements in the competitiveness as well as the nitrogen-fixation efficiency of inoculants would be desirable.

Recent discoveries concerning regulation of *nod* genes may provide new approaches to the problem of competition. In particular, it has been shown that genes involved in the earliest stages of nodulation are not highly expressed in the culture conditions traditionally used for preparation of commercial inocula. The inclusion of specific inducers (flavones or isoflavones) in the preparation of inocula might enhance nodulation and competitiveness. Alternatively, manipulations of *nod* promoters and regulatory genes might also improve nodulation. Comparisons of more competitive and less competitive natural isolates for differences in the expression and activity of specific *nod* gene products may also reveal clues as to the genetic basis of competition.

Improvements in the nitrogen-fixation abilities of some strains have reportedly been achieved by mutagenesis and screening for increased nitrogenase activity on plants or under defined laboratory conditions. However, the genetic basis for these improvements in unknown.[169-171] The identification of structural and regulatory genes for nitrogenase and for other nodule functions (such as carbon and energy supply) provides opportunities for enhancing the expression of gene products that might be limiting in field situations.[172] The identification and interstrain transfer of a cluster of hydrogen uptake (*hup*) genes that enable *Bradyrhizobium* to recycle hydrogen gas (a by-product of nitrogenase activity) may provide a means for improving the efficiency of inoculum strains that are naturally Hup⁻.[173]

The identification of host-range genes might make it possible for a strain with desirable characteristics (competitiveness or efficient nitrogen fixation) on its normal host to be modified so that it can be used with other hosts. The identification of genes for nodule formation on *Parasponia* might provide clues as to how symbiosis can be extended to other nonlegumes.

More generally, elucidation of the molecular mechanisms involved in *Rhizobium/Bradyrhizobium*-legume symbioses may provide insights into plant-microbe interactions and regulation of plant development that will form the basis for selecting crops with disease resistance or other desirable traits.

VI. FUTURE PROSPECTS

Many genetic loci that influence symbiosis have now been identified in a variety of *Rhizobium* and *Bradyrhizobium* strains. In many cases, the genes have been cloned and sequenced, and the products have been associated with a particular stage of infection, nodule development, or nitrogen fixation. The list now includes at least 15 different *nod* genes, 10 or more *exo/ndv* genes, and 15 or more *nif* and *fix* genes.

However, the biochemical activities of the proteins encoded by symbiotic genes have not been defined, except in the case of some *nif* genes (i.e., *nifHDK*, encoding the nitrogenase polypeptides). Specific biochemical assays are needed to examine the roles of genes that may be involved in synthesis of polysaccharides (*exo, ndv, psi*, some *nod* genes), synthesis of plant growth regulators (some *nod* genes?), transport of substances into or out of the cell (common *nod* genes?), gene regulation (*nodD, nifA*), and nitrogenase activity (*nif* and *fix* genes). Genetic techniques may yet provide further hints as to biochemical functions. For example, selection of mutations in a gene which enable it to suppress the effect of mutation in another gene can reveal interactions between gene products. The phenotypes resulting from changes in specific nucleotides or from construction of hybrid genes can suggest ways in which specific proteins interact with DNA or plant signals. In the end, however, careful biochemical analysis and "inspired guesswork"[174] (see also Chapter 1) will be required for further progress.

The possibility that other symbiotic genes remain to be discovered should not be overlooked. The recent discovery of *exo* genes is a case in point. Random Tn5 mutagenesis and screening for symbiotic defects on plants did not reveal most of these genes; they were discovered only when cell-surface changes were specifically looked for, using Calcofluor, phage, and monoclonal antibodies. Screenings of mutagenized populations for other specific biochemical defects may yet disclosse additional genes with a role in symbiosis. Lac fusions can be used to select mutations in previously unrecognized loci that regulate the fusion promoters. Other *nod, ndv*, and *fix* genes that are specific to particular strain/host combinations probably remain to be discovered, especially in the broad host-range rhizobia and bradyrhizobia which have not been studied as intensively as *R. meliloti, R. trifolii*, and *R. leguminosarum*. Also, the interactions of symbiotic genes with metabolic genes that affect symbiosis may be revealing. For example, examination of the structure and function of genes involved in dicarboxylic acid transport[175,176] has identified important symbiotic functions, as well as interesting mechanisms of gene regulation. Finally, the identification of genes that may affect the survival and competitiveness of *Rhizobium* and *Bradyrhizobium* in nature is just beginning.[11,177-181]

ACKNOWLEDGMENT

I thank Eric Johansen, Janice Kimpel, Renee Kosslak, and Don Merlo for helpful suggestions on the manuscript, and Jo Adang for assistance with preparation of illustrations. This is Agrigenetics Advanced Science Company Manuscript No. 79.

REFERENCES

1. **Gussin, G. N., Ronson, C. W., and Ausubel, F. M.,** Regulation of nitrogen fixation genes, *Annu. Rev. Genet.*, 20, 567, 1986.
2. **Long, S. R.,** Genetic advances in the study of *Rhizobium* nodulation, *Genet. Eng.*, 8, 135, 1986.
3. **Hodgson, A. L. M. and Stacey, G.,** Potential for *Rhizobium* improvement, *CRC Crit. Rev. Biotechnol.*, 4, 1, 1986.
4. **Long, S. R.,** Genetics of *Rhizobium* nodulation, in *Plant-Microbe Interactions,* Kosuge, T. and Nester, E. W., Eds., Macmillan, New York, 1984.
5. **Verma, D. P. S. and Long, S.,** The molecular biology of *Rhizobium*-legume symbiosis, *Int. Rev. Cytol.*, Suppl. 14, 211, 1983.
6. **Rossen, L., Johnston, A. W. B., Firmin, J. L., Shearman, C. A., Evans, I. J., and Downie, J. A.,** Structure, function and regulation of nodulation genes of *Rhizobium, Oxford Surv. Plant Mol. Cell Biol.*, 3, 441, 1986.
7. **Downie, J. A. and Johnston, A. W. B.,** Nodulation of legumes by *Rhizobium:* the recognized root?, *Cell,* 47, 153, 1986.
8. **Halverson, L. J. and Stacey, G.,** Signal exchange in plant-microbe interactions, *Microbiol. Rev.,* 50, 193, 1986.
9. **Sprent, J. I.,** Benefits of *Rhizobium* to agriculture, *Trends Biotechnol.,* 4, 124, 1986.
10. **Stacey, G. and Upchurch, R. G.,** *Rhizobium* inoculation of legumes, *Trends Biotechnol.,* 2, 65, 1984.
11. **Dowling, D. N. and Broughton, W. J.,** Competition for nodulation of legumes, *Annu. Rev. Microbiol.,* 40, 131, 1986.
12. **Prakash, R. K. and Atherly, A. G.,** Plasmids of *Rhizobium* and their role in symbiotic nitrogen fixation, *Int. Rev. Cytol.,* 104, 1, 1986.
13. **Kondorosi, E. and Kondorosi, A.,** Nodule induction of plant roots by *Rhizobium, Trends Biochem. Sci.,* 11, 296, 1986.
14. **Lambert, G. R., Harker, A. R., Zuber, M., Dalton, D. A., Hanus, F. J., Russell, S. A., and Evans, H. J.,** Characterization, significance and transfer of hydrogen uptake genes from *Rhizobium japonicum, Nucleic Acids Res.,* 14, 209, 1986.
15. **Jordan, D. C.,** Rhizobiacea CONN 1938, in *Bergey's Manual of Systemic Bacteriology,* Krieg, N. R. and Holt, J. G., Eds., Williams & Wilkins, Baltimore, 1984, 234.
16. **Jordan, D. C.,** Transfer of *Rhizobium japonicum* Buchana 1980 to *Bradyrhizobium* gen. nov., a genus of slow-growing, root nodule bacteria from leguminous plants, *Int. J. Syst. Bacteriol.,* 32, 136, 1982.
17. **Hennecke, H., Kaluza, K., Thöny, B., Fuhrmann, M., Ludwig, W., and Stackebrandt, E.,** Concurrent evolution of nitrogenase genes and 16S rRNA in *Rhizobium* species and other nitrogen fixing bacteria, *Arch. Microbiol.,* 142, 342, 1985.
18. **Jarvis, B. D. W., Gillis, M., and De Ley, J.,** Intra- and intergeneric similarities between the ribosomal ribonucleic acid cistrons of *Rhizobium* and *Bradyrhizobium* species and some related bacteria, *Int. J. Syst. Bacteriol.,* 36, 129, 1986.
19. **Wedlock, N. D. and Jarvis, B. D. W.,** DNA homologies between *Rhizobium fredii,* rhizobia that nodulate *Galega* sp., and other *Rhizobium* and *Bradyrhizobium* species, *Int. J. Syst. Bacteriol.,* 36, 550, 1986.
19A. **Hollis, A. B., Kloos, W. E., and Elkan, G. H.,** DNA: DNA hybridization studies of *Rhizobium japonicum* and related *Rhizobiaceae, J. Gen. Microbiol.,* 123, 215, 1981.
20. **Holsters, M., Van Den Eede, G., Goethals, K., Van Montagu, M., and Dreyfus, B.,** Nodulation genes of the stem nodulating *Sesbania rostrata* symbiont, strain ORS571, *Molecular Genetics of Plant-Microbe Interactions,* Verma, D. P. S. and Brisson, N., Eds., Martinus Nijhoff, Dordrecht, 1987, 208.
21. **Djordjevic, M. A., Innes, R. W., Wijffelman, C. A., Schofield, P. R., and Rolfe, B. G.,** Nodulation of specific legumes is controlled by several distinct loci in *Rhizobium trifolii, Plant Mol. Biol.,* 6, 389, 1986.
22. **Djordjevic, M. A., Schofield, P. R., and Rolfe, B. G.,** Tn5 mutagenesis of *Rhizobium trifolii* host-specific nodulation genes results in mutants with altered host-range ability, *Mol . Gen. Genet.,* 200, 463, 1985.
23. **Simon, R. and Priefer, U.,** Vector technology of relevance to nitrogen fixation research, in *The Molecular Biology of Symbiotic Nitrogen Fixation,* Gresshoff, P., Ed., CRC Press, Boca Raton, FL, 1989.

24. **Vincent, J. M.**, Factors controlling the legume-*Rhizobium* symbiosis, in *Nitrogen Fixation*, Vol. II, Newton, W. E. and Orme-Johnson, W. H., Eds., University Park Press, Baltimore, 1980, 103.

25. **Dart, P. J.**, Infection and development of leguminous nodules, in *A Treatise on Dinitrogen Fixation*, Hardy, R. W. F. and Silver, W. S., Eds., John Wiley & Sons, New York, 1977, 367.

26. **Bauer, W. D.**, Infection of legumes by rhizobia, *Annu. Rev. Plant Physiol.*, 32, 407, 1981.

27. **Lancelle, S. A. and Torrey, J. G.**, Early development of *Rhizobium*-induced root nodules of *Parasponia rigida*. I. Infection and early nodule initiation, *Protoplasma*, 123, 26, 1984.

28. **Lancelle, S. A. and Torrey, J. G.**, Early development of *Rhizobium*-induced root nodules of *Parasponia rigida*. II. Nodule morphogenesis and symbiotic development, *Can. J. Bot.*, 63, 25, 1985.

29. **Price, G. D., Mohapatra, S. S., and Gresshoff, P. M.**, Structure of nodules formed by rhizobium strain ANU289 in the nonlegume parasponia and the legume siratro *(Macroptilium atropurpureum)*, *Bot. Gaz.*, 145, 444, 1984.

30. **Calvert, H. E., Pence, M. K., Pierce, M., Malik, N. S. A., and Bauer, W. D.**, Anatomical analysis of the development and distribution of *Rhizobium* infections in soybean roots, *Can. J. Bot.*, 62, 2375, 1984.

31. **Finan, T. M., Hirsch, A. M., Leigh, J. A., Johansen, E., Kuldau, G. A., Deegan, S., Walker, G. C., and Signer, E. R.**, Symbiotic mutants of *Rhizobium meliloti* that uncouple plant from bacterial differentiation, *Cell*, 40, 869, 1985.

32. **Leigh, J. A., Signer, E. R., and Walker, G. C.**, Exopolysaccharide-deficient mutants of *Rhizobium meliloti* that form ineffective nodules, *Proc. Natl. Acad. Sci. U.S.A.*, 82, 6231, 1985.

33. **VandenBosch, K. A., Noel, K. D., Kaneko, Y., and Newcomb, E. H.**, Nodule initiation elicited by noninfective mutants of *Rhizobium phaseoli*, *J. Bacteriol.*, 162, 950, 1985.

34. **Hirsch, A. M., Wilson, K. J., Jones, J. D. G., Bang, M., Walker, V. V., and Ausubel, F. M.**, *Rhizobium meliloti* genes allow *Agrobacterium tumefaciens* and *Escherichia coli* to form pseudonodules on alfalfa, *J. Bacteriol.*, 158, 1133, 1984.

35. **Hirsch, A. M., Drake, D., Jacobs, T. W., and Long, S. R.**, Nodules are induced on alfalfa roots by *Agrobacterium tumefaciens* and *Rhizobium trifolii* containing small segments of the *Rhizobium meliloti* nodulation region, *J. Bacteriol.*, 161, 223, 1985.

36. **Truchet, G., Debelle, F., Vasse, J., Terzaghi, B., Garnerone, A., Rosenberg, C., Batut, J., Maillet, F., and Denarie, J.**, Identification of *Rhizobium meliloti* pSym2011 region controlling the host specificity of root hair curling and nodulation, *J. Bacteriol.*, 164, 1200, 1985.

37. **Degenhardt, T. L., LaRue, T. A., and Paul, E. A.**, Investigation of a non-nodulating cultivar of *Pisum sativum*, *Can J. Bot.*, 54, 1633, 1976.

38. **Ruvkun, G. B. and Ausubel, F. M.**, Interspecies homology of nitrogenase genes, *Proc. Natl. Acad. Sci. U.S.A.*, 77, 191, 1980.

39. **Dylan, T., Ielpi, L., Stanfield, S., Kachyap, L., Douglas, C., Yanofsky, M., Nester, E., Helinski, D. R., and Ditta, G,** *Rhizobium meliloti* genes required for nodule development are related to chromosomal virulence genes in *Agrobacterium tumefaciens*, *Proc. Natl. Acad. Sci. U.S.A.*, 83, 4403, 1986.

40. **Fred, E. B., Baldwin, I. L., and McCoy, E.**, Root nodule bacteria and leguminous plants, *University of Wisconsin Studies in Science No. 5*, University of Wisconsin, Madison, 1932.

41. **Debellé, F., Rosenberg, C., Vasse, J., Maillet, F., Martinez, E., Denariae, J., and Truchet, G.**, Assignment of symbiotic developmental phenotypes to common and specific nodulation (*nod*) genetic loci of *Rhizobium meliloti*, *J. Bacteriol.*, 168, 1075, 1986.

42. **Downie, J. A., Knight, C. D., Johnston, A. W. B., and Rossen, L.**, Identification of genes and gene products involved in the nodulation of peas by *Rhizobium leguminosarum*, *Mol. Gen. Genet.*, 198, 225, 1985.

43. **Egelhoff, T. T., Fischer, R. F., Jacobs, T. W., Mulligan, J. T., and Long, S. R.**, Nucleotide sequence of *Rhizobium meliloti* 1021 nodulation genes: *nodD* is read divergently from *nodABC*, *DNA*, 4, 241, 1985.

44. **Wijffelman, C. A., Pees, E., van Brussel, A. A. N., Okker, R. J. H., and Lugtenberg, B. J. J.**, Genetic and functional analysis of the nodulation region of the *Rhizobium leguminosarum* Sym plasmid pRL1JI, *Arch. Microbiol.*, 143, 225, 1985.

45. **Kondorosi, E., Banfalvi, Z., and Kondorosi, A.**, Physical and genetic analysis of a symbiotic region of *Rhizobium meliloti*: identification of nodulation genes, *Mol. Gen. Genet.*, 193, 445, 1984.

46. **Evans, I. J. and Downie, J. A.**, The *nodI* gene product of *Rhizobium leguminosarum* is closely related to ATP-binding bacterial transport proteins; nucleotide sequence analysis of the *nodI* and *nodJ* genes, *Gene*, 43, 95, 1986.

47. **Scott, K. F.**, Conserved nodulation genes from the non-legume symbiont *Bradyrhizobium* sp. *(Parasponia)*, *Nucleic Acids Res.*, 14, 2905, 1986.

48. **Nieuwkoop, A. J., Banfalvi, Z., Deshmande, N., Gerhold, D., Schell, M. G., Sirotkin, K. M., and Stacey, G.**, A locus encoding host range is linked to the common nodulations genes of *Bradyrhizobium japonicum*, *J. Bacteriol.*, 169, 2631, 1987.

49. **Appelbaum, E., Thompson, D., Barkei, J., Chartrain, N., and Maroney, M.**, unpublished data.

49a. **Appelbaum, E., Thompson, D. V., Idler, K., and Chartrain, N.**, *Rhizobium japnicum* USDA 191 has two *nodD* genes that differ in primary structure and function, *J. Bacteriol.*, 170, 12, 1988.

50. **Appelbaum, E., Chartrain, N., Thomson, D., Johansen, E., Idler, K., O'Connell, M., and Mc-Loughlin, T.**, Genes of *Rhizobium japonicum* involved in development of nodules, *Nitrogen Fixation Research Progress*, Evans, H. J., Bottomley, P. J., and Newton, W. E., Eds., Martinus Nijhoff, Dordrecht, 1985, 101.

51. **Lewin, A., Rosenberg, C., Meyer, H., Wong, Z. A. C. H., Nelson, L., Manen, J.-F., Stanley, J., Dowling, D. N., Dénarie, J., and Broughton, W. J.**, Multiple host-specificity loci of the broad host-range *Rhizobium* sp. NGR234 selected using the widely compatible legume *Vigna unguiculata*, *Plant Mol. Biol.*, 8, 447, 1987.

52. **Bassam, B. J., Rolfe, B. G., and Djordjevic, M. A.**, *Macroptilium atropurpureum* (siratro) host specificity genes are linked to a *nodD*-like gene in the broad host range *Rhizobium* strain NGR234, *Mol. Gen. Genet.*, 203, 49, 1986.

53. **Bachem, C. W. B., Banfalvi, Z., Kondorosi, E., Schell, J., and Kondorosi, A.**, Identification of host range determinants in the *Rhizobium* species MPIK3030, *Mol. Gen. Genet.*, 203, 42, 1986.

54. **Pankhurst, C. E., Broughton, W. J, and Wieneke, U.**, Transfer of indigenous plasmid of *Rhizobium loti* to other rhizobia and *Agrobacterium tumefaciens*, *J. Gen Microbiol.*, 129, 2535, 1983.

55. **Appelbaum, E. R., Johansen, E., and Chartrain, N.**, Symbiotic mutants of USDA191, a fast-growing *Rhizobium* that nodulates soybeans, *Mol. Gen. Genet.*, 201, 454, 1985.

56. **Masterson, R. V., Prakash, R. K., and Atherly, A. G.**, Conservation of symbiotic nitrogen fixation gene sequences in *Rhizobium japonicum* and *Bradyrhizobium japonicum*, *J. Bacteriol.*, 163, 21, 1985.

57. **Haugland, R. and Verma, D. P. S.**, Interspecific plasmid and genomic DNA sequence homologies and localization of *nif* genes in effective and ineffective strains of *Rhizobium japonicum*, *J. Mol. Appl. Sci.*, 1, 205, 1981.

58. **Schofield, P. R. and Watson, J. M.**, DNA sequence of *Rhizobium trifolii* nodulation genes reveals a reiterated and potentially regulatory sequence preceding *nod*ABC and *nod*FE, *Nucleic Acids Res.*, 14, 2891, 1986.

59. **Shearman, C. A., Rossen, L., Johnston, A. W. B., and Downie, J. A.**, The *Rhizobium leguminosarum* nodulation gene *nod*F encoded a polypeptide similar to acyl carrier protein and is regulated by *nodD* plus a factor in pea root exudate, *EMBO J.*, 5, 647, 1986.

60. **Rossen, L., Johnston, A. W. B., and Downie, J. A.**, DNA sequence of the *Rhizobium leguminosarum* nodulation genes *nodA, B,* and *C* required for root hair curling, *Nucleic Acids Res.*, 12, 9497, 1984.

61. **Göttfert, M., Horvath, B., Kondorosi, E., Putnoky, P., Rodriguez-Quiñones, F., and Kondorosi, A.**, At least two *nodD* genes are necessary for efficient nodulation of alfalfa by *Rhizobium meliloti*, *J. Mol. Biol.*, 191, 411, 1986.

62. **Török, I., Kondorosi, E., Stepkowski, T., Pósfai, J., and Kondorosi, A.**, Nucleotide sequence of *Rhizobium meliloti* nodulation genes, *Nucleic Acids Res.*, 12, 9509, 1984.

63. **Jacobs, T. W., Egelhoff, T. T., and Long, S. R.**, Physical and genetic map of *Rhizobium meliloti* nodulation gene region and nucleotide sequence of *nodC*, *J. Bacteriol.*, 162, 469, 1985.

64. **Lamb, J. W. and Hennecke, H.**, In *Bradyrhizobium japonicum* the common nodulation genes, *nodABC*, are linked to *nifA* and *fixA*, *Mol. Gen. Genet.*, 202, 512, 1986.

65. **Rodriguez-Quiñones, F., Balfalvi, Z., Murphy, P., and Kondorosi, A.**, Interspecies homology of nodulation genes in *Rhizobium*, *Plant Mol. Biol.*, 8, 61, 1987.

66. **Djordjevic, M. A., Schofield, P. R., Ridge, R. W., Morrison, N. A., Bassam, B. J., Plazinski, J., Watson, J. M., and Rolfe, B. G.**, *Rhizobium* nodulation genes involved in root hair curling (Hac) are functionally conserved, *Plant Mol. Biol.*, 4, 147, 1985.

67. **Fisher, R. F., Tu, J. K., and Long, S. R.**, Conserved nodulation genes in *Rhizobium meliloti* and *Rhizobium trifolii*, *Appl. Environ. Microbiol.*, 49, 1432, 1985.

68. **Noti, J. D., Dudas, B., and Szalay, A. A.**, Isolation and characterization of nodulation genes from *Bradyrhizobium* sp. (*Vigna*) strain IRc 78, *Proc. Natl. Acad. Sci. U.S.A.*, 82, 7379, 1985.

69. **Marvel, D. J., Kuldau, G., Hirsch, A., Richards, E., Torrey, J. G., and Ausubel, F. M.**, Conservation of nodulation genes between *Rhizobium meliloti* and a slow-growing *Rhizobium* strain that nodulates a nonlegume host, *Proc. Natl. Acad. Sci. U.S.A.*, 82, 5841, 1985.

70. **Wilson, K. J., Anjaiah, V., Nambiar, P. T. C., and Ausubel, F. M.**, Isolation and characterization of symbiotic mutants of *Bradyrhizobium* sp. (*Arachis*) strain NC92: mutants with host-specific defects in nodulation and nitrogen fixation, *J. Bacteriol.*, 169, 2177, 1987.

71. **Horvath, B., Kondorosi, E., John, M., Schmidt, J., Török, I., Györgypal, Z., Barabas, I., Wieneke, U., Schell, J., and Kondorosi, A.**, Organization, structure and symbiotic function of *Rhizobium meliloti* nodulation genes determining host specificity for alfalfa, *Cell*, 46, 335, 1986.

72. **Zaat, S. A. J., VanBrussel, A. A. N., Tak, T., Pees, E., and Lugtenberg, B. J. J.**, Flavonoids induce *Rhizobium leguminosarum* to produce *nodDABC* gene-related factors that cause thick, short roots and root hair responses on common vetch, *J. Bacteriol.*, 169, 3388, 1987.

73. **Dazzo, F. B., Hollingsworth, R. I., Philip, S, Smith, K. B., Welsch, M. A., Salzwedel, J., Morris, P., and McLoughlin, L.**, Involvement of pSYM nodulation genes in production of surface and extracellular components of *Rhizobium trifolii* which interact with white clover root hairs, in *Molecular Genetics of Plant-Microbe Interactions*, Verma, D. P. S. and Brisson, N., Eds., Martinus, Nijhoff, Dordrecht, 1987, 171.

74. **Bauer, W. G., Bhuvaneswari, T. V., Calvert, H. E., Law, I. J., Malik, N. S. A., and Vesper, S. J.**, Recognition and infection by slow-growing rhizobia, in *Nitrogen Fixation Research Progress*, Evans, H. J., Bottomley, P. J., and Newton, W. E., Eds., Martinus Nijhoff, Dordrecht, 1985, 247.

75. **Dazzo, F. B., Hollingsworth, R. I., Sherwood, J. E., Abe, M., Hrabak, E. M., Gardiol, A. E., Pankratz, H. S., Smith, K. B., and Yang, H.**, Recognition and infection of clover root hairs by *Rhizobium trifolii*, in *Nitrogen Fixation Research Progress*, Evans, H. J., Bottomley, P. J., and Newton, W. E., Eds., Martinus Nijhoff, Dordrecht, 239, 1985.

76. **Marvel, D. J., Torrey, J. G., and Ausubel, F. M.**, *Rhizobium* symbiotic genes required for nodulation of legume and nonlegume hosts, *Proc. Natl. Acad. Sci. U.S.A.*, 84, 1319, 1987.

77. **Scott, K. F., Saad, M., Price, G. D., Gresshoff, P. M., Kane, H., and Chua, K. Y.**, Conserved nodulation genes are obligatory for nonlegume nodulation, in *Molecular Genetics of Plant-Microbe Interactions*, Verma, D. P. S. and Brisson, N., Eds., Martinus Nijhoff, Dordrecht, 1987, 238.

78. **Scott, K. F. and Bender, G.**, The *Parasponia-Bradyrhizobium* symbiosis, in *The Molecular Biology of Symbiotic Nitrogen Fixation*, Gresshoff, P., Ed., CRC Press, Boca Raton, FL, 1989.

78A. **Better, M. and Helinski, D. R.**, Isolation and characterization of the *recA* gene of *Rhizobium meliloti*, *J. Bacteriol.*, 155, 311, 1983.

79. **Downie, J. A., Surin, B. P., Evans, I. J., Rossen, L., Firmin, J. L., Shearman, C. A., and Johnston, A. W. B.**, Nodulation genes of *Rhizobium leguminosarum*, in *Molecular Genetics of Plant-Microbe Interactions*, Verma, D. P. S. and Brisson, N., Eds., Martinus Nijhoff, Dordrecht, 1987, 225.

80. **Horvath, B., Bachem, C. W. B., Schell, J., and Kondorosi, A.**, Host-specific regulation of nodulation genes in *Rhizobium* is mediated by a plant-signal, interacting with the *nodD* gene product, *EMBO J.*, 6, 841, 1987.

81. **Downie, J. A., Hombrecher, G., Ma, Q.-S, Knight, C. D., Wells, B., and Johnston, A. W. B.**, Cloned nodulation genes of *Rhizobium leguminosarum* determine host-range specificity, *Mol. Gen. Genet.*, 190, 395, 1983.

82. **Bánfalvi, Z., Randhawa, G. S., Kondorosi, E., Kiss, A., and Kondorosi, A.**, Construction and characterization of R-prime plasmids carrying symbiotic genes of *R. meliloti*, *Mol. Gen. Genet.*, 189, 129, 1983.

83. **Appelbaum, E. R., McLoughlin, T. J., O'Connell, M., and Chartrain, N.**, Expression of symbiotic genes of *Rhizobium japonicum* USDA 191 in other rhizobia, *J. Bacteriol.*, 163, 385, 1985.

84. **Djordjevic, S. P., Chen, H., Batley, M., Redmond, J. W., and Rolfe, B. G.**, Nitrogen fixation ability of exopolysaccharide synthesis mutants of *Rhizobium* sp. strain NGR234 and *Rhizobium trifolii* is restored by the addition of homologous exopolysaccharides, *J. Bacteriol.*, 169, 53, 1987.

85. **Debellé, F. and Sharma, S. B.**, Nucleotide sequence of *Rhizobium meliloti* RCR2011 genes involved in host specificity of nodulation, *Nucleic Acids Res.*, 14, 7453, 1986.

86. **Ramakrishnan, N., Prakash, R. K., Shantharam, S., Duteau, N. M., and Atherly, A. G.**, Molecular cloning and expression of *Rhizobium fredii* USDA 193 nodulation genes: extension of host range for nodulation, *J. Bacteriol.*, 168, 1087, 1986.

87. **Hombrecher, G., Brewin, N. J., and Johnston, A. W. B.**, Cloning and mutagenesis of nodulation genes from *Rhizobium leguminosarum* TOM, a strain with extended host range, *Mol. Gen. Genet.*, 182, 767, 1984.

88. **Keyser, H. H., Bohlool, B. B., Hu, T. S., and Weber, D. F.**, Fast-growing rhizobia isolated from root nodules of soybean, *Science*, 215, 1631, 1982.

89. **Heron, D. S. and Pueppke, S. G.**, Mode of infection, nodulation specificity, and indigenous plasmids of 11 fast-growing *Rhizobium japonicum* strains, *J. Bacteriol.*, 160, 1061, 1984.

90. **John, M., Schmidt, J., Wieneke, U., Kondorosi, E., Kondorosi, A., and Schell, J.**, Expression of the nodulation gene *nod C* of *Rhizobium meliloti* in *Escherichia coli*: role of the *nod C* gene product in nodulation, *EMBO J.*, 4, 2425, 1985.

91. **Schmidt, J., John, M., Wieneke, U., Krüssmann, H.-D., and Schell, J.**, Expression of the nodulation gene *nodA* in *Rhizobium meliloti* and localization of the gene product in the cytosol, *Proc. Natl. Acad. Sci. U.S.A.*, 83, 9581, 1986.

92. **Innes, R. W., Kuempel, P. L., Plazinski, J., Canter-Cremers, H., Rolfe, B. G., and Djordjevic, M. A.**, Plant factors induce expression of nodulation and host range genes in *Rhizobium trifolii*, *Mol. Gen. Genet.*, 201, 420, 1985.

93. **Mulligan, J. T. and Long, S. R.**, Induction of *Rhizobium meliloti nodC* expression by plant exudate requires *nodD*, *Proc. Natl. Acad. Sci. U.S.A.*, 82, 6609, 1985.

94. **Rossen, L., Johnston, A. W. B., and Downie, J. A.**, The *nodD* gene of *Rhizobium leguminosarum* is autoregulatory and in the presence of plant exudates induces the *nod*A, B, C genes, *EMBO J.*, 4, 3369, 1985.

95. **Djordjevic, M. A., Redmond, J. W., Batley, M., and Rolfe, B. G.**, Clovers secrete specific phenolic compounds which either stimulate or repress *nod* gene expression in *Rhizobium trifolii*, *EMBO J.*, 6, 1173, 1987.

96. **Zaat, S. A. J., Wijffelman, C. A., Spaink, H. P., van Brussel, A. A. N., Okker, R. J. H., and Lugtenberg, B. J. J.**, Induction of the *nod*A promoter of *Rhizobium leguminosarum* Sym plasmid pRL1J1 by plant flavenones and flavones, *J. Bacteriol.*, 169, 198, 1987.

97. **Kosslak, R. M., Bookland, R., Barkei, J., Paaren, H., and Appelbaum, E. R.**, Induction of *Bradyrhizobium japonicum* common *nod* genes by isoflavanones isolated from *Glycine max*, *Proc. Natl. Acad. Sci. U.S.A.*, 84, 7428, 1987.

98. **Yelton, M. M., Mulligan, J. T., and Long, S. R.**, Expression of *Rhizobium meliloti nod* genes in *Rhizobium* and *Agrobacterium* backgrounds, *J. Bacteriol.*, 169, 3094, 1987.

99. **Egelhoff, T. T. and Long, S. R.**, *Rhizobium meliloti* nodulation genes: identification of *nodDABC* gene products, purification of *nodA* protein, and expression of *nodA* in *Rhizobium meliloti*, *J. Bacteriol.*, 164, 591, 1985.

100. **Fischer, R. F., Brierley, H. L., Mulligan, J. T., and Long, S. R.**, Transcription of *Rhizobium meliloti* nodulation genes, *J. Biol. Chem.*, 262, 6849, 1986.

101. **Rostas, K., Kondorosi, E., Horvath, B., Simoncsits, A., and Kondorosi, A.**, Conservation of extended promoter regions of nodulation genes in *Rhizobium*, *Proc. Natl. Acad. Sci. U.S.A.*, 83, 1757, 1986.

102. **Spaink, H. P., Okker, R. J. H., Wijffelman, C. A., Pees, E., and Lugtenberg, B. J. J.**, Regulation of the promoters in the nodulation region of the symbiosis plasmid pRL1JI of *Rhizobium leguminosarum*, in *Molecular Genetics of Plant-Microbe Interactions*, Verma, D. P. S. and Brisson, N., Eds., Martinus Nijhoff, Dordrecht, 1987, 244.

103. **Peters, N. K., Frost, J. W., and Long, S. R.**, A plant flavone, luteolin, induces expression of *Rhizobium meliloti* nodulation genes, *Science*, 233, 977, 1986.

104. **Redmond, J. W., Batley, M., Djordjevic, M. A., Innes, R. W., Kuempel, P. L., and Rolfe, B. G.**, Flavones induce expression of nodulation genes in *Rhizobium*, *Nature (London)*, 323, 632, 1986.

105. **Firmin, J. L., Wilson, K. E., Rossen, L., and Johnston, A. W. B.**, Flavonoid activation of nodulation genes in *Rhizobium* reversed by other compounds present in plants, *Nature (London)*, 324, 90, 1986.

106. **Dewick, P. M.**, Isoflavonoids, in *The Flavonoids: Advances in Research*, Harborne, J. B. and Mabry, T. J., Eds., Chapman and Hall, New York, 1982, 620.

107. **Ingham, J.**, Phytoalexins from the *Leguminosae*, in *Phytoalexins*, Bailey, J. A. and Mansfield, J. W., Eds., John Wiley & Sons, New York, 1982, 21.

108. **Werner, D., Mellor, R. B., Hahn, M. G., and Grisebach, H.**, Soybean root response to symbiotic infection glyceollin I accumulation in an ineffective type of soybean nodules with an early loss of the peribacteroid membrane, *Z. Naturforsch.*, 40c, 179, 1984.

108a. **Markham, K.**, *Techniques of Flavonoid Identification*, Academic Press, New York, 1982.

108b. **Pankhurst, C. E. and Biggs, D. R.**, Sensitivity of *Rhizobium* to selected isoflavonoids, *Can. J. Microbiol.*, 26, 542, 1980.

108c. **Spaink, H. P., Wijffelman, C. A., Pees, E., Okker, R. J. H., and Lugtenberg, B. J. J.**, *Rhizobium* nodulation gene *nodD* as a determinant of host specificity, *Nature (London)*, 328, 337, 1987.

109. **Albersheim, P. and Darvill, A. G.**, Oligosaccharins, *Sci. Am.*, 253, 58, 1985.

110. **Nap, J. P. and Bisseling, T.**, Nodulin function and nodulin gene regulation in root nodule development, in *The Molecular Biology of Symbiotic Nitrogen Fixation*, Gresshoff, P., Ed., CRC Press, Boca Raton, FL, 1989.

111. **Stachel, S. E., Messens, E., Van Montagu, M., and Zambryski, P. C.**, Identification of the signal molecules produced by wounded plant cells that activate T-DNA transfer in *Agrobacterium tumefaciens*, *Nature (London)*, 318, 624, 1985.

112. **Bolton, G. W., Nester, E. W., and Gordon, M. P.**, Plant phenolic compounds induce expression of the *Agrobacterium tumefaciens* loci needed for virulence, *Science*, 232, 983, 1986.

112a. **Murphy, P. J., Heycke, N., Banfalu, Z., Tate, M. E., de Brujin, F., Kondorosi, A., Tempé, J., and Schell, J.**, Genes for the catabolism and synthesis of an opine-like compound in *Rhizobium meliloti* are closely linked and on the Sym plasmid, *Proc. Natl. Acad. Sci. U.S.A.*, 84, 493, 1987.

112b. **Scott, B. D., Wilson, R., Shaw, G. J., Petit, A., and Tempé, J.**, Biosynthesis and degradation of nodule-specific *Rhizobium loti* compounds in *Lotus* nodules, *J. Bacteriol.*, 169, 278, 1987.

113. **Chakravorty, A. V., Zurkowski, W., Shine, J., and Rolfe, B. G.**, Symbiotic nitrogen fixation: molecular cloning of *Rhizobium* genes involved in exopolysaccharide synthesis and effective nodulation, *J. Mol. Appl. Genet.*, 1, 585, 1982.

114. **Chen, H., Batley, M., Redmond, J., and Rolfe, B. G.**, Alteration of the effective nodulation properties of a fast-growing broad host range *Rhizobium* due to changes in exopolysaccharide synthesis, *J. Plant Physiol.*, 120, 331, 1985.

115. **Sanders, R., Carlson, R. W., and Albersheim, P.,** A *Rhizobium* mutant incapable of nodulation and normal polysaccharide secretion, *Nature (London)*, 271, 240, 1978.

116. **Sanders, R., Raleigh, E., and Signer, E.,** Lack of correlation between extracellular polysaccharide and nodulation ability in *Rhizobium*, *Nature (London)*, 292, 241, 1981.

117. **Noel, K. D., Vandenbosch, K. A., and Kulpaca, B,** Mutations in *Rhizobium phaseoli* that lead to arrested development of infection threads, *J. Bacteriol.*, 168, 1392, 1986.

118. **Maier, R. J. and Brill, W. J.,** Involvement of *Rhizobium japonicum* O-antigen in soybean nodulation, *J. Bacteriol.*, 133, 1295, 1978.

119. **Cangelosi, G. A., Hung, L., Puvanesarajah, V., Stacey, G., Ozga, D. A., Leigh, J. A., and Nester, E. W.,** Common loci for *Agrobacterium tumefaciens* and *Rhizobium meliloti* exopolysaccharide synthesis and their roles in plant interactions, *J. Bacteriol.*, 169, 2086, 1987.

120. **Douglas, C. J., Halperin, W., and Nester, E. W.,** *Agrobacterium tumefaciens* mutants affected in attachment to plant cells, *J. Bacteriol.*, 152, 1265, 1982.

121. **Douglas, C. J., Staneloni, R. J., Rubin, R. A., and Nester, E. W.,** Identification and genetic analysis of an *Agrobacterium tumefaciens* chromosomal virulence region, *J. Bacteriol.*, 161, 850, 1985.

122. **Puvanesarajah, V., Schell, F. M., Stacey, G., Douglas, C. J., and Nester, E. W.,** A role for 2-linked-β-D-glucan in the virulence of *Agrobacterium tumefaciens*, *J. Bacteriol.*, 164, 102, 1985.

123. **Borthakur, D., Downie, J. A., Johnston, A. W. B., and Lamb, J. W.,** *psi*, a plasmid-linked *Rhizobium phaseoli* gene that inhibits exopolysaccharide production and which is required for symbiotic nitrogen fixation, *Mol. Gen. Genet.*, 200, 278, 1985.

124. **Borthakur, D., Barber, C. E., Lamb, J. W., Daniels, M. J., Downie, J. A., and Johston, A. W. B.,** A mutation that blocks exopolysaccharide synthesis prevents nodulation of peas by *Rhizobium leguminosarum* but not of beans by *R. phaseoli* and is corrected by cloned DNA from *Rhizobium* or the phytopathogen *Xanthomonas*, *Mol. Gen. Genet.*, 203, 320, 1986.

125. **Borthakur, D. and Johnston, A. W. B.,** Sequence of *psi*, a gene on the symbiotic plasmid of *Rhizobium phaseoli* which inhibits exopolysaccharide synthesis and nodulation and demonstration that its transcription is inhibited by *psr*, another gene on the symbiotic plasmid, *Mol. Gen. Genet.*, 207, 149, 1987.

126. **Noel, K. D., Pachori, P., Kulpaca, B., Vandenbosch, K. A., Brink, B. A., and Cava, J. R.,** *Rhizobium* mutants defective in lipopolysaccharide and infection, in *Molecular Genetics of Plant-Microbe Interactions*, Verma, D. P. S. and Brisson, N., Eds., Marinus, Nijhoff, Dordrecht, 1987, 167.

127. **Stacey, G., Pocratsky, L. A., and Puvanesarajah, V.,** Bacteriophage that can distinguish between wild-type *Rhizobium japonicum* and a non-nodulating mutant, *Appl. Environ. Microbiol.*, 48, 68, 1984.

128. **Bishop, P. E., Dazzo, F. B., Appelbaum, E. R., Maier, R. J., and Brill, W. J.,** Intergeneric transfer of genes involved in the *Rhizobium*-legume symbiosis, *Science*, 198, 938, 1977.

129. **Maier, R. J., Bishop, P. E., and Brill, W. J.,** Transfer from *Rhizobium japonicum* to *Azotobacter, vinelandii* of genes required for nodulation, *J. Bacteriol.*, 134, 1199, 1978.

130. **Dereylo, M., Skorupska, A., Bednara, J., and Lorkiewicz, Z.,** *Rhizobium trifolii* mutants deficient in exopolysaccharide production, *Physiol. Plant*, 66, 699, 1986.

131. **Geremia, R. A., Cavaignac, S., Zorreguieta, A., Toro, N., Olivares, J., and Ugalde, R. A.,** A *Rhizobium meliloti* mutant that forms ineffective pseudonodules in alfalfa produces exopolysaccharide but fail to form β-(1→2) Glucan, *J. Bacteriol.*, 169, 880, 1987.

132. **Ugalde, R. A., Handelsman, J., and Brill, W. J.,** Role of galactosyltransferase activity in phage sensitivity and nodulation competitiveness of *Rhizobium meliloti*, *J. Bacteriol.*, 166, 148, 1986.

133. **Roberts, G. P., MacNeil, T., MacNeil, D., and Brill, W. J.,** Regulation and characterization of protein products coded by the *nif* (nitrogen fixation) genes of *Klebsiella pneumoniae*, *J. Bacteriol.*, 136, 267, 1978.

134. **Ausubel, F. M.,** Regulation of nitrogen fixation genes, *Cell*, 37, 5, 1984.

135. **Dixon, R. A.,** Genetic complexity of nitrogen fixation, *J. Gen. Microbiol.*, 130, 2745, 1984.

136. **Roberts, G. P. and Brill, W. J.,** Genetics and regulation of nitrogen fixation, *Annu. Rev. Microbiol.*, 35, 207, 1981.

137. **Brill, W. J.,** Biochemical genetics of nitrogen fixation, *Microbiol. Rev.*, 44, 449, 1980.

138. **Hawkes, T. R., McLean, P. A., and Smith, B. E.,** Nitrogenase from *nifV* mutants of *Klebsiella pneumoniae* contains an altered form of the iron-molybdenum cofactor, *Biochem. J.*, 217, 317, 1984.

139. **Roberts, G. P. and Brill, W. J.,** Gene-product relationship of the *nif* regulon of *Klebsiella pneumoniae*, *J. Bacteriol.*, 144, 210, 1980.

140. **Imperial, J., Ugalde, R. A., Shah, V. K., and Brill, W. J.,** Role of the *nifQ* gene product in the incorporation of molybdenum into nitrogenase in *Klebsiella pneumoniae*, *J. Bacteriol.*, 158, 187, 1984.

141. **Ruvkun, G. B., Sundaresan, V., and Ausubel, F. M.,** Directed transposon Tn5 mutagenesis and complementation analysis of *Rhizobium meliloti* symbiotic nitrogen fixation genes, *Cell*, 29, 551, 1982.

142. **Corbin, D., Barran, L, and Ditta, G.,** Organization and expression of *Rhizobium meliloti* nitrogen fixation genes, *Proc. Natl. Acad. Sci. U.S.A.*, 80, 3005, 1983.

143. **Schetgens, T. M. P., Bakkeren, G., van Dun, C., Hontelez, J. G. J., van den Box, R. C., and van Kammen, A.**, Molecular cloning and functional characterization of *Rhizobium leguminosarum* structural *nif* genes by site-directed transposon mutagenesis and expression in *Escherichia coli* minicells, *J. Mol. Appl. Genet.*, 2, 406, 1984.

144. **Scott, K. F., Rolfe, B. G., and Shine, J.**, Biological nitrogen fixation: primary structure of the *Rhizobium trifolii* iron protein gene, *DNA*, 2, 149, 1983.

145. **Quinto, C., De La Vega, H., Flores, M., Leemans, J., Cevallos, M. A., Pardo, M. A., Azpiroz, R., De Lourdes Girard, M., Calva, E., and Palacios, R.**, Nitrogenase reductase: a functional multigene family in *Rhizobium phaseoli*, *Proc. Natl. Acad. Sci. U.S.A.*, 82, 1170, 1985.

146. **Better, M., Lewis, B., Corbin, D., Ditta, G., and Helinski, D. R.**, Structural relationship among *Rhizobium meliloti* symbiotic promoters, *Cell*, 35, 479, 1983.

147. **Scott, K. F., Rolfe, B. G., and Shine, J.**, Nitrogenase structural genes are unlinked in the nonlegume symbiont *Parasponia Rhizobium*, *DNA*, 2, 141, 1983.

148. **Adams, T. H., McClung, C. R., and Chelm, B. K.**, Physical organization of the *Bradyrhizobium japonicum* nitrogenase gene region, *J. Bacteriol.*, 159, 857, 1984.

149. **Fischer, H.-M. and Hennecke, H.**, Linkage map of the *Rhizobium japonicum nifH* and *nifDK* operons encoding the polypeptides of the nitrogenase enzyme complex, *Mol. Gen. Genet.*, 196, 537, 1984.

150. **Hennecke, H., Fischer, H.-M., Ebeling, S., Gubler, M., Thöny, B., Göttfert, M., Lamb, J., Hahn, M., Ramseier, T., Regensburger, B., Alvarez-Morales, A., and Studer, D.**, Nif, Fix and Nod gene clusters in *Bradyrhizobium japonicum*, and *Nifa*-mediated control of symbiotic nitrogen fixation, in *Molecular Genetics of Plant-Microbe Interactions*, Verma, D. P. S. and Brisson, N., Eds., Martinus, Nijhoff, Dordrecht, 1987, 191.

151. **Fuhrmann, M., Fischer, H.-M., and Hennecke, H.**, Mapping of *Rhizobium japonicum nifB-*, *fixBC-* and *fixA*-like genes and identification of the *fixA* promoter, *Mol. Gen. Genet.*, 199, 315, 1985.

152. **Ebeling, S., Hahn, M., Fischer, H.-M., and Hennecke, H.**, Identifcation of *nifE-*, *nifN-* and *nifS*-like genes in *Bradyrhizobium japonicum*, *Mol. Gen. Genet.*, 207, 503, 1987.

153. **Fischer, H.-M., Alvarez-Morales, A., and Hennecke, H.**, The pleiotropic nature of symbiotic regulatory mutants: *Bradyrhizobium japonicum nifA* gene is involved in control of *nif* gene expression and formation of determinate symbiosis, *EMBO J.*, 5, 1165, 1986.

154. **Buikema, W. J., Klingensmith, J. A., Gibbons, S. L., and Ausubel, F. M.**, Conservation of structure and location of *Rhizobium meliloti* and *Klebsiella pneumoniae nifB* genes, *J. Bacteriol.*, 169, 1120, 1987.

155. **Szeto, W. W., Zimmerman, J. L., Sundaresan, V., and Ausubel, F. M.**, A *Rhizobium meliloti* symbiotic regulatory gene, *Cell*, 36, 1035, 1984.

156. **Rossen, L., Ma, Q.-S., Mudd, E. A., Johnston, A. W. B., and Downie, J. A.**, Identification of DNA sequence of *fixZ*, a *nifB*-like gene from *Rhizobium leguminosarum*, *Nucleic Acids Res.*, 12, 7123, 1984.

157. **Hontelez, J., Lankhorst, R. K., Jansma, J.-D., Jacobsen, E., van den Bos, R. C., and van Kammen, A.**, Characterization of symbiotic genes and regulation of their expression in *Rhizobium leguminosarum* PRE, in *Molecular Genetics of Plant-Microbe Interactions*, Verma, D. P. S. and Brisson, N., Eds., Martinus Nijhoff, Dordrecht, 1987, 241.

158. **Earl, C. D., Ronson, C. W., and Ausubel, F. M.**, Genetic and structural analysis of the *Rhizobium meliloti fixA*, *fixB*, *fixC*, and *fixX* genes, *J. Bacteriol.*, 169, 1127, 1987.

159. **Kahn, D., Batut, J., Boistard, P., Daveran, M. L., David, M., Domergue, O., Garnerone, A. M., Ghai, J., Hertig, C., Infante, D., and Renalier, M. H.**, Molecular analysis of a *fix* cluster from *Rhizobium meliloti*, in *Molecular Genetics of Plant-Microbe Interactions*, Verma, D.P. S. and Brisson, N., Eds., Martinus Nijhoff, 1987, 258.

160. **Forrai, T., Vincze, E., Banfalvi, Z., Kiss, G. B., Randhawa, G. S., and Kondorosi, A.**, Localization of symbiotic mutations in *Rhizobium meliloti*, *J. Bacteriol.*, 153, 635, 1983.

161. **Stacey, G., Paau, A. S., Noel, K. D., Maier, R. J., Silver, L. E., and Brill, W. J.**, Mutants of *Rhizobium japonicum* defective in nodulation, *Arch. Microbiol.*, 132, 219, 1982.

162. **Regensburger, B., Meyer, L., Filser, M., Weber, J., Studer, D., Lamb, J. W., Fischer, H.-M., Hahn, M., and Hennecke, H.**, *Bradyrhizobium japonicum* mutants defective in root-nodule bacteroid development and nitrogen fixation, *Arch. Microbiol.*, 144, 355, 1986.

163. **Hirsch, A. M., Bang, M., and Ausubel, F. M.**, Ultrastructural analysis of ineffective alfalfa nodules formed by *nif*::Tn5 mutants of *Rhizobium meliloti*, *J. Bacteriol.*, 155, 367, 1983.

164. **Hirsch, A. M. and Smith, C. A.**, Effects of *Rhizobium meliloti nif* and *fix* mutants of alfalfa root nodule development, *J. Bacteriol.*, 169, 1137, 1987.

165. **Gussin, G. N., Ronson, C. W., and Ausubel, F. M.**, Regulation of nitrogen fixation genes, *Annu. Rev. Genet.*, 20, 567, 1986.

166. **Ditta, G., Virts, E., Palomares, A., and Kim, C.-H.**, The *nifA* gene of *Rhizobium meliloti* is oxygen regulated, *J. Bacteriol.*, 169, 3217, 1987.

167. **Hennecke, H., Fischer, H.-M., Ebeling, S., Gubler, M., Thöny, B., Göttfert, M., Lamb, J., Hahn, M., Ramseier, T., Regensburger, B., Alvarez-Morales, A., and Studer, D.,** *nif, fix,* and *nod* gene clusters in *Bradyrhizobium japonicum,* and *nifa*-mediated control of symbiotic nitrogen fixation, in *Molecular Genetics of Plant Microbe Interactions,* Verma, D. P. S. and Brisson, N., Eds., Martinus Nijhoff, Dordrecht, 1987, 191.

168. **Kamicker, B. J. and Brill, W. J.,** Identification of *Bradyrhizobium japonicum* nodule isolates from Wisconsin soybean farms, *Appl. Environ. Microbiol.,* 51, 487, 1986.

169. **Hua, S. S. T., Scott, D. B., and Lim, S. T.,** *Genetic Engineering of Symbiotic Nitrogen Fixation,* Plenum Press, New York, 1981, 95.

170. **Williams, L. E. and Phillips, D. A.,** Increased soybean productivity with a *Rhizobium japonicum* mutant, *Crop Sci.,* 23, 246, 1983.

171. **Maier, R. J. and Brill, W. J.,** Mutant strains on *Rhizobium japonicum* with increased ability to fix nitrogen for soybeans, *Science,* 201, 448, 1978.

172. **Ezzell, C.,** EPA clears the way for release of nitrogen-fixing microbe, *Nature (London),* 327, 90, 1987.

173. **Lambert, G. R., Cantrell, M. A., Hanus, F. J., Russell, S. A., Haddad, K. R., and Evans, H. J.,** Intra- and interspecies transfer and expression of *Rhizobium japonicum* hydrogen uptake genes and autotrophic growth capability, *Proc. Natl. Acad. Sci. U.S.A.,* 82, 3232, 1985.

174. **Hayes, W.,** *The Genetics of Bacteria and Their Viruses,* Blackwell Scientific, Oxford, 1968, 91.

175. **Ronson, C. W., Astwood, P. M., and Downie, J. A.,** Molecular cloning and genetic organization of C4-dicarboxylate transport genes from *Rhizobium leguminosarum, J. Bacteriol.,* 160, 903, 1984.

176. **Ronson, C. W. and Astwood, P. M.,** Genes involved in the carbon metabolism of bacteroids, in *Nitrogen Fixation Research Progress,* Martinus Nijhoff, Dordrecht, The Netherlands, 1985, 201.

177. **Dowling, D. N., Samrey, U., Stanley, J., and Broughton, W. J.,** Cloning of *Rhizobium leguminosarum* genes for competitive nodulation blocking on peas, *J. Bacteriol.,* 169, 1345, 1987.

178. **Pankhurst, C. E., MacDonald, P. E., and Reeves, J. M.,** Enhanced nitrogen fixation and competitiveness for nodulation of *Lotus pedunculatus* by a plasmid-cured derivative of *Rhizobium loti, J. Gen. Microbiol.,* 132, 2321, 1986.

179. **Brewin, N. J., Wood, E. A., and Young, J. P. W.,** Contribution of the symbiotic plasmid to competitiveness of *Rhizobium leguminosarum, J. Gen. Microbiol.,* 129, 2973, 1983.

180. **McLoughlin, T. J., Merlo, A. O., Satola, S. W., nd Johansen, E.,** Isolation of competition-defective mutants of *Rhizobium fredii, J. Bacteriol.,* 169, 410, 1987.

181. **Handelsman, J., Ugalde, R. A., and Brill, W. J.,** *Rhizobium meliloti* competitiveness and the alfalfa agglutinin, *J. Bacteriol.,* 157, 703, 1985.

Chapter 7

NITRATE INHIBITION OF NODULATION IN LEGUMES

Bernard J. Carroll and Anne Mathews

TABLE OF CONTENTS

I. INTRODUCTION

Nitrate is usually the principle form of mineral nitrogen available to plants in the soil.[1,2] Many legume species also have direct access to biologically fixed nitrogen, but preferentially utilize nitrate rather than develop a root nodule symbiosis with *Rhizobium* or *Bradyrhizobium*.[3-6] Many other edaphic factors such as pH, nutrient deficiencies and toxicities, water, and temperature affect nodulation,[7] but nitrate is unique in that it is generally not inhibitory to plant growth. It is, therefore, interesting to speculate on the evolutionary significance of nitrate inhibition of nodulation. Estimates of energy costs are generally greater for nitrogen fixation than for nitrate assimilation.[8-11] However, provided a small amount of combined nitrogen is available, the yield of some legumes, such as soybean, is not related to the proportion of nitrogen derived from the soil[12,13] and is optimal when nitrogen is obtained from both nitrate and symbiotic nitrogen fixation.[14] Harper[11] suggested that the difference in energy costs between nitrate utilization and nitrogen fixation were insufficient to influence soybean yield under field conditions. There are other developmental and ecological considerations which may have resulted in natural selection for nitrate inhibition of nodulation. Nitrate can be assimilated in either, or both, the root and shoot tissue of the plant,[2,15] whereas nitrogen fixation requires the development of a specialized organ, the root or stem nodule.[16] Thus, there is developmental restriction on nitrogen fixation in that it can only occur after a nodule has been formed. In young white clover seedlings, for example, maximum activities of nitrate reductase (the principal enzyme in the nitrate assimilation pathway) precedes the highest rates of nitrogenase activity by a matter of weeks.[17] Indeed, plants that are dependent on nitrogen fixation as the sole nitrogen source do not grow as well as those which are supplemented with low noninhibitory or larger levels of nitrate.[14] In the ecological context, it can be assumed that nitrate utilization by legumes decreases the amount of soil nitrate available to adjacent nonsymbiotic plants that are competing for other nutrients. Thus, preferential utilization of nitrate may be advantageous for legume species by decreasing the competitive ability of other plants that are unable to form a nitrogen-fixing symbiosis. In monoculture, however, this proposed competitive advantage is negated, and inhibition of nitrogen fixation by soil nitrate increases the need for nitrogen fertilizers in crop rotation systems involving legumes.

All the stages of symbiotic development that have been investigated are inhibited by nitrate; that is, root hair formation,[4,18] attachment of rhizobia to the root,[19] root hair curling,[4,18] infection thread formation,[18,20,21] and the level of immunologically detectable lectin on the root surface.[19,22] Host-derived lectins are considered to be important in the attachment of the microsymbiont to the root[23,24] and/or in conditioning the microsymbiont to nodulate its host.[25-27] Additionally, nitrate delays the appearance of nodules,[28] inhibits the number of nodules that are formed, nodule development and specific nitrogenase activity (activity per unit of nodule mass), and it induces premature nodule senescence.[17,29-33] As far as nodule formation in soybean is concerned, the initial stages in the infection process[34] appear to be the most sensitive steps. Inhibition is substantially alleviated by delaying exposure to nitrate for 18 h after the time of inoculation.[21] The first instances of root hair curling and subepidermal cortical cell divisions in soybean occur within this time after inoculation.[34] The degree of inhibition of nodule formation is also seen in the number of actual infections (centers of subepidermal cell division with infection threads) and pseudoinfections (centers without infection threads) that are formed in the presence of nitrate.[21]

The carbohydrate deprivation hypothesis has been put forward as an explanation for nitrate inhibition and argues that nitrate assimilation decreases the amount of reducing equivalents available to symbiotic processes.[30,35] However, this hypothesis lacks precise definition, particularly for the early stages of symbiosis and nodule ontogeny. It has been demonstrated that exposure of nitrogen-fixing plants to nitrate results in less photosynthate being translocated to the nodule,[36-38] but this may be explained by a loss in the sink capacity of the nodules rather than by a lowered level of reducing equivalents being available in the plant. The metabolites from nitrate assimilation, particularly nitrite, have also been implicated as mediators of nitrate inhibition. Nitrite, the product of nitrate reductase (NR) activity and the first intermediate in the nitrate assimilation pathway, has been postulated to destroy hormones that may be involved in nodulation.[39] It has also been demonstrated to inactivate nitrogenase[40] and convert leghemoglobin to an inactive form.[41]

The emphasis of this review will be on the recent advances in the understanding of the effect of nitrate on nodule ontogeny and its relevance to nodulation in general. The effects of nitrate on nitrogenase activity will also be briefly considered. Both plant and bacterial attributes have been studied extensively, but the most fruitful research has been undertaken on the host domain of the symbiosis. Much of the discussion is speculative, in the hope that it will stimulate further research. The potential for enhancing nodulation, nitrogen fixation, and yield in legumes will also be addressed.

II. NITRATE INHIBITION OF NODULE FORMATION IS A LOCALIZED EFFECT

Wilson[3] showed that nodule formation in soybean was only inhibited on those root parts that were in direct contact with nitrate, and he concluded that the effect of nitrate was localized. Since then, the localized effect of nitrate on nodulation has been demonstrated in several legume species using split-root systems, where half of the root system is exposed to inhibitory levels of nitrate and the other half is not.[17,42,43] Nodule development is affected on the nitrate-free portion of the root, provided the level of nitrate is sufficiently high on the other half of the root.[17,43] However, this phenomenon may have been due to enhanced metabolite diversion to the nitrate portion of the root, which was generally larger than the nitrate-free side. Root growth does respond to nutrient supply,[44] and this is an important consideration when interpreting data from split-root culture systems. By altering the placement of combined nitrogen in the growth medium, Harper and Cooper[45] showed that inhibition of soybean nodule development by ammonium nitrate is dependent on the combined nitrogen being in the direct vicinity of the root. It is, therefore, likely that nitrate inhibition of nodule development, as well as nodule formation, is at least in part a localized phenomenon. In a white clover split-root system it has been demonstrated that nitrate inhibition of nitrogenase activity is systemic, but as mentioned above, this may have been due to preferential allocation of assimilates to the nitrate moiety rather than to carbohydrate deprivation of the nodules per se.[17]

Further support for a localized effect of nitrate on nodule formation was reported by Gibson and Harper.[28] Soybean plants were grown in water culture at various nitrate levels, and the external concentration of nitrate, rather than the rate of nitrate uptake, determined the degree of inhibition of nodule appearance as monitored by the time required for the plants to nodulate. Clearly, nitrate itself has a regulatory role in nodule appearance, and this will be discussed in more detail later in this review. Contrary to this conclusion, Malik et al.[21] suggested that nitrate may not directly inhibit infection initiation, since the efficiency of nodule formation was still inhibited when separate portions of a soybean root were exposed to nitrate and *Bradyrhizobium*. In these experiments,[21] plastic growth pouches containing a paper towel were modified in such a manner that the roots were exposed to nitrate either above or below the inoculated zone. There is a problem in interpreting their data in that the root portion exposed to nitrate may be responding to the nutrient supply, thereby diverting resources away from the inoculated region of the root (as referred to above for split-root systems). Another problem with the pouch method in general for studying nitrate effects, is that all surfaces of the nitrate-exposed root portions may not be subjected to identical nitrate concentrations. For example, those surfaces in direct contact with the paper towel (reservoir) may have a more constant, and therefore higher supply of nitrate.

III. THE MICROSYMBIONT AND NITRATE INHIBITION OF NODULATION AND NITROGEN FIXATION

Several investigations have been directed toward assessing variability among *Bradyrhizobium* or *Rhizobium* strains for nodulation tolerance to nitrate. While variation does exist between strains of *R. meliloti*,[46] *R. leguminosarum*,[47] and *B. japonicum*[28,48] at low and medium levels of nitrate, no striking differences have been reported at higher levels of nitrate that severely limit nodulation on the host plant. Furthermore, attempts to select nitrate-tolerant variants of *B. japonicum* by successive inoculation and reisolation of nodule occupants in soybean plants grown on nitrate did not prove fruitful.[48]

The capacity of the microsymbiont to assimilate nitrate has no bearing on its ability to nodulate its host in the presence of nitrate. Mutants that lack nitrate reductase activity and/ or the ability to grow on nitrate as the major nitrogen source have been isolated in the cowpea *Bradyrhizobium* strain 32H1, *R. trifolii* strain TA1,[30] and *B. japonicum* strain USDA 110,[49] but all of these variants when used as inoculant on host plants respond in the same fashion as the wild-type parental strains.[30,50-53] This observation applies to both nitrate inhibition of nodulation and nitrogenase activity. While inhibition of nitrogenase (acetylene reduction) activity by isolated soybean bacteroids is dependent on the bacteroids possessing NR activity,[54] recently it has been reported that nitrate does not enter the infected region of the nodule[2] where nitrite has been postulated to interfere with leghemoglobin[41] and nitrogenase.[40] Clearly, NR activity in the microsymbiont is not important in nitrate inhibition of the symbiosis. Even when NR-deficient or wild-type strains of *B. japonicum* are inoculated onto nitrate-tolerant soybean mutants,[55-57] no strain-dependent differences were observed for nodulation in the presence of nitrate.[53] These soybean mutants, described in detail later, nodulate profusely in the presence of nitrate, and the lack of an effect of bacterial NR activity in association with these deregulated hosts suggests that the potential effect of bacterial nitrate metabolism on nodulation of the wild-type host is not masked by other regulatory phenomena.[53]

A slight, but significant, bacterial effect on the regulation of nodulation by nitrate is the observation that high dose inoculation partially alleviates the degree of inhibition in soybean.[58,59] This effect was not dependent on the ability of the inoculant to utilize nitrate in the rhizosphere, which may have been a plausible explanation.[59] The effect of nitrate on colonization has not been reported, but the response to high levels of inoculum may be partly counteracting an inhibitory effect of nitrate on the proliferation of rhizobia around the roots, and nitrate-tolerant soybean mutants[56] do have higher numbers of rhizobia in the rhizosphere.[60,176] In analogy to the nitrate effect, nonnodulating soybean mutants do occasionally nodulate if the inoculant dose is sufficiently high.[61,62] La Favre and Eaglesham[61] suggested that nodulation of nonnodulating rj_1 soybeans at high *B. japonicum* cell numbers may be due to the enhanced production of a bacterial by-product(s) required to induce nodule formation. This is also a credible hypothesis for explaining the response of wild-type soybeans to increased rates of inoculant in the presence of inhibitory nitrate.

The research that has concentrated on the microsymbiont has been useful in discounting the role of bacterial nitrate metabolism in the inhibition of nodulation. While slight bacterial effects have been observed, these investigations have implied a major role of the legume host in the regulation of nodulation and nitrogen fixation in the presence of nitrate.

IV. THE LEGUME HOST AND NITRATE INHIBITION OF NODULATION

A. VARIATION BETWEEN LEGUME SPECIES, CULTIVARS, AND ACCESSIONS

Although nitrate inhibition of nodulation is a common phenomenon among legume species, there is considerable variation in the degree of regulation between different plant-bacteria combinations. Harper and Gibson[63] reported more severe effects of nitrate on nodule formation in barrel medic, soybean, and lablab bean than in siratro, lupin, chick-pea, pea, and subterranean clover. These authors employed a water culture system for the comparisons. Differences have also been noted between species within the *Trifolium* genus.[63,64] Nodulation studies using the plate method[17,65] have demonstrated that *T. subterranean* is more sensitive than *T. repens*, and both of these clover species are more inhibited than *T. dubium*.[64] In the comparison of eight legume species mentioned earlier, there was no obvious correlation between sensitivity and nitrate uptake, NR activity, or the shoot vs. root location of NR activity.[63]

Natural variants with increased nodulation have been identified in several legume species,[66-72] but many of these were not selected specifically for nitrate-tolerant nodulation. Specific variation in nitrate tolerance between cultivars and accessions has been reported in soybean.[28,73-75] During early development, soybean cultivars Elf and Avoyelles when inoculated with *B. japonicum* strain USDA 110 displayed a higher degree of nitrate tolerant nodulation than did ten other cultivars.[28] In field studies, Hardarson et al.[73] showed that nitrogen fixation in cultivar Dunadja was less sensitive to higher rates of N fertilizer application than in several other cultivars. More recently, Herridge and Betts[74,75] screened 489 soybean genotypes (cultivars or geographical accessions) in sand pots for differences in nitrate tolerance, and 32 of the most promising lines were later screened in a high-nitrate soil. Several superior accessions of Korean origin were identified[75] as having up to 17 times more nodules, 15 times greater nodule mass, and higher rates of nitrogen fixation than the commercial cultivar Bragg in a high nitrate soil.[74] Based on the relative ureide technique to estimate nitrogen fixation,[76-78] the Korean lines had up to 20 times the level of nitrogen fixation measured in Bragg.[74] While nitrate tolerance in these Korean lines is associated with a higher nodule number per plant,[74] the mechanism of the pronounced nitrate tolerance in these lines has not been determined. Increased nodulation was not observed in the absence of nitrate,[75] and no significant differences in the location and activity of NR were detected between cultivar Bragg and two of the Korean genotypes.[75,177]

The distinct variation between legume species and cultivars or accessions within a species and the lack of any clear relationship to nitrate metabolism indicates that the mechanism of nitrate inhibition of nodulation may be multifaceted.[63] The considerable diversity between the genomes of legume species and cultivars emphasized the need for genetic analysis and isolation of host mutants that are specifically altered in either nitrate metabolism or the regulation of nodulation by nitrate.

B. NITRATE REDUCTASE (NR) MUTANTS

NR-deficient mutants have been isolated in pea[79,80] and soybean.[81-83] Inducible NR activity in pea mutant E_1 is about 20% of the parent cultivar Rondo, and when nitrate serves as the sole source of mineral nitrogen, E_1 grows poorly and accumulates nitrate.[80] In the presence or absence of nitrate, mutant E_1 has the same nodule number and nodule mass as the wild type.[84] Jacobsen[84] concluded that the nitrate concentration in the plant and/or in the growth medium was responsible for inhibition of nodulation, rather than the ability of the plants to assimilate nitrate.

NR in soybean is more complicated and there is noninducible or constitutive (cNR) as well as inducible (iNR) activities.[11,14,85,86] cNR is expressed in developmentally young leaf tissue, independent of nitrate being present, whereas iNR is found in root and shoot tissue provided the plants are grown with nitrate.[87] There are two constitutive NRs, c_1 and c_2, and a single iNR.[11] Mutants have been identified that lack both forms of cNR activity; namely, nr_1[87] and NR345.[82] Both of these are normal in iNR activity,[81,88] as are LNR-5 and LNR-6, which lack only c_2NR.[83] NR328 is a leaky cNR mutant[82] and is unique from the other soybean mutants in that it is partly defective in iNR activity.[88] For this reason, it grows poorly on high levels of nitrate and develops necrosis on the distal margins of leaves, probably due to excessive nitrate accumulation.[88] Mutant nr_1 (formerly LNR-2), which is totally deficient in cNR, maintains its ability to utilize nitrate, thus indicating that sufficient nitrate can be assimilated through iNR alone.[89] Nodulation in nr_1, NR345, and NR328 is equally inhibited by nitrate as compared to the wild type.[82,89]

The nodulation response of pea mutant E_1[84] and soybean mutant NR328[53,82,88] supports the theory that nitrate itself has a major regulatory role, independent of its assimilation. These genetic studies substantiate the observed correlation between the degree of inhibition and the nitrate level in the growth medium, rather than the rate of nitrate uptake.[28] It has been reported that azide causes the accumulation of labeled sulfate in the region exterior to the endodermis of maize roots, thus indicating that the site of active uptake of sulfate by the root is probably in the endodermal cells interior to the root cortex.[90] Should this be the case for nitrate uptake, which is also considered to be an active process,[91] it would mean that early events in nodule ontogeny that occur in the cortex, root hairs, and rhizosphere are external to the site of active uptake. In these distal regions of the root where nodule initiation occurs, the nitrate concentration is probably directly related to the levels exterior to the root. Such an explanation is consistent with the observed correlation between inhibition and the nitrate level in the nutrient media and the response of NR-deficient legume mutants. This possible compartmentalization in the root is of relevance to the effect of other nutrients on nodule initiation in general and would be an interesting area of investigation. The endodermal cells may also act as a regulatory barrier governing the exudation of host signals to the microsymbiont that are required for nodule formation (e.g., host inducers for *Rhizobium* or *Bradyrhizobium nod* gene expression; see Chapter 6).

C. NITRATE-TOLERANT NODULATION MUTANTS

While nonnodulation variants exist in several legume species,[92-103] mutants that nodulate profusely in the presence of nitrate have only been identified in mutagenized populations of pea[84,104,105] soybean,[55-57,106] and bean.[184] These mutants nodulate more than their parent cultivars in the presence and absence of nitrate and therefore have been termed nts (*nitrate tolerant symbiosis*) and supernodulating.[55,57] The pea mutant nod₃ and soybean mutant nts382[56,57,107,108] are not defective in their ability to utilize nitrate, but have lost the ability to control the number of nodules that are formed. In soybean, an additional 14 nts mutants were isolated from cv Bragg,[56] and all of these are supernodulating and have normal nitrate metabolism as assessed by NR activity.[107,178,183] There are both intermediate and extreme supernodulators in soybean, and these two classifications are illustrated in Figures 1 and 2. Mutant nts1116 (Figure 1; Table 1) is the sole representative of the former category whereas nts382 (Figure 2; Table 1) and the other selections are extreme supernodulators.[56,183] The symbiotic characteristics of nts1116 and nts382 in the presence and absence of nitrate are demonstrated in Table 1. The salient features are that nts1116 resembles the parent cv Bragg more closely than the extreme supernodulators, such as nts382, and nitrate accentuates the discrepancy in nodulation and nitrogenase activity observed between the mutant lines and the wild type. Depending on the inoculant dose, which affects the degree of supernodulation,[110] the extreme mutants generally have lower rates of specific nitrogenase activity and lower plant dry weights, particularly in the absence of combined nitrogen.[56,108] The differ-

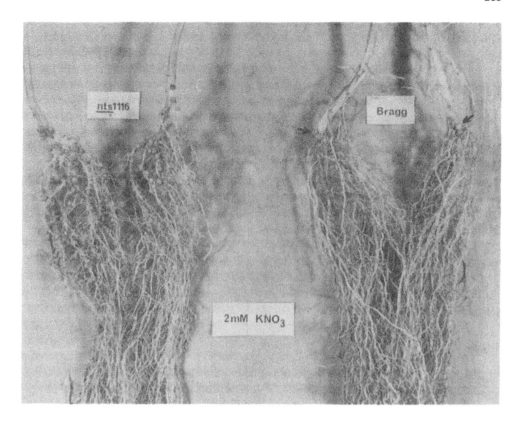

FIGURE 1. The nodulation characteristics of parent cv Bragg and the intermediate supernodulator nts1116 after culture on nutrients supplemented with 2 mM KNO$_3$. Nodule formation on Bragg plants is largely restricted to the upper region of the tap root (see arrows), whereas nodulation occurs at a greater frequency in nts1116 on both the tap and lateral roots. Plants were grown as described by Carroll et al.[56] and harvested 45 d after planting and inoculation with *B. japonicum* CB1809 (= USDA136) (2 × 10^8 cells per pot).

ential growth of wild type and extreme supernodulators is largely due to the enhanced nodulation, or another feature of the symbiotic interaction, since uninoculated Bragg and nts382 grow at similar rates during early development when nitrate is the sole source of nitrogen.[108] Under greenhouse or growth-cabinet conditions, the growth rates of nts1116 are similar to the parent cv Bragg, but it is often slightly smaller during early development. While extreme supernodulation is presently associated with reduced growth, a thorough assessment of the effect of these mutations on plant vigor cannot be made until back-crossed material is available. This material would be devoid of other superfluous mutations that exist in the original selections.

Pea mutant nod$_3$[84] and soybean mutant nts382[108] are also characterized by increased lateral root formation, thus highlighting a pleiotrophic effect of these mutations and a likely hormonal imbalance in these plants. Reciprocal grafts between mutants and the wild type demonstrated that supernodulation is largely controlled by the shoot in soybean[60,109,111] as is the case in the pea mutant isolated by Messager.[105] In contrast, supernodulation in pea mutant nod$_3$ is controlled by the root tissue.[112,113] Thus, the research on the pea mutants suggests that anomalies in gene expression in either the shoot or the root can result in supernodulation. Enhanced nodulation in all of the soybean mutants was determined by the shoot,[111] and this may reflect a difference in the control of nodulation between pea and soybean. Alternatively, the genetic constitution of soybean may have made it impossible to detect root-expressed anomalies that effect the control of nodulation. Soybean is generally

FIGURE 2.   The nodulation characteristics of wild-type Bragg and the extreme supernodulator nts382 after culture on 2.75 mM KNO$_3$. Plants were cultured as outlined by Carroll et al.[56,57] and the pots received 10^8 cells of *B. japonicum* CB1809 (= USDA136) at day 0 and again 4 d after planting. The plants were harvested 28 d after planting. Under these conditions prolific nodulation occurs all over the roots in nts382, such that nodules often occupy entire portions of the root tissue. Note the difference in the degree of supernodulation between nts382 and the intermediate nts1116 (Figure 1 and Table 1).

considered to be a diploidized tetraploid,[114] and there is cytological evidence for genome duplication in this species.[115] Perhaps there are genes in wild-type soybean that are expressed in root tissue and are responsible for limiting the number of nodules formed, but anomalies in these genes were not detected due to the presence of a duplicated wild-type locus.[116] The focus of the subsequent section on the relationship between nitrate inhibition and autoregulation of nodulation will be on soybean, which has been characterized in the most detail.

D. NITRATE INHIBITION AND AUTOREGULATION OF NODULATION

In the absence of nitrate, nodulation is tightly controlled and the number of infections is much greater than the number of nodules formed (for review see Reference 117). In wild-type soybean, the majority of infections are blocked at a very early stage of nodule ontogeny.[34,185] The points of blockage occur prior to the formation of visible lumps on the root, such that most infections on the root are comprised of an infection thread(s) associated with limited cell divisions in the outer cortical tissue.[34] Preliminary characterization indicated that the supernodulating soybean mutants were defective in this internal control of nodulation, called autoregulation, such that prolific nodulation occurred all over the tap and lateral roots.[56,57,116] This has since been confirmed by anatomical analysis of nodule ontogeny in nts382 and the wild-type cv Bragg.[185] While the majority of infections are blocked at an early stage in Bragg, as it is in wild-type cv Williams,[34] infection events are not arrested in nts382, and many advanced stages were observed along the root.[60] Mutant nts382 had

TABLE 1

The Symbiotic Characteristics of Parent cv Bragg, nts1116, and nts382 Grown in the Presence and Absence of Nitrate

Genotype	Nodulation phenotype	Nodule no. per plant		Nodule dry wt. (mg) per plant			Nitrogenase activity per plant[a]		
		N-free	5 mM KNO₃	N-free	5 mM KNO₃	% inhibition[b]	N-free	5 mM KNO₃	% inhibition[b]
Bragg	Wild-type	26 ± 6	19 ± 7	31 ± 10	5 ± 3	84	71 ± 13	1 ± 1	99
nts1116	Intermediate supernodulation nitrate-tolerance	101 ± 26	74 ± 45	66 ± 12	30 ± 12	55	85 ± 17	23 ± 10	73
nts382	Extreme supernodulation nitrate-tolerance	576 ± 77	1007 ± 154	166 ± 9	193 ± 35	−16[c]	119 ± 35	69 ± 11	42

Note: Data are taken from Delves et al. (1986), with permission from *Plant Physiology*. Pots of sand:vermiculite (2:1) were inoculated with *B. japonicum* USDA110, and the plants received either N-free nutrients or nutrients supplemented with 5 m*M* KNO₃ until harvest at 45 d after planting. Nitrogenase (acetylene reduction) activity was measured as described by Carroll et al.[56] Data are the mean ± SD of 3 to 6 plants.

[a] nmol C₂H₄ produced · per minute.
[b] Percent inhibition caused by the presence of nitrate.
[c] Nitrate-stimulated nodule dry weight per plant in nts382.

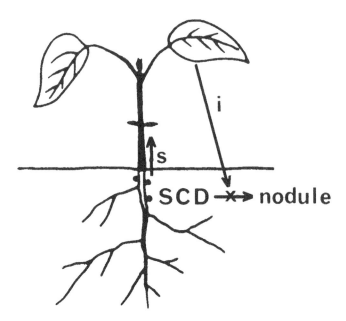

FIGURE 3.   A simple model for autoregulation of nodulation in the wild type. Early (and late) infection events produce a signal "s", which is translocated to the shoot. The response of the shoot is to produce an inhibitor "i" which impedes the rate of progress of subepidermal cell divisions (SCD) with associated infection threads into visible nodules. The supernodulating soybean mutants are wild type for signal "s", but are defective in the synthesis of inhibitor "i". A more detailed discussion of the model is presented in the text.

about the same number of infections as Bragg,[185] such that the number of infection events could not explain its altered nodulation capacity.

The density of nodules on wild-type roots is often,[118] but not always,[119] highest on root tissue in the region of the root tip at the time of inoculation. Pierce and Bauer[118] suggested that a rapid regulatory response occurs such that early infections inhibit subsequent nodulation. This regulation occurs prior to nodule formation and nitrogenase activity. Using a split-root system, Kosslak and Bohlool[120] showed that prior inoculation of one side resulted in suppression of nodulation on the second side, and this again was not related to nodule formation or nitrogenase activity. The clear demonstration that this suppression was due to autoregulation was shown by Olsson et al. using a split-root system with several of the supernodulating soybean mutants.[121] This work also emphasized the requirement for a small noninhibitory level of combined nitrogen (as nitrate in this case) in the divided growth media, since the original observation of suppression by Kosslak and Bohlool[120] could have been explained by the first side responding to inoculation in an analogous fashion to a response of the root to nutrient supply. Indeed, when nts382 was tested in a split-root system without any nitrate or combined nitrogen, some suppression of nodulation did occur on the side that was inoculated after a delay in comparison to the side that was inoculated earlier.[121] With a low level of starter nitrate available, suppression of nodulation on the root moiety that received inoculant after a delay was evident in the parent cv Bragg but not in nts382.[121] The response of nts1116 was intermediate between Bragg and the extreme supernodulator nts382, thereby reflecting its nodulation phenotype and leaky defect in autoregulation. Thus, autoregulation in wild-type soybeans not only involves the shoot, but it acts in a systemic fashion as demonstrated by split-root experiments.

A model of autoregulation in wild-type soybean is shown in Figure 3. Early infection events result in the production of a signal (designated "s" in Figure 3) which is translocated to the shoot. The response of the shoot is to produce an autoregulation inhibitor ("i" in Figure 3) that is systematically translocated to the root with the function to arrest the development of infected subepidermal cell divisions (SCD) into nodules. The signal emanating from early infections is unlikely to be derived from the rhizobia inside the infection thread because supernodulating rhizobia have not been identified, even though the genetic contribution of the microsymbiont to nodulation has been extensively characterized.[122-125] The proviso of this assertion is that the genetic attributes of the rhizobia required for nodule formation are not also needed for the production of the signal "s" (Figure 3). A *R. fredii* strain, that forms only juvenile cell divisions[119] analogous to pseudoinfections, did not detectably suppress subsequent nodulation, and, therefore, it is not likely that sufficient synthesis of the autoregulation signal emanates from pseudoinfections. Thus, a plausible candidate for the source of signal "s" is the plant moiety of actual infections. Grafts involving wild-type scions on mutant roots indicate that with a wild-type shoot, the supernodulating soybean mutants have essentially normal roots[109] and are capable of producing the presumptive autoregulation signal emanating from early (and late) infection events. These mutants, however, lack the autoregulation inhibitor produced in the shoots, and infection events progress through to nodules. Strong evidence for the synthesis of an autoregulation inhibitor in soybean (designated "i" in Figure 3) is that shoot extracts from inoculated, but not uninoculated, wild-type plants suppress supernodulation when injected into nts382 plants.[126,179]

The simple model for autoregulation in soybean, illustrated in Figure 3, may need some modification for other legume species. Shoot-controlled enhanced nodulation indicates that autoregulation also involves the shoot in pea.[105] On the other hand, root control in pea mutant nod$_3$ emphasizes the importance of the root tissue.[112,113] Perhaps nod$_3$ is incapable of producing the presumptive signal that emanates from infection events, or alternatively, the autoregulation inhibitor produced in the shoot has its effect on nodulation via a root process that is anomalous in nod$_3$.

The supernodulators lack autoregulation and are also nitrate tolerant, and therefore these processes must be coordinately regulated in wild-type plants. Based on the discussions earlier in this chapter, the simplest explanation for nitrate inhibition of nodule formation is that it is dependent on the synthesis of the autoregulation inhibitor in the shoot and on nitrate itself in contact with the roots, as illustrated in Figure 4. The supernodulating mutants are nitrate tolerant for nodule formation and nodule growth,[56,104] thus indicating dual control of these processes and the likelihood that the autoregulation inhibitor in the wild type is involved in the regulation of nodule growth, as well as nodule formation. The control of nodule function appears separate from nodulation in that these plants do not support high levels of specific nitrogenase activity.[56,110]

An important feature of the supernodulating soybean mutants is that they nodulate quicker than the parent cv Bragg,[127,180] and they also form more nodules above the root tip, as marked at the time of inoculation.[185] Thus, the rate of progress of the early events in nodule ontogeny is accelerated in these mutants. This suggests that both autoregulation and nitrate may act by simply slowing down the rate of progress of infections in the immature zone of the root. The immature tissue of the soybean root is the region where the early events in nodule ontogeny are initiated.[128] It is proposed that infection events in supernodulators progress quickly, such that they are sufficiently advanced and developmentally committed by the time they are located in mature root tissue. Conversely, in the wild type, the majority of infections are not sufficiently advanced on maturation of the root tissue and therefore do not develop into nodules. According to this model, the autoregulation inhibitor and nitrate itself impede the rate of progress of wild-type infections. These two inhibitors act synergistically in the wild type, and the presence of nitrate may further decrease the rate of progress of preinfection events to such an extent that the number of actual infection threads

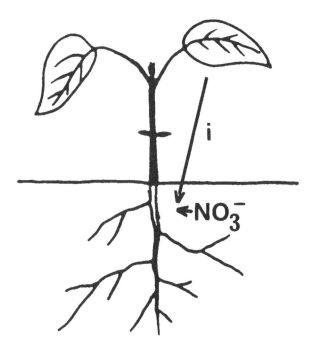

FIGURE 4.   A simple model illustrating the relationship between autoregulation and nitrate inhibition of nodulation. Nitrate itself, rather than its assimilation, is a major regulatory factor affecting nodulation. Inhibition of nodulation by nitrate is, however, dependent on the synthesis of the autoregulation inhibitor "i" in the shoot tissue. In wild-type legumes, nitrate and the inhibitor "i" act synergistically to slow down the rate of progress of infections into nodules. See text for further details.

(or infection centers) is decreased or prevented. Nitrate, however, requires the presence of the autoregulation inhibitor in order to exert its effect, and because the supernodulators lack (or have diminished levels of) the inhibitor, they are able to nodulate in the presence of nitrate. Such an explanation is consistent with the observations that infection thread formation does not solely limit nodulation in the presence of nitrate[17,18] and that delaying exposure to nitrate for 18 h after inoculation significantly reduced the inhibition of nodule formation on root tissue above the root tip mark at the time of inoculation.[21] Once nodules are formed or infections are developmentally committed to nodule formation, the regulation of nodule growth is probably dependent on the autoregulation inhibitor, since nodule growth is not inhibited by nitrate in the supernodulators.[56]

Of course, considerably more research needs to be done to confirm the validity of these models for explaining autoregulation and nitrate inhibition of nodulation. The nature of the autoregulation inhibitor is still under investigation and it may involve a common phytohormone.[179] Several phytohormones, such as abscisic acid,[129] gibberellins,[130] and ethylene[131] are known to inhibit nodulation when exogenously applied. Inoculated Bragg plants have higher levels of abscisic acid than inoculated nts382 plants, and this may be related to their nodulation phenotypes.[179] Ligero et al.[132] also observed a positive correlation between the nitrate concentration in the growth medium and the quantity of ethylene released from inoculated alfalfa roots. They suggested that inhibition of nodulation by nitrate may be mediated through the synthesis of this phytohormone. The characterization of the synthesis of ethylene by the supernodulating pea and soybean mutants has not been reported in the literature. Mutant nts382 has increased numbers of *Bradyrhizobium* in the rhizosphere,[60,176] and the autoregulation mechanism in the wild type, whatever its nature, may act by directly inhibiting metabolism in the microsymbiont.

V. NITRATE INHIBITION OF NITROGENASE (ACETYLENE REDUCTION) ACTIVITY

As mentioned earlier, defects in the microsymbiont's ability to utilize nitrate does not alleviate nitrate inhibition of nitrogenase activity.[30,52] Nevertheless, the role of nitrite in nitrate inhibition has not been thoroughly discounted.[2,40,41,133-136] Nitrogenase activity is often less inhibited (in proportional terms) when plants are infected with a less effective bacteria,[53,137] but this effect is probably a result of the nitrogen status of the host plant and not a direct effect of the inoculant's effectiveness.

NR-deficient legume mutants have been characterized for nitrate inhibition of nitrogenase activity. Nitrogenase activity is less affected in pea mutant E_1 than in its wild-type parent, and the extent of inhibition is proportional to its lowered level of NR activity.[138] E_1 is an iNR mutant processing 20% of the wild-type *in vivo* NR activity, and treatment of N_2-dependent plants for 2 d with nitrate resulted in a 47% inhibition of nitrogenase (acetylene reduction) activity in the wild type, whereas nodule activity in mutant E_1 was only inhibited by 19%.[138] These results indicated that either carbohydrate deprivation or products derived from nitrate reduction, and not nitrate per se, was responsible for nitrate inhibition of specific nitrogenase activity in wild-type pea.[84,138] Analogous experiments with soybean mutant NR328 and its parental wild type indicated that alterations in the plants' ability to utilize nitrate did not substantially affect nitrate inhibition of specific nitrogenase activity.[53] NR328 accumulated nitrate and had approximately 70% of the wild-type iNR activity in its root tissue.[88] These data suggest that the mechanism of nitrate inhibition of nitrogenase activity may vary between pea and soybean, and that nitrate itself may play a role in controlling activity in the latter species. This is in addition to its regulatory effect on nodulation.

There is also evidence that specific nitrogenase activity in soybean and pea respond differently to elevation in the ambient CO_2; increased specific rates were observed in the short term for pea,[139] but enhanced nitrogenase activity in soybean with increasing CO_2 only occurred with the ensuing increment in nodule growth.[140] Differences between soybean[141] and *Phaseolus vulgaris*[136,142] have also been described in terms of the effect of nitrate on carbohydrate composition of nodules. Nodule carbohydrate concentrations decline with the onset of nitrate inhibition of nitrogenase activity in *P. vulgaris*, but increased in soybean between 6 and 24 h after exposure to nitrate, during which time nitrogenase activity was progressively inhibited.[136] Microsymbiont integrity in soybean is not affected during the early stages of nitrate-induced nodule senescence as assessed by nitrogenase activity of isolated bacteroids,[33,143,144] whereas bacteroids isolated from similarly-treated *P. vulgaris* plants display only limited ability in reducing acetylene to ethylene.[133] Soybean bacteroids are eventually affected by nitrate in that they cannot reduce acetylene *ex planta*,[33,143,144] but this is only evident several days after the initial decline in nitrogenase activity. Detectable changes in bacteroid proteins,[134] total leghemoglobin[33,143,144] and other host proteins and enzymes[33,135] also occur subsequent to the inhibition of nitrogenase activity.

Much of the research discussed above has been based on the acetylene reduction assay, and recently it was shown that acetylene often induces a rapid decline in nitrogenase activity.[145,146] This decline is due to an increased diffusive resistance to oxygen and may lead to erroneous interpretation when quantitative comparisons, such as those reported above, are being made. The implications and applications of the phenomenon have been thoroughly reviewed,[147] but it is important to emphasize that this research demonstrated that legume nodules have the capacity to rapidly adjust the supply of oxygen to the nitrogen-fixing bacteroids.[148] Earlier, it was shown that the oxygen delivery in soybean nodules is subject to regulation in that the nodules adjust to changes in rhizosphere pO_2, such that the pO_2 inside the nodule is optimal for nitrogen fixation.[149,150]

The lack of an unequivocal explanation for nitrate inhibition of nitrogenase activity by carbohydrate deprivation or the products of nitrate reduction in legume species indicated that some other control mechanism may be operative. In view of the variable O_2 diffusion barrier probably operative in legume nodule,[146,148,151] a plausible explanation for inhibition was that O_2 supply to the bacteroids was limited in the presence of nitrate. Evidence in support of this hypothesis has been obtained for soybean[107,152] and white clover.[147,153] Nitrate inhibition of nitrogenase (acetylene reduction) activity was alleviated when soybean root systems were assayed at 60% O_2.[152] While this was not observed in nitrate-inhibited white clover plants,[147] the diffusion resistance to O_2 increased considerably as estimated by respiratory CO_2 production.[153] This discrepancy may reflect a difference between soybean and white clover in the rate of adjustment to the oxygen diffusion barrier, and the latter may be able to respond more quickly.

It had been shown that oxygen supply also limits acetylene reduction activity in water-stressed soybeans,[154-158] and there is also evidence for oxygen limitation in nodules of plants that have been subjected to prolonged darkness,[107,152,159] defoliation,[160,161] or simply shaking prior to the assay.[160] Thus, the O_2 limitation phenomenon is a common regulatory mechanism for controlling nitrogenase activity in symbiotically stressed plants and probably in plants grown under optimal conditions during their development. The nature of the oxygen diffusion barrier is not known, but it is likely to reside, at least partly, in the cortex external to the infected region of the nodule.[147,152,162]

O_2 deprivation appears to preempt other metabolic changes in the nitrate-affected nodules that could affect nitrogenase activity,[2,153] and it may be directly responsible for the reported accumulation of nitrite and the eventual decline in transport of carbohydrates into the nodule by decreasing its sink capacity. Nitrite content of tissues is O_2-dependent, and Heckmann and Drevon[163] showed that nitrite accumulation in soybean nodules decreases with increasing external pressure of O_2. Research on the iNR pea mutant E_1 indicates that the assimilation of nitrate is required for inhibition of nitrogenase activity.[138] In soybean, on the other hand, nitrogenase activity by iNR mutant NR328 is inhibited by nitrate, and perhaps in this species nitrate itself is capable of regulating the diffusion of oxygen into the nodule. This hypothesis has also been suggested by Becana and Sprent.[2] Further characterization of mutant NR328 and the isolation and investigation of additional iNR-deficient variants could substantiate the hypothesis that nitrate per se is capable of regulating nitrogenase activity in soybean. The metabolic rate of the shoot tissue directly regulates nitrogenase (acetylene reduction) activity in soybean, since short decreases in the temperature around the shoots has been demonstrated to cause a decline in nodule activity.[164] Perhaps the level of certain carbohydrates or other metabolites in the shoot are also important in controlling nitrogenase activity.

It has been reported that exposure of soybean roots to low O_2 tension caused a curtailment of nodule development, but not nodule appearance.[165] It is therefore plausible that an O_2 deficit in the nodules of nitrate-grown plants restricts nodule growth as well as specific nitrogenase activity. However, supra-ambient O_2 concentrations, even as high as 80%, do not alleviate nitrate inhibition of nodule growth in soybean.[181]

VI. NITROGEN METABOLISM AND THE PROSPECT OF ENHANCING YIELD IN GRAIN LEGUMES

Breeding for increased yield in cereals has taken precedence over grain legumes. This subject has been reviewed by Summerfield,[166] where it was cited that in 1976 there were about 400 corn breeders in the U.S., compared to 25 soybean breeders, even though these two crops are of similar economic magnitude. Major reasons for this bias in plant breeding were described.[166] With regard to nitrogen assimilation, the prospects are encouraging for increasing nitrogen fixation and yield by genetic manipulation of the host. The strict control of nodulation exercised by the legume host limits the possibility of selecting superior inoculant strains, although uptake hydrogenase in *B. japonicum* may be a valuable agronomic attribute for the symbiosis and grain yield in soybean.[167]

Seasonal profiles of nitrogen assimilation in soybean[11,168-171] and *P. vulgaris*[172] have been investigated. These studies have indicated that nitrate assimilation precedes nitrogen fixation during plant development. This can be partly explained in terms of nitrate inhibition of nodulation and the depletion of nitrate in the soil.[11] In addition, the developmental regulation of nitrate assimilation and nitrogen fixation in soybean do not appear to be the same. Imsande[171] has shown that soybean is ineffective in utilizing nitrate during pod fill, even when abundant nitrate was supplied in the water culture medium. Plants that received a less inhibitory form of combined nitrogen and were able to nodulate during pod fill assimilated considerably more nitrogen and produced 50% more seed weight than plants receiving nitrate alone.[171] The enhanced yield showed a strong positive correlation to nitrogenase (acetylene reduction) activity during pod fill.[171] These data indicated that the ability of the soybean plant to utilize nitrate is lost before the capacity to assimilate symbiotically fixed nitrogen is eroded and that improved nitrogen fixation late in crop development should enhance seed yield.

The capacity of plants to fix nitrogen is dependent on prior formation of nodules, and therefore nitrate inhibition of nodulation limits the amount of nitrogen that can be fixed later in crop development by wild-type plants. On the other hand, nitrate-tolerant nodulation mutants have considerable potential for enhanced nitrogen fixation as the crop matures. Preliminary field trials have shown that soybean mutant nts1116, an intermediate supernodulator, yields at least as well as its parent cv Bragg.[182] The extreme supernodulating soybean mutants do not yield as well as the wild type under optimal field conditions. However, as mentioned earlier, these lines probably harbor other mutations that may be affecting vigor, and therefore, improved agronomic performance is anticipated on the introduction of the intermediate and extreme supernodulation characters into superior genetic backgrounds. Genetic analysis indicates that the mutants arose from single mutation events,[116,173] and therefore, transfer of the characters to other genetic backgrounds should not be difficult.

Under greenhouse conditions, the supernodulating soybeans nodulate better than the wild type at low pH and in the presence of toxic levels of aluminum,[127] and therefore, there is considerable potential for this material in soils that are not conducive to nodulation. Another possible benefit is that supernodulating plants will leave higher levels of nitrate in the soil, but this may have limited potential at present because these lines have retained the ability to utilize nitrate.[56,84,108]

VII. CONCLUSION

While nitrate inhibition of nodulation is a complex phenomenon, recent advances have been made by complementing host genetics with other areas of research. There is good evidence that nitrate, separate from its assimilation, contributes to its regulatory properties. Genetic studies on the fungus *Aspergillus nidulans* have also indicated that nitrate itself can suppress the catabolism of other nitrogen sources, notably the uptake and assimilation of various amino acids.[174] The nitrate effect on legume nodulation gives credence to the notion that nitrate may act as a plant hormone.[175]

Nitrate inhibition is a common regulatory feature of legume root nodule symbioses, and it is primarily controlled by the host. The supernodulation and nitrate-tolerant mutants in pea and soybean illustrate the coordinate regulation on nitrate inhibition and autoregulation of nodulation and, thus, accentuate the importance of nitrate inhibition to our understanding of these symbioses. The regulation of nodule formation and nodule growth are also related.

The supernodulating mutants may also be useful in studying earlier processes in nodulation. Autoregulation represents a barrier to thorough genetic analysis of the events before nodule appearance, because only null mutants will be reliably recovered in a mutagenesis program. Leaky mutants would be masked by autoregulation, since the number of infection events in wild-type symbioses greatly exceeds the number of nodules that are formed. This latter class of mutants would be more easily detected in a supernodulation background, where a greater proportion of infections progress through to nodules.[185] Future research on nitrate effects will continue to contribute to the knowledge of nodulation in general. It will be useful in differentiating between the basic host and microbial requirements for nodulation and the regulation of these interactions. For example, compounds that are exuded by legume roots and involved in *nod* gene expression in the microsymbiont are basic requirements for nodulation,[122,123] but are they subject to regulation such as to mediate the control of nodule number and the amount of nodule tissue that is formed? The success of attempts to enhance nitrogen fixation and plant productivity is not only dependent on an understanding of the basic attributes of the symbionts, but on how the interaction is regulated. The prospect of achieving these aims by modifying the regulation of nodulation in the legume host is extremely promising.

ACKNOWLEDGMENTS

We thank Diana Quiggin, Charles Lawson, Angela Delves, Alan Gibson, Arno Krotzky, Jane Olsson, David Day, David Herridge, John Betts, Peter Gresshoff, Alexander Hansen, and John Brockwell for discussion or editorial comments.

REFERENCES

1. **Beringer, J. E.,** The significance of symbiotic nitrogen fixation in plant production, *CRC Crit. Rev. Plant. Sci.,* 1, 269, 1984.
2. **Becana, M. and Sprent, J. I.,** Nitrogen fixation and nitrate reduction in the root nodules of legumes, *Physiol. Plant.,* 70, 757, 1987.
3. **Wilson, J. K.,** Physiological studies of *Bacillus radicicola* or soybean (*Soja max* Piper) and factors influencing nodule production, *Cornell Univ. Agric. Exp. Stn. Bull.,* 386, 369, 1917.
4. **Thorton, H. G.,** Action of Na-nitrate of infection of lucerne root hairs by nodule bacteria, *Proc. R. Soc. London Ser. B,* 199, 47, 1936.
5. **Harper, J. E.,** Contribution of dinitrogen and soil or fertilizer nitrogen to soybean (*Glycine max* L. Merr.) production, in *World Soybean Research,* Hill, L. D., Ed., Interstate Printers and Publishers, Danville, IL, 1976, 101.
6. **Gibson, A. H.,** Recovery and compensation by nodulated legumes to environmental stress, in *Symbiotic Nitrogen Fixation in Plants, IBP Synth.,* Vol. 7. Cambridge University Press, London, 1976, 385.
7. **Lie, T. A.,** Environmental effects on nodulation and symbiotic nitrogen fixation, in *The Biology of Nitrogen Fixation,* Quispel, A., Ed., North-Holland, Amsterdam, 1974, 555.
8. **Ryle, G. J. A., Powell, C. E., and Gordon, A. J.,** The respiratory costs of nitrogen fixation in soya bean, cowpea and white clover. II. Comparisons of the cost of nitrogen fixation and the utilization of combined nitrogen, *J. Exp. Bot.,* 30, 145, 1979.
9. **Finke, R. L., Harper, J. E., Hageman, R. H.,** Efficiency of nitrogen assimilation by N_2-fixing and nitrate-grown soybean plants (*Glycine max* [L.] Merr.), *Plant Physiol.,* 70, 1178, 1982.
10. **Atkins, C. A.,** Efficiencies and inefficiencies in the legume/*Rhizobium* symbiosis — a review, *Plant Soil,* 82, 273, 1984.
11. **Harper, J. E.,** Nitrogen metabolism, in *Soybeans: Improvement, Production, and Uses,* 2nd ed., Wilcox, J. R., Ed., American Society of Agronomy, Crop Sciences Society of America, and Soil Sciences Society of America, Madison, WI, 1987, 497.

12. **Welch, L. F., Boone, V., Chambliss, C. G., Christiansen, A. T., Mulvaney, D. L., Oldham, M. G., and Pendleton, J. W.**, Soybean yields with direct and residual nitrogen fertilization, *Agron. J.*, 65, 547, 1973.

13. **McNeil, D. L. and La Rue, T. A.**, Effect of nitrogen sources on ureides in soybeans, *Plant Physiol.*, 74, 227, 1984.

14. **Harper, J. E.**, Soil and symbiotic nitrogen requirements for optimum soybean production, *Crop. Sci.*, 14, 255, 1974.

15. **Beevers, I. and Hageman, R. H.**, Uptake and reduction of nitrate: bacteria and higher plants, in *Inorganic Plant Nutrition*, Lauchli, A. and Bieleski, R. L., Eds., Springer-Verlag, Berlin, 1983, 351.

16. **Bergersen, F. J.**, *Root Nodules of Legumes: Structure and Functions*, Research Studies Press, Chichester, U.K., 1982.

17. **Carroll, B. J. and Gresshoff, P. M.**, Nitrate inhibition of nodulation and nitrogen fixation in white clover, *Z. Pflanzenphysiol.*, 110, 77, 1983.

18. **Munns, D. N.**, Nodulation of *Medicago sativa* in solution culture. III. Effects of nitrate on root hairs and infection, *Plant Soil*, 29, 33, 1968.

19. **Dazzo, F. B. and Brill, W. J.**, Regulation by fixed nitrogen of host-symbiont recognition in the *Rhizobium*-clover symbiosis, *Plant Physiol.*, 62, 18, 1978.

20. **Darbyshire, J. F.**, Studies on the physiology of nodule formation. IX. The influence of combined nitrogen, glucose, light intensity and day length on root hair infection in clover, *Ann. Bot.*, 30, 623, 1966.

21. **Malik, N. S. A., Calvert, H. E., and Bauer, W. D.**, Nitrate induced regulation of nodule formation in soybean, *Plant Physiol.*, 84, 266, 1987.

22. **Sherwood, J. E., Truchet, G. L., and Dazzo, F. B.**, Effect of nitrate supply on the in-vivo synthesis and distribution of trifoliin A, a *Rhizobium trifolii*-binding lectin, in *Trifolium repens* seedlings, *Planta*, 162, 540, 1984.

23. **Bohlool, B. and Schmidt, E.**, Lectins: a possible basis for specificity in the *Rhizobium*-legume root nodule symbiosis, *Science*, 185, 269, 1974.

24. **Dazzo, F. B. and Hollingsworth, R. E.**, Trifoliin A and carbohydrate receptors as mediators of cellular recognition in the *Rhizobium trifolii*-clover symbiosis, *Biol. Cell.*, 51, 267, 1984.

25. **Halverson, L. J. and Stacey, G.**, Host recognition in the *Rhizobium*-soybean symbiosis. Detection of a protein factor in soybean root exudate which is involved in the nodulation process, *Plant Physiol.*, 74, 84, 1984.

26. **Halverson, L. J. and Stacey, G.**, Host recognition in the *Rhizobium*-soybean symbiosis. Evidence for the involvement of lectin in nodulation, *Plant Physiol.*, 77, 621, 1985.

27. **Halverson, L. J. and Stacey, G.**, Effect of lectin on nodulation by wild-type *Bradyrhizobium japonicum* and a nodulation-defective mutant, *Appl. Environ. Microbiol.*, 4, 753, 1986.

28. **Gibson, A. H. and Harper, J. E.**, Nitrate effect on nodulation of soybean by *Rhizobium japonicum*, *Crop Sci.*, 25, 497, 1985.

29. **Chen, P. and Phillips, D. A.**, Induction of root nodule senescence by combined nitrogen in *Pisum sativum* L., *Plant Physiol.*, 59, 440, 1977.

30. **Gibson, A. H. and Pagan, J. D.**, Nitrate effects on the nodulation of legumes inoculated with nitrate-reductase deficient mutants of *Rhizobium*, *Planta*, 134, 17, 1977.

31. **Wong, P. P.**, Nitrate and carbohydrate effects on nodulation and nitrogen fixation (acetylene reduction) activity of lentil (*Lens esculenta* Moench.), *Plant Physiol.*, 66, 78, 1980.

32. **Houwaard, F.**, Influence of ammonium and nitrate nitrogen on nitrogenase activity of pea plants as affected by light intensity and sugar addition, *Plant Soil*, 54, 271, 1980.

33. **Schuller, K. A., Day, D. A., Gibson, A. H., and Gresshoff, P. M.**, Enzymes of ammonia assimilation and ureide biosynthesis in soybean nodules: effect of nitrate, *Plant Physiol.*, 80, 646, 1986.

34. **Calvert, H. E., Pence, M. K., Pierce, M., Malik, N. S. A., and Bauer, W. D.**, Anatomical analysis of the development and distribution of *Rhizobium* infections in soybean roots, *Can. J. Bot.*, 62, 2375, 1984.

35. **Oghoghorie, C. G. O. and Pate, J. S.**, The nitrate stress syndrome of the nodulated field pea (*Pisum arvense* L.). Techniques for measurement and evaluation in physiological terms, in *Biological Nitrogen Fixation in Natural and Agricultural Habitats*, Plant and Soil Spec. Vol., Lie, T. A. and Muller, E. G., Eds., Martinus Nijhoff, The Hague, 1971, 185.

36. **Small, J. G. C. and Leonard, O. A.**, Translocation of C14-labelled photosynthate in nodulated legumes as influenced by nitrate nitrogen, *Am. J. Bot.*, 56, 187, 1969.

37. **Latimore, M., Giddens, J., and Ashley, D. A.**, Effect of ammonium and nitrate nitrogen upon photosynthate supply and nitrogen fixation by soybeans, *Crop Sci.*, 17, 399, 1977.

38. **Ursino, D. J., Hunter, D. M., Laing, R. D., and Keighley, J. L. S.**, Nitrate modification of photosynthesis and photoassimilate export in young nodulated soybean plants, *Can. J. Bot.*, 60, 2665, 1982.

39. **Tanner, J. W. and Anderson, I. C.**, External effect of combined nitrogen on nodulation, *Plant Physiol.*, 39, 1039, 1964.

40. **Trinchant, J. C. and Rigaud, J.,** Nitrate inhibition of nitrogenase from soybean bacteroids, *Arch. Microbiol.,* 124, 49, 1980.

41. **Rigaud, J. and Puppo, A.,** Effect of nitrate upon leghemoglobin and interaction with nitrogen fixation, *Biochim. Biophys. Acta,* 497, 702, 1977.

42. **Virtanen, A. I., Jorma, J., Linkola, H., and Linnasalmi, A.,** On the relation between nitrogen fixation and leghemoglobin content of leguminous root nodules, *Acta Chem. Scand.,* 1, 90, 1947.

43. **Hinson, K.,** Nodulation responses from nitrogen applied to soybean half-root systems, *Agron. J.,* 67, 799, 1975.

44. **Drew, M. C., Saker, L. R., and Ashley, T. W.,** Nutrient supply and the growth of the seminal root system in barley. I. The effect of nitrate concentration on the growth of the axes and laterals, *J. Exp. Bot.,* 24, 1189, 1973.

45. **Harper, J. E. and Cooper, R. L.,** Nodulation response of soybeans (*Glycine max* [L]. Merr.) to application rate and placement of combined nitrogen, *Crop Sci.,* 11, 438, 1971.

46. **Heichel, G. H. and Vance, C. P.,** Nitrate-N and *Rhizobium* strain roles in alfalfa seedling nodulation and growth, *Crop Sci.,* 19, 512, 1979.

47. **Nelson, L. M.,** Variation in the response of N_2-fixing *Rhizobium leguminosarum* isolates to application of NH_4NO_3, in *Advances in Nitrogen Fixation Research,* Veeger, C. and Newton, W. E., Eds., Martinus Nijhoff/Junk, The Hague, 1984, 542.

48. **McNeil, D. L.,** Variations in ability of *Rhizobium japonicum* strains to nodulate soybeans and maintain fixation in the presence of nitrate, *Appl. Environ. Microbiol.* 44, 647, 1982.

49. **Streeter, J. G. and Devine, P. J.,** Evaluation of nitrate reductase activity in *Rhizobium japonicum, Appl. Environ. Microbiol.,* 46, 521, 1983.

50. **Streeter, J. G.,** Synthesis and accumulation of nitrite in soybean nodule supplied with nitrate, *Plant Physiol.,* 69, 1429, 1982.

51. **Streeter, J. G.,** Nitrate inhibition of legume nodule growth and activity. Long term studies with a continuous supply of nitrate, *Plant Physiol.,* 77, 321, 1985.

52. **Streeter, J. G.,** Nitrate inhibition of legume nodule growth and activity. Short term studies with high nitrate supply, *Plant Physiol.,* 77, 325, 1985.

53. **Lawson, C. G. R., Gresshoff, P. M., and Carroll, B. J.,** Contribution of plant and bacterial nitrate metabolism to the inhibition of the soybean-*Bradyrhizobium* symbiosis by nitrate, Honours thesis, Botany Department, Australian National University, Canberra, 1986.

54. **Stephens, B. D. and Neyra, C. A.,** Nitrate and nitrite reduction in relation to nitrogenase activity in soybean nodules and *Rhizobium japonicum* bacteroids, *Plant Physiol.,* 71, 731, 1983.

55. **Carroll, B. J., McNeil, D. L., and Gresshoff, P. M.,** Breeding soybeans for increased nodulation in the presence of external nitrate, in *Symbiotic Nitrogen Fixation,* Vol. 1, Ghai, B. S., Ed., USG Publishers, Ludhiana, India, 1984, 43.

56. **Carroll, B. J.,, McNeil, D. L., and Gresshoff, P. M.,** Isolation and properties of soybean (*Glycine max* [L.] Merr) mutants that nodulate in the presence of high nitrate concentrations, *Proc. Natl. Acad. Sci. U.S.A.,* 82, 4162, 1985.

57. **Carroll, B. J., McNeil, D. L., and Gresshoff, P. M.,** A supernodulation and nitrate tolerant symbiotic (nts) soybean mutant, *Plant Physiol.,* 78, 34, 1985.

58. **Herridge, D. F., Roughley, R. J., and Brockwell, J.,** Effect of rhizobia and soil nitrate on the establishment and functioning of the soybean symbiosis in the field, *Aust. J. Agric. Res.,* 35, 149, 1984.

59. **Lawson, C. G. R., Carroll, B. J., and Gresshoff, P. M.,** Alleviation of nitrate inhibition of soybean nodulation by high inoculum does not involve bacterial nitrate metabolism, *Plant Soil,* 110, 123, 1988.

60. **Mathews, A.,** Host Involvement in Nodule Initiation in the Soybean- *Bradyrhizobium* Symbiosis, Ph.D. thesis, Department of Botany, Australian National University, Canberra, 1987.

61. **La Favre, J. S. and Eaglesham, A. R. J.,** Increased nodulation of 'non-nodulating' (rj_1, rj_1) soybeans by high dose inoculation, *Plant Soil,* 80, 297, 1984.

62. **Mathews, A., Carroll, B. J., and Gresshoff, P. M.,** Characterization of non-nodulating mutants of soybean [*Glycine max* (L.) Merr.]: *Bradyrhizobium* effects and absence of root hair curling, *J. Plant Physiol.,* 113, 349, 1987.

63. **Harper, J. E. and Gibson, A. H.,** Differential nodulation tolerance to nitrate among legume species, *Crop Sci.,* 24, 797, 1984.

64. **Schuller, K. A.,** Nitrate Effect on Nodulation and Nitrogen Fixation by Three Clover Species, Honours thesis, Botany Department, Australian National University, Canberra, 1982.

65. **Rolfe, B. G., Gresshoff, P. M., and Shine, J.,** Rapid screening for symbiotic mutants of *Rhizobium* of nodulation in white clover, *Plant Sci. Letters,* 19, 277, 1980.

66. **Nutman, P. S.,** Symbiotic effectiveness in nodulated red clover. I. Variation in host and in bacteria, *Heredity,* 8, 35, 1953.

67. **Nutman, P. S., Mareckova, H., and Raicheva, L.,** Selection for increased nitrogen fixation in red clover, in *Biological Nitrogen Fixation in Natural and Agricultural Habitats,* Plant and Soil Spec. Vol., Lie, T. A. and Mulder, E. G., Eds., Martinus Nijhoff, The Hague, 1971, 27.

68. **Mytton, L. R. and Jones, D. G.,** The response to selection for increased nodule tissue in white clover (*Trifolium repens* L.), in *Biological Nitrogen Fixation in Natural and Agricultural Habitats,* Plant and Soil Spec. Vol., Lie, T. A. and Mulder, E. G., Eds., Martinus Nijhoff, The Hague, 1971, 17.

69. **Gelin, O. and Blixt, S.,** Root nodulation in peas, *Agric. Hortic. Genet.,* 22, 149, 1964.

70. **Duhigg, P., Melton, B., and Baltensperger, A.,** Selection for acetylene reduction rates in 'Mesilla' alfalfa, *Crop Sci.,* 18, 813, 1978.

71. **Heichel, G. H., Barnes, D. K., Vance, C. P., and Hardarson, G.,** Environmental and genotypic effects on dinitrogen fixation of contrasting alfalfa clones, in *Advances in Nitrogen Fixation Research,* Veeger, C. and Newton, W. E., Eds., Martinus Nijhoff/Junk, The Hague, 1984, 595.

72. **Imsande, J.,** Plant genotype and the control of nitrogen fixation in *Advances in Nitrogen Fixation Research,* Veeger, C. and Newton, W. E., Eds., Martinus Nijhoff/Junk, The Hague, 1984, 596.

73. **Hardarson, G., Zapata, F., and Danso, S. K. A.,** Effect of plant genotype and nitrogen fertilizer on symbiotic nitrogen fixation by soybean cultivars, *Plant Soil,* 82, 397, 1984.

74. **Herridge, D. F. and Betts, J. H.,** Nitrate tolerance in soybean: variation between genotypes, in *Nitrogen Fixation Research Progress,* Evans, H. J., Bottomley, P. J., and Newton, W. E., Eds., Martinus Nijhoff, Dordrecht, 1985, 32.

75. **Betts, J. H. and Herridge, D. F.,** Isolation of soybean lines capable of nodulation and nitrogen fixation under high levels of nitrate supply, *Crop Sci.,* 27, 1156, 1987.

76. **Herridge, D. F.,** Relative abundance of ureides and nitrate in plant tissues of soybean as a quantitative assay of nitrogen fixation, *Plant Physiol.,* 70, 1, 1982.

77. **Herridge, D. F.,** Use of the ureide technique to describe the nitrogen economy of field-grown soybeans, *Plant Physiol.,* 70, 7, 1982.

78. **Herridge, D. F.,** Effects of nitrate and plant development on the abundance of nitrogenous solutes in root-bleeding and vacuum-extracted exudates of soybean, *Crop Sci.,* 24, 173, 1984.

79. **Kleinhofs, A., Warner, R. L., Muehlbauer, F. J., and Nilan, R. A.,** Induction and selection of specific gene mutations in *Hordeum* and *Pisum, Mutat. Res.,* 51, 29, 1978.

80. **Feenstra, W. J. and Jacobsen, E.,** Isolation of a nitrate reductase deficient mutant of *Pisum sativum* by means of selection for chlorate resistance, *Theor. Appl. Genet.,* 58, 39, 1980.

81. **Nelson, R. S., Ryan, S. A., and Harper, J. E.,** Soybean mutants lacking constitutive nitrate reductase activity. I. Selection and initial plant characterization, *Plant Physiol.,* 72, 503, 1983.

82. **Carroll, B. J. and Gresshoff, P. M.,** Isolation and initial characterization of constitutive nitrate reductase-deficient mutants NR328 and NR345 of soybean *(Glycine max), Plant Physiol.,* 81, 572, 1986.

83. **Streit, L. and Harper, J. E.,** Biochemical characterization of soybean mutants lacking constitutive NADH:nitrate reductase, *Plant Physiol.,* 81, 593, 1986.

84. **Jacobsen, E.,** Modification of symbiotic interaction of pea *(Pisum sativum* L.) and *Rhizobium leguminosarum* by induced mutations, *Plant Soil,* 82, 427, 1984.

85. **Lahav, E., Harper, J. E., and Hageman, R. H.,** Improved soybean growth in urea with pH buffered by a carboxy resin, *Crop Sci.,* 16, 325, 1976.

86. **Harper, J. E., Nelson, R. S., and Streit, L.,** Nitrate metabolism of soybean — physiology and genetics, in *World Soybean Research Conference. III. Proceedings,* Shibles, R., Ed., Westview Press, Boulder, CO, 1985, 476.

87. **Streit, L., Nelson, R. S., and Harper, J. E.,** Nitrate reductase from wild-type and nr_1-mutant soybean *(Glycine max* [L.] Merr.) leaves. I. Purification, kinetics and physical properties, *Plant Physiol.,* 78, 80, 1985.

88. **Whitmore Smith, D.,** Nitrate metabolism in nitrate reductase deficient soybean mutants. Honours thesis, Botany Department, Australian National University, Canberra, 1985.

89. **Ryan, S. A., Nelson, R. S., and Harper, J. E.,** Soybean mutants lacking constitutive nitrate reductase activity. II. Nitrogen assimilation, chlorate resistance and inheritance, *Plant Physiol.,* 72, 510, 1983.

90. **Luttge, U.,** Import and export of mineral nutrients in plant roots, in *Encyclopedia of Plant Physiology,* N.S., Vol. 15A, Inorganic Plant Nutrition, Lauchli, A. and Bieleski, R. L., Eds., Springer-Verlag, Berlin, 1983, 181.

91. **Butz, R. G. and Jackson, W. A.,** A mechanism for nitrate transport and reduction, *Phytochemistry,* 16, 409, 1977.

92. **Nutman, P. S.,** Genetical factors concerned in the symbiosis of clover and nodule bacteria, *Nature (London),* 157, 463, 1946.

93. **Williams, L. F. and Lynch, D. L.,** Inheritance of a non-nodulating character in the soybean, *Agron. J.,* 46, 28, 1954.

94. **Lie, T. A.,** Temperature-dependent root-nodule formation in pea cv. Iran, *Plant Soil,* 34, 751, 1971.

95. **Holl, F. B.,** Host plant control of the inheritance of dinitrogen fixation in the *Pisum-Rhizobium* symbiosis, *Euphytica,* 24, 767, 1975.

96. **Gorbet, D. W. and Burton, J. C.,** A non-nodulating peanut, *Crop. Sci.,* 19, 727, 1979.

97. **Peterson, M. A. and Barnes, D. K.,** Inheritance of ineffective nodulation and non-nodulation traits in alfalfa, *Crop Sci.,* 21, 611, 1981.

98. **Ohlendorf, H.**, Selektion auf Resistenz von *Pisum sativum* gegen *Rhizobium leguminosarum* Stamm 311d, *Z. Pflanzenzuecht.*, 90, 204, 1983.

99. **Ohlendorf, H.**, Untersuchungen zur Vererbung der Resistenz von *Pisum sativum* gegen *Rhizobium leguminosarum* Stamm 311d, *Z. Pflanzenzuecht.*, 91, 13, 1983.

100. **Nambiar, P. R. C., Nigan, S. N., Dart, P. J., and Gibbon, R. W.**, Absence of root hair in non-nodulating groundnut, *Arachis hypogeae* L., *J. Exp. Bot.*, 34, 484, 1983.

101. **Kneen, B. E. and La Rue, T. A.**, Nodulation resistant mutant of *Pisum sativum* (L)., *J. Hered.*, 75, 238, 1984.

102. **Carroll, B. J., McNeil, D. L., and Gresshoff, P. M.**, Mutagenesis of soybean (*Glycine max* [L.] Merr.) and the isolation of non-nodulating mutants, *Plant Sci.*, 47, 109, 1986.

103. **Davis, T. M., Foster, K. W., and Phillips, D. A.**, Inheritance and expression of three genes controlling root nodule formation in chickpea, *Crop Sci.*, 26, 719, 1986.

104. **Jacobsen, E. and Feenstra, W. J.**, A new pea mutant with efficient nodulation in the presence of nitrate, *Plant Sci. Lett.*, 33, 337, 1984.

105. **Messager, A.**, Selection of pea mutants for nodulation and nitrogen fixation, in *Analysis of the Plant Genes Involved in the Legume-Rhizobium Symbiosis*, Organization for Economic Cooperation and Development, Paris, 1985, 52.

106. **Gremaud, M. F. and Harper, J. E.**, Selection and initial characterisation of partially nitrate tolerant nodulation mutants of soybean, *Plant Physiol.* 89, 169, 1989.

107. **Carroll, B. J., McNeil, D. L., Whitmore Smith, D., and Gresshoff, P. M.**, Host genetics and physiological studies on nitrate inhibition of nodulation and nitrogen fixation in soybean, in *Nitrogen Fixation Research Progress*, Evans, H. J., Bottomley, P. J., and Newton, W. E., Eds., Martinus Nijhoff, Dordrecht, 1985, 39.

108. **Day, D. A., Lambers, H., Bateman, J., Carroll, B. J., and Gresshoff, P. M.**, Growth comparisons of a supernodulating soybean (*Glycine max* L.) mutant and its wildtype parent, *Physiol. Plant.*, 68, 375, 1986.

109. **Delves, A. C., Mathews, A., Day, D. A., Carter, A. S., Carroll, B. J., and Gresshoff, P. M.**, Regulation of the soybean-*Rhizobium* symbiosis by shoot and root factors, *Plant Physiol.*, 82, 588, 1986.

110. **Day, D. A., Price, G. D., Schuller, K. A., and Gresshoff, P. M.**, Nodule physiology of a supernodulating soybean (*Glycine max*) mutant, *Aust. J. Plant Physiol.*, 14, 527, 1987.

111. **Delves, A. C., Higgins, A. V., and Gresshoff, P. M**, Shoot control of supernodulation in a number of mutant soybeans, *Glycine max* (L.) Merr., *Aust. J. Pla t Physiol.*, 14, 689, 1987.

112. **Jacobsen, E., Postma, J. G., Nijdam, H., and Feenstra, W. J.**, Genetical and grafting experiments with pea mutants in studies on symbiosis, in *Nitrogen Fixation Research Progress*, Evans, H. J., Bottomley, P. J. and Newton, W. E., Eds., Martinus Nijhoff, Dordrecht, 1985, 43.

113. **Postma, J. G., Jacobsen, E., and Feenstra, W. J.**, Three pea mutants with an altered nodulation studied by genetic analysis and grafting, *J. Plant Physiol.*, 132, 424, 1988.

114. **Lackey, J. A.**, Phaseoleae DC, in *Advances in Legume Systematics*, Polhill, R. M. and Raven, P. H., Eds., Royal Botanical Gardens, Kew, U.K., 1981, 301.

115. **Crane, C. F., Beversdorf, W. D., and Bingham, E. T.**, Chromosome pairing and associations at meiosis in haploid soybean (*Glycine max*), Can. *J. Genet. Cytol.*, 24, 293, 1982.

116. **Carroll, B. J., Gresshoff, P. M., and Delves, A. C.**, Inheritance of supernodulation in soybean and estimation of the genetically effective cell number, *Theor. Appl. Genet.*, 76, 54, 1988.

117. **Bauer, W. D.**, Infection of legumes by rhizobia, *Annu. Rev. Plant Physiol.*, 32, 407, 1981.

118. **Pierce, M. and Bauer, W. D.**, A rapid regulatory response governing nodulation in soybean, *Plant Physiol.*, 73, 286, 1983.

119. **Heron, D. S. and Pueppke, S. G.**, Regulation of nodulation in the soybean-*Rhizobium* symbiosis: strain and cultivar variability, *Plant Physiol.*, 84, 1391, 1987.

120. **Kosslak, R. M. and Bohlool, B. B.**, Suppression of nodule development of one side of a split-root system of soybeans caused by prior inoculation of the other side, *Plant Physiol.*, 75, 125, 1984.

121. **Olsson, J. E., Nakao, P., Bohlool, B. B., and Gresshoff, P. M.**, Lack of systemic suppression of nodulation in split root system of supernodulating soybean (*Glycine max* [L.] Merr.) mutants, *Plant Physiol.*, in press.

122. **Peters, N. K., Fros , J. W., and Long, S. R.**, A plant flavone, luteolin, induces expression of *Rhizobium meliloti* nodulation genes, *Science*, 233, 977, 1986.

123. **Redmond, J. W., Batle , M., Djordjevic, M. A., Innes, R. W., Kuempel, P. L., and Rolfe, B. G.**, Flavones induce expression of nodulation genes in *Rhizobium, Nature (London)*, 323, 632, 1986.

124. **Firman, J. L., Wilson, K. E., Rossen, L., and Johnston, A. W. B.**, Flavonoid activation of nodulation genes in *Rhizobium* reversed by other compounds present in plants, *Nature (London)*, 324, 90, 1986.

125. **Kosslak, R. M., Bookland, R., Barkei, J., Paaren, H., and Appelbaum, E. R. A.**, Induction of *Bradyrhizobium japonicum* common *nod* genes by isoflavones isolated from *Glycine max*, *Proc. Natl. Acad. Sci. U.S.A.*, 84, 7428, 1987.

126. **Gresshoff, P. M., Krotzky, A., Mathews, A., Day, D. A., Schuller, K. A., Olsson, J., Delves, A. C., and Carroll, B. J.**, Suppression of the symbiotic supernodulation symptoms of soybean, *J. Plant Physiol.*, 132, 417, 1988.

127. **Alva, A. K., Edwards, D. G., Carroll, B. J., Asher,, C. J., and Gresshoff, P. M.,** Nodulation and early growth of soybean mutants with increased nodulation capacity under acid soil infertility factors, *Agron. J.,* 80, 836, 1988.

128. **Turgeon, B. G. and Bauer, W. D.,** Spot inoculation of soybean roots with *Rhizobium japonicum, Protoplasma,* 115, 122, 1983.

129. **Phillips, D. A.,** Abscisic acid inhibition of root nodule initiation in *Pisum sativum, Planta,* 100, 181, 1971.

130. **Williams, P. M. and Sicardi de Mallorca, M.,** Effect of gibberellins and the growth retardant CCC on the nodulation of soya, *Plant Soil,* 77, 53, 1984.

131. **Grobbelaar, N., Clarke, B., and Hough, M. C.,** The nodulation and nitrogen fixation of isolated roots of *Phaseolus vulgaris* L. III. The effect of carbon dioxide and ethylene, in *Biological Nitrogen Fixation in Natural and Agricultural Habitats,* Plant Soil Spec. Vol., Lie, T. A. and Mulder, E. G., Eds., Martinus Nijhoff, The Hague, 1971, 215.

132. **Ligero, F., Lluch, C., and Olivares, J.,** Evolution of ethylene from roots and nodulation rate of alfalfa (*Medicago sativa* L.) plants inoculated with *Rhizobium meliloti* as affected by the presence of nitrate, *J. Plant Physiol.,* 129, 461, 1987.

133. **Trinchant, J. C. and Rigaud, J.,** Nitrogen fixation in French-beans in the presence of nitrate. Effect on bacteroid respiration and comparison with nitrate, *J. Plant Physiol.,* 116, 209, 1984.

134. **Becana, M., Aparicio-Tejo, P. M., and Sanchez-Diaz, M.,** Levels of ammonia, nitrite and nitrate in alfalfa root nodules supplied with nitrate, *J. Plant Physiol.,* 199, 359, 1985.

135. **Becana, M., Aparicio-Tejo, P. M., and Sanchez-Diaz, M.,** Nitrate and nitrate reduction by alfalfa root nodules: accumulation of nitrate in *Rhizobium meliloti* bacteroids and senescence of nodules, *Physiol. Plant.,* 64, 353, 1985.

136. **Wasfi, M. and Prioul J.-L.,** A comparison of inhibition of French-bean and soybean nitrogen fixation by nitrate, 1% oxygen or direct assimilate deprivation, *Physiol. Plant.,* 66, 481, 1986.

137. **Manhart, J. R. and Wong, P. P.,** Nitrate effect on nitrogen fixation (acetylene reduction). Activities of legume root nodules induced by rhizobia with varied nitrate reductase activities, *Plant Physiol.,* 65, 502. 1980.

138. **Feenstra, W . J., Jacobsen, E., van Swaay, A. C. P. M., and de Visser, A. J. C.,** Effect of nitrate on acetylene reduction in a nitrate reductase deficient mutant of pea (*Pisum sativum* L.), *Z. Pflanzenphysiol.,* 105, 471, 1982.

139. **Phillips, D. A., Newell, K. D., Hassell, S. A., and Felling, C. E.,** The effect of CO$_2$ enrichment on root nodule development and symbiotic N$_2$ reduction in *Pisum sativum* L., *Am. J. Bot.,* 63(3), 356, 1976.

140. **Williams, L. E., DeJong, T. M., and Phillips, D. A.,** Effect of changes in shoot carbon-exchange rate on soybean root nodule activity, *Plant Physiol.,* 69, 432, 1982.

141. **Streeter, J. G.,** Effect of nitrate in the rooting medium on carbohydrate composition of soybean nodules, *Plant Physiol.,* 68, 840, 1981.

142. **Streeter, J. G.,** Effect on nitrate on acetylene reduction activity and carbohydrate composition of *Phaseolus vulgaris* nodules, *Physiol. Plant.,* 68, 294, 1986.

143. **McNeil, D. L., Carroll, B. J., and Gresshoff, P. M.,** The nitrogen fixation capacity of bacteroids extracted from soybean nodules inhibited by nitrate, ammonia or dark treatment, in *Symbiotic Nitrogen Fixation,* Vol. 1, Ghai, B. S., Ed., USG Publishers, Ludhiana, India, 1984, 79.

144. **Carroll, B. J.,** The Plant Contribution to the Soybean-*Rhizobium* Symbiosis, Ph.D. thesis, Australian National University, Canberra, 1985.

145. **Minchin, F. R., Witty, J. F., Sheehy, J. E., and Muller, M.,** A major error in the acetylene reduction assay. Decreases in nodular nitrogenase activity under assay conditions, *J. Exp. Bot.,* 34, 641, 1983.

146. **Witty, J. F., Minchin, F. R., Sheehy, J. E., and Minguez, M. I.,** Acetylene induced changes in the oxygen diffusion resistance and nitrogenase activity of legume root nodules, *Ann. Bot.,* 53, 13, 1984.

147. **Witty, J. F., Minchin, F. R., Skot, L., and Sheehy, J. E.,** Nitrogen fixation and oxygen in legume root nodules, *Oxford Surv. Plant Mol. Cell Biol.,* 3, 276, 1986.

148. **Sheehy,, J. E., Minchin, F. R.,and Witty, J. F.,** Biological control of the resistance to oxygen flux in nodules, *Ann. Bot.,* 52, 565, 1983.

149. **Criswell, J. G., Havelka, U. D., Quebedeaux, B., and Hardy, R. W. F.,** Adaptation of nitrogen fixation by intact soybean nodules to altered rhizosphere pO$_2$, *Plant Physiol.,* 58, 622, 1976.

150. **Criswell, J. G., Havelka, U. D., Quebedeaux, B., and Hardy, R. W. F.,** Effect of rhizosphere pO$_2$ on nitrogen fixation by excised and intact nodulated soybean roots, *Crop Sci.,* 17, 39, 1977.

151. **Witty, J. F., Minchin, F. R., and Sheehy, J. E.,** Carbon costs of nitrogenase activity in legume root nodules determined using acetylene and oxygen, *J. Exp. Bot.,* 34, 951, 1983.

152. **Carroll, B. J., Hansen, A. P., McNeil, D. L., and Gresshoff, P. M.,** Effect of oxygen supply on nitrogenase activity of nitrate- and dark-stressed soybean (*Glycine max* [L.] Merr.) plants, *Aust. J. Plant Physiol.,* 14, 679, 1987.

153. **Minchin, F. R., Minguez, M., Sheehy, J. E., Witty J. F., and Skot, L.,** Relationship between nitrate and oxygen supply in symbiotic nitrogen fixation by white clover, *J. Exp. Bot.,* 36, 1103, 1986.

154. **Pankhurst, C. E. and Sprent, J. I.,** Effects of water stress on the respiratory and nitrogen-fixing activity of soybean root nodules, *J. Exp. Bot.,* 26, 287, 1975.

155. **Ralston, E. J. and Imsande, J.,** Entry of oxygen and nitrogen into intact soybean nodules, *J. Exp. Bot.,* 33, 208, 1982.

156. **Sinclair, T. R., Weisz, P. R., and Denison, R. F.,** Oxygen limitation to nitrogen fixation in soybean nodules, in *World Soybean Research Conference. III. Proceedings,* Shibles, R., Ed., Westview Press, Boulder, CO, 1985, 797.

157. **Weisz, P. R., Denison, R. F., and Sinclair, T. R.,** Response to drought stress of nitrogen fixation (acetylene reduction) rates by field-grown soybeans, *Plant Physiol.,* 78, 525, 1985.

158. **Durand, J.-L., Sheehy, J. E., and Minchin, F. R.,** Nitrogenase activity, photosynthesis and nodule water potential in soybean plants experiencing water deprivation, *J. Exp. Bot.,* 38, 311, 1987.

159. **Minchin, F. R., Sheehy, J. E., and Minguez, M.,** Characterization of the resistance to oxygen diffusion in legume nodules, *Ann. Bot.,* 55, 53, 1985.

160. **Minchin, F. R., Sheehy, J. E., and Witty, J. F.,** Further errors in the acetylene reduction assay: effects of plant disturbance,, *J. Exp. Bot.,* 37, 1581, 1986.

161. **Hartwig, U., Boller, B., and Nosberger, J.,** Oxygen supply limits nitrogenase activity of clover nodules after defoliation, *Ann. Bot.,* 59, 285, 1987.

162. **Sinclair, T. R. and Goudriaan, J.,** Physical and morphological constraints on transport in nodules, *Plant Physiol.,* 67, 143, 1981.

163. **Heckmann, M. O. and Drevon, J. J.,** Nitrate metabolism in soybean root nodules, *Physiol. Plant.,* 68, 721, 1987.

164. **Schweitzer, L. E. and Harper, J. E.,** Effect of light, dark and temperature on root nodule activity (acetylene reduction) of soybeans, *Plant Physiol.,* 65, 51, 1980.

165. **Bond, G.,** Symbiosis of leguminous plants and nodule bacteria. IV. The importance of the oxygen factor in nodule formation and function. *Ann. Bot.,* 15, 95, 1950.

166. **Summerfield, R. J.,** The contribution of physiology to breeding for increased yields in grain legume crops, in *Opportunities for Increasing Crop Yields,* Hurd, R. G., Biscoe, P. V., and Dennis, C., Eds., Pitman Publishing, Boston, 1980, 51.

167. **Evans, H. R., Hanus, F. J., Haugland, R. A., Cantrell, M. A., Xu, L. S., Russell S. A., Lambert, G. R., and Harker, A. R.,** Hydrogen recycling in nodules affects nitrogen fixation and growth of soybeans, in *World Soybean Research Conference. III. Proceedings,* Shibles, R., Ed., Westview Press, Boulder, CO, 1985, 935.

168. **Harper, J. E. and Hageman, R. H.,** Canopy and seasonal profiles of nitrate reductase in soybean (*Glycine max* [L.] Merr.), *Plant Physiol.,* 49, 146, 1972.

169. **Streeter, J. G.,** Nitrogen nutrition of field-grown soybean plants. II. Seasonal variations in nitrate reductase, glutamate dehydrogenase, and nitrogen constituents of plant parts, *Agron. J.,* 64, 315, 1972.

170. **Thibodeau, P. S. and Jaworski, E. G.,** Patterns of nitrogen utilization in the soybean, *Planta,* 127, 133, 1975.

171. **Imsande, J.,** Ineffective utilization of nitrate by soybean during pod fill, *Physiol. Plant.,* 68, 689, 1986.

172. **Attewell, J. and Bliss, F. A.,** Host plant characteristics of common bean lines selected using indirect measures of N_2 fixation, in *Nitrogen Fixation Research Progress,* Evans, H. J., Bottomley, P. J., and Newton, W E., Eds., Martinus Nijhoff, Dordrecht, 1985, 3.

173. **Delves, A. C., Carroll, B. J., and Gresshoff, P. 'A.,** Genetics analysis and complementation studies on a number of mutant supernodulating soybean lines, *J. Genet.,* 67, 1, 1988.

174. **Cove, D. J.,** Chlorate toxicity in *Aspergillus nidulans:* studies of mutants altered in nitrate assimilation, *Mol. Gen. Genet.,* 146, 147, 1976.

175. **Trewavas, A. J.,** Nitrate as a plant hormone, in *Interactions Between Nitrogen and Growth Regulators in the Control of Plant Development* Monogr. 9, Jackson, M. B., Ed., British Growth Regulator Group, Wantage, U.K., 1983, 97.

176. **Brockwell, J.,** personal communication.

177. **Betts, J. H. and Carroll, B. J.,** unpublished data.

178. **Carroll, B. J. and Gresshoff, P. M.,** unpublished data.

179. **Krotzky, A. J.,** personal communication.

180. **Carroll, B. J., Gresshoff, P. M., and Gibson, A. H.,** unpublished data.

181. **Hansen, A. P., Gresshoff, P. M., and Carroll, B. J.,** unpublished data.

182. **Carroll, B. J., Boerma, H. K., Ashley, D. A., and Gresshoff, P. M.,** unpublished data.

183. **Day, D. A., Carroll, B. J., Delves, A. C., and Gresshoff, P. M.,** Relationship between autoregulation and nitrate inhibition of nodulation in soybeans, *Physiol. Plant.,* 75, 37, 1989.

184. **Park, S. J. and Buttery, B. R.,** Nodulation mutants of white bean (*Phaseolus vulgaris* L.) induced by ethyl-methane sulphonate, *Can. J. Plant Sci.* 68, 199, 1988.

185. **Mathews A., Carroll, B. J., and Gresshoff, P. M.,** Development of *Bradyrhizobium* infections in a supernodulating and non-nodulating mutants of soybean (*Glycine max* [L.] Merr.), *Protoplasma,* in press, 1989.

Chapter 8

NODULIN FUNCTION AND NODULIN GENE REGULATION IN ROOT NODULE DEVELOPMENT

Jan-Peter Nap and Ton Bisseling

TABLE OF CONTENTS

I. INTRODUCTION

The interactions leading to a nitrogen-fixation nodule have been well analyzed from a morphological point of view.[1-5] In general, the sequence of events is as follows. Rhizobia interact with epidermal cells in the region where root hairs are beginning to emerge.[6] The root hairs respond by a marked curling due to uneven growth of the root hair, thereby entrapping bacteria within a "pocket".[7,8] Bacteria penetrate the plant cell wall through partial dissolution of the host cell wall.[8-11] Subsequently, bacteria invade the root hair cell and then the root cortex through a tube-like structure, the infection thread. As the infection threads ramify, the bacteria proliferate within the thread and become surrounded by mucopolysaccharide.[12] Meanwhile, but independently from the infection process,[13,14] cells of the root cortex enter the new developmental program of root nodule formation. At several places, cortical cells start dividing. From these centers of mitotic activity, the nodule primordia are formed.[15] Infection threads grow toward these meristematic centers, and upon contact, rhizobia bud off from the tips of the infection threads into the cytoplasm of the plant cells. This release is an endocytotic process[16] in which the bacteria become enclosed by a membrane, the so-called peribacteroid membrane, that is derived from the plasmalemma of the host cell.[17] After release, bacteria and peribacteroid membranes divide in a coordinated fashion to fill the host cell cytoplasm.[18] The bacteria differentiate into pleiomorphic bacteroids which synthesize the nitrogen-fixing system. Not all nodule cells are invaded by rhizobia. About half of the nodule cells remain uninfected. These uninfected cells occur among the infected cells.[5]

Numerous variations on nodule development have been described.[2] For example, not all leguminous plants show infection threads. In peanut (*Arachis hypogea*), infection by rhizobia is not via root hairs but by inter- and intracellular invasion.[19] Differences also exist in the susceptibility of root hairs to become infected. In alfalfa (*Medicago sativa*), the epidermal cells in the region of rapid root elongation are susceptible to infection, whereas in white clover (*Trifolium repens*) some of the mature root hairs are also susceptible to infection.[20] The way *Rhizobium* invades its nonlegume host *Parasponia* differs substantially from the infection pathways of most legume hosts[21,22] (see Chapter 9). The initial infection involves intercellular penetration of the epidermis, frequently accompanied by degradation of cortical cells. Eventually infection threads develop. Rhizobia are not released from this infection thread, do not differentiate into bacteroids, and fix nitrogen while retained within the infection threads. Persistent infection threads without bacterial release are also observed in certain tree legume nodules (*Andira* species). These nodules may represent a primitive stage in the evolution of root nodules.[23]

By their morphology, two main categories of leguminous nodules can be recognized, determinate and indeterminate nodules, although more refined classifications have been proposed.[24] In general, temperate legumes, such as *Pisum, Vicia, Trifolium,* and *Medicago*

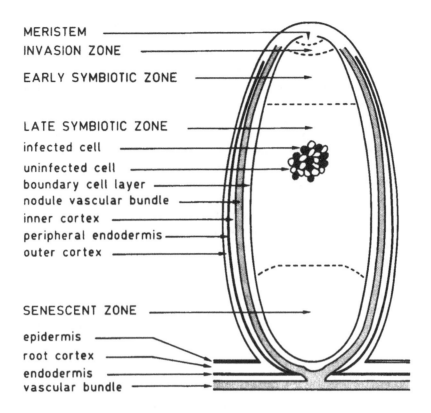

FIGURE 1.   Overall organization of an indeterminate nodule.

species, develop indeterminate nodules, while determinate nodules occur on the roots of tropical legumes such as *Glycine, Phaseolus, Vigna,* and *Arachis* species. The *Parasponia* nodule is indeterminate, but differs from legume nodules in that it has a central vascular bundle.[22,25] Nodule morphology is the result of a developmental program under control of the host plant, because for a number of *Rhizobium* strains it has been demonstrated that one and the same *Rhizobium* strain can induce determinate nodules on one host, indeterminate nodules on another,[2] and can, moreover, nodulate *Parasponia*.[26]

A schematic representation of a longitudinal section of an indeterminate nodule, modified after Sutton,[27] is shown in Figure 1. Uninfected tissue, known as the nodule cortex, surrounds the central infected zone of the nodule. A peripheral endodermis divides this nodule cortex into an outer and inner cortex. Within the latter tissue, the vascular bundles are located. Two cell layers composed of small uninfected cells, termed the boundary cell layer,[28] separate the inner cortex from the central tissue of the nodule. In the mature nodule, the central tissue is divided in distinct zones which differ in developmental stage.[29] The most distal zone is the apical meristem, which is adjacent to enlarging cells that may become infected by *Rhizobium* (the invasion zone). In the early symbiotic zone, host cells differentiate into infected and uninfected cells, while in the late symbiotic zone uninfected cells and fully packed infected cells are found. In this late symbiotic zone, nitrogen fixation and ammonia assimilation occur. The most proximal zone in the nodule is the senescent region where both plant cells and bacteroids degenerate.

In contrast to indeterminate nodules, determinate nodules exhibit a different orientation and duration of meristematic activity within the nodule. Determinate nodules do not have a persistent meristem.[14] After release of rhizobia from the infection thread, the infected cells continue to divide till about 1 week after the onset of nitrogen fixation. When mitotic activity has ceased, increase in nodule size is caused by cell expansion rather than by cell division. As a consequence, the developmental phases in a determinate nodule are separated in time rather than in space. All cells of the central tissue within a single nodule are progressing through the same stage of development.

The different nodule tissues described above are all initiated from the nodule meristem. In an orderly fashion one stage of development follows another in the proper sequence, and at each stage, cells which hitherto had shown a common lineage diverge into alternative pathways of differentiation. Unlike the developmental program of a lateral root which is entirely pericycle-derived, the legume nodule originates predominantly from cells of the root cortex. The vascular strands are positioned peripherally and not centrally. These two arguments do not apply to the *Parasponia* nodule, in which nodule growth starts in the pericycle[21] and the vascular bundle is central.[22] For this reason, the *Parasponia* nodule seems similar to a modified lateral root. The nodule meristem differentiates into nodule cells in one direction only. Also, the nodule does not form a root cap. Therefore, the nodule can be considered a unique organ, different from lateral roots.[1]

If the legume nodule is considered a unique organ, the developmental program leading to root nodules will be as complex as other developmental programs in plants[30,31] and will involve numerous genes. Central to the problem of development is the mechanism whereby the right genes are activated in the right cells at the right time. Therefore, a useful first approach to understand the nature of root nodule development seems to be the identification of the genes, the expression of which distinguishes a root nodule from other plant organs. In many *Rhizobium*-legume symbioses, the occurrence of plant-encoded, nodule-specific proteins, the so-called nodulins,[32] has now been firmly established.[33] It has become evident that differential expression of nodulin genes accompanies the development of root nodules. Considering nodulin gene expression as the most specific aspect of nodule differentiation, the study of nodulin gene expression may provide a path toward an understanding of root-nodule development.

In this chapter we will discuss nodulins, nodulin genes, the relationship between nodulin gene expression and nodule development, and the communication between the two partners in the symbiosis correlated with regulation of nodulin gene expression.

II. NODULINS AND NODULIN GENES

A. NODULIN NOMENCLATURE

By definition, nodulins are plant gene-encoded proteins which are found only in root nodules and not in uninfected roots nor in other parts of the host plant.[34] Nodulin genes are, by consequence, plant genes exclusively expressed during the development of the symbiosis. Until a defined biochemical function has been assigned to the protein, the identified nodulins are indicated by the letter N and the molecular weight as determined by SDS polyacrylamide gel electrophoresis (SDS-PAGE). In addition, we propose to add the plant genus and species initials in lower case to the N in order to facilitate the discrimination between nodulins of different plant species. If the protein turns out to be a nodule-specific form of an enzyme that also occurs elsewhere in the host plant, like glutamine synthetase (see Section II.C.3), addition of the prefix "n" to the name of the protein is recommended. *In vivo* nodule proteins, that by analysis with, e.g., antisera are shown to be nodule specific, should be indicated by Nsp and the molecular weight, until their nodulin nature has been ascertained (for discussion, see Govers et al.)[35]. In this chapter we will adopt this nomenclature, implying that some nodulins will be named differently from what previously has been published.

Nodulin genes are differentially expressed during nodule development.[36-42] The majority of nodulin genes is expressed around the onset of nitrogen fixation. Typical representatives of these nodulin genes are the leghemoglobin genes. Few nodulins are detectable at earlier stages of development, at the stage in which the nodule structure is being formed.[43-45] To account for the apparent difference in timing in expression, nodulin genes have been classified in class I and class II,[35] and in class A and class B nodulin genes,[40] but for class I/class A nodulins the term "early nodulins" is frequently used. Therefore, we will adopt the term "late nodulin" for class II/class B nodulins. Early nodulin cDNA clones are designated by ENOD,[45] while late nodulin cDNA clones re best designated by NOD,[42] to avoid confusion with the bacterial *nod* genes. The names of both NOD and ENOD clones should be preceded by the plant genus and species initials.

It should be noted that the factor "time" as discriminator between nodulins may prove to be inadequate when steps of development that are successive in one plant species, are synchronized in another species. Furthermore, there may be early nodulins yet undetected that are involved in stages earlier than the early nodulins identified so far. Some nodulins may also be involved in nodule senescence, so play a role later than the identified late nodulins. Bearing all this in mind, the proposed terminology of early and late nodulins is an operational one, one that serves our purposes for the time being, but will have to be changed as more data become available.

Histological analyses of nodule development in combination with studies on the timing of nodulin gene expression have shown that the complete nodule structure with all its defining characteristics is formed when only early nodulin genes are expressed and no late nodulin gene expression is yet detectable.[40,43] This applies to both determinate (soybean) and indeterminate (pea) nodules. Therefore, early nodulins can be involved in nodule organogenesis and the infection process, but late nodulins are not. The expression of late nodulin genes during wild-type nodule development is correlated with the onset of nitrogen fixation.[37,38] Thus, late nodulins will most probably function in establishing and maintaining a proper environment within the nodule that allows nitrogen fixation and ammonium assimilation to occur. In the following paragraphs, we will discuss in more detail the functions that have been assigned to early and late nodulins.

B. EARLY NODULINS

The only early nodulin that has been characterized in sufficient detail to warrant discussion about its function is the soybean (*Glycine max*) early nodulin Ngm-75. This nodulin was identified after analysis by two-dimensional gel electrophoresis of the polypeptides obtained upon *in vitro* translation of nodule and root RNA, respectively.[40] Hybrid released-translation experiments showed that the early nodulin cDNA clone pGmENOD2 encodes Ngm-75.[45] This cDNA clone was isolated from a soybean nodule cDNA library by differential hybridization. Using pGmENOD2 as probe on RNA transfer blots, ENOD2-homologous nodulin RNA has been demonstrated in pea (*Pisum sativum*),[43] vetch (*Vicia sativa* subsp. *nigra*),[47] alfalfa,[48] and clover.[49] The ENOD2-like sequences of pea and soybean appear to be better conserved than the leghemoglobin sequences.[41] The widespread occurrence of the ENOD2-like gene among legumes and its apparent conservation suggest a similar function of the encoded protein in the various symbioses. Interestingly, there is a significant difference in ENOD2 gene expression during nodule development between soybean and pea. In soybean, the Ngm-75 genes are transiently expressed during nodule development,[40] whereas in pea the ENOD2-like transcript accumulates during development.[41] This different pattern of expression could be related to the different developmental program of the determinate soybean nodule, in which mitotic activity decreases during development, as compared to the indeterminate pea nodule with a persistent meristem.

The function of the Ngm-75 nodulins seems not related to the infection process, because Ngm-75 gene expression is detected in nodule-like structures on soybean roots which are devoid of intracellular bacteria and infection threads.[45] Also, in alfalfa nodules induced by *R. meliloti exo* mutants in which infection threads are missing,[48] a homologous gene is expressed. Therefore, a role for Ngm-75 in the formation of a nodule structure is far more

```
MTSVLHYSLLLLLLGVVLITTPVLAN
  LKPRFFYEP
            PPIEK     PPTYE P                              PPFYK PPYYP PPVHH
                  P  PPEYQ      PPHEK  T  PPEYL P
            PPHEK P PPEYL      PPHEK  P  PPEYQ
            PPHEK             PPHEN  P  PPEHQ
            PPHEK    PPEHQ    PPHEK  P  PPEYE
            PPHEK    PPEYQ    PPHEK  P  PPEYQ
            PPHEK P PPEYQ     PPHEK  P  PPEHQ
            PPHEK    PPEHQ    PPHEK  P  PPEYQ
            PPHEK P PPEYQ     PPQEK
            PPHEK P PPEYQ     PPHEK  P  PPEHQ
            PPHEK P                                        PPVYP
            PPYEK P PPVYE     PPYEK  P                     PPVVY P
            PPHEK    PPIYE P PPLEK                         PPVYN P PPYGR Y PPSKK N
```

FIGURE 2. Amino acid sequence of the soybean early nodulin Ngm-75. The sequence should be read left to right from top to bottom. Spacing is arranged to stress the repetitive structure and pentapeptide building blocks of this protein.

likely than its functioning in the infection process. This is also suggested by *in situ* hybridization analyses that showed Ngm-75 gene expression to be confined to the inner cortex of the nodule.[325]

DNA sequence analysis of the partial cDNA clone pGmENOD2[45] and the corresponding complete genomic sequence[49] revealed that proline is the major amino acid of the Ngm-75 nodulin. The amino acid sequence is organized in highly repetitive units, composed of 17 times the pentapeptide repeat Pro-Pro-His-Glu-Lys, 6 times this repeat with only a single amino acid substitution, in addition to 13 times the repeat Pro-Pro-Glu-Tyr/His-Gln (Figure 2). These repeats are found at the core of the amino acid sequence. At the amino and carboxy terminal ends, the pentapeptide repetitive structure is maintained, although individual amino acids deviate. A putative signal peptide is present at the amino terminal sequence,[49,50] suggesting that the Ngm-75 protein is transported across a membrane. As no proline-rich proteins have been described in plants, and as soybean nodule tissue is known to be extremely hydroxyproline rich,[51] it seems likely that the prolines become hydroxylated *in vivo* to yield a hydroxyproline-rich protein.

Several classes of hydroxyproline-rich (glyco) proteins have been described to occur in plants.[52-54] These include the cell wall structural hydroxyproline-rich glycoproteins (HRGPs) known as extensins,[55-57] the arabinogalactan proteins,[58-60] solanaceous lectins,[61] and hydroxyproline-rich agglutinins.[62] These classes are, in the first place, characterized by their overall amino acid composition. Arabinogalactan proteins are very Ala-rich and are acidic; potato lectin is very rich in Cys, whereas in the extensins the amino acids Pro, Ser, His, Tyr, Val, and Lys comprise 95% of the polypeptide backbone. In addition, the extensins have a characteristic repeating pentapeptide, Ser-Hyp-Hyp-Hyp-Hyp, and a less prominent repeat Thr-Hyp-Val-Tyr-Lys.

A small family of developmentally regulated genes coding for proline-rich proteins has been identified in soybean.[63] A cDNA clone for one of these genes hybridizes with two mRNAs that differ in size. One mRNA is predominant in meristematic and elongating tissue; the other occurs primarily in the quiescent, mature region of the hypocotyl. The latter mRNA codes for a protein, designated SbPRP1, that contains a signal peptide and 43 repeats of a sequence consisting primarily of Pro-Pro-Val-Tyr-Lys,[63] whereas the mRNA from the meristematic tissue encodes a peptide that contains several Pro-Pro-Val-Glu-Lys repeats.[64] A very similar cDNA clone, 1A10, representing a gene expressed in the axis of germinating soybean seedlings, was actually proven to encode a hydroxyproline-rich protein that is part of the cell wall. In view of the structural similarities with the SbPRP proteins, we assume 1A10 to represent a gene of the SbPRP gene family.

Although the notably high glutamic acid content and the absence of Ser and Ala indicates that Ngm-75 represents a novel type of (hydroxy)proline-rich proteins, the presence of a signal peptide and the repetitive nature of the amino acid sequence suggest a relationship with the extensins and SbPRPs. Comparison of the hydropathy profiles of Ngm-75, SbPRP1, and extensin indicates indeed a remarkable similarity of the primary amino acid sequence of these proteins with respect to their hydrophilic nature and repeat structure.[49,63] The best-studied protein of these three is extensin. The extensins constitute up to 15% of the cell walls of most dicotyledonous plants.[57,65] With time, they become insolubilized in cell walls, probably due to the formation of diisotyrosine bonds.[66] The side-chain functional groups of the nonproline amino acids Lys, Tyr, His, and Glu could be used to form inter- and intracellular electrovalent or (ir)reversible covalent links with other cell wall components, indicating an important role for extensin in determining cell wall architecture.[54] Extensins have been shown to accumulate upon wounding,[67] pathogen attack,[68-70] oligosaccharide elicitors,[70,71] and ethylene treatment.[71,72] Thus, extensins are assumed to be involved in host defense responses, while a differential expression during seed development[73] indicates a role in developmental processes as well. Two thirds of the *in vivo* extensin molecule is carbohydrate,[54] illustrating the large *in vivo* modifications this protein undergoes. Preliminary studies using an extensin cDNA clone did not indicate an enhanced extensin gene expression during root-nodule formation.[49] Therefore, it is likely that different mechanisms regulate the induction of ENOD2 and extensin gene expression.

The structural features of the Ngm-75 nodulin and the similarities found with extensin, the SbPRPs, and the protein encoded by the cDNA clone 1A10[326] do not prove, but strongly suggest, that Ngm-75 is a cell wall protein closely related to the extensin and the SbPRPs. This suggestion is furthermore supported by the finding that cortical cell walls of the soybean root nodule are one of the richest sources of hydroxyproline yet found.[54]

In our laboratory, several other early nodulin cDNA clones from soybean and pea have been isolated.[49] Sequence analysis and comparison of the deduced amino acid sequence with protein data bases have shown that some other early nodulins might also be associated with the cell wall matrix, like Ngm-75,[325] while the sequence of other early nodulins does not resemble any other sequence present in DNA and protein data bases. Hence, at this time the role of the latter early nodulins in nodule development cannot be deduced.

C. LATE NODULINS
1. Leghemoglobin

The only nodulin the presence of which can be observed by the eye is leghemoglobin. This nodulin constitutes up to 25% of the total soluble protein in a nitrogen-fixing nodule, and due to its high concentration nodules acquire their characteristic, reddish color. Leghemoglobin is a hemoprotein resembling the vertebrate globins, and it has been the subject of many investigations (for review, see Appleby).[74] Leghemoglobin has a high oxygen affinity and controls as oxygen carrier the concentration of free oxygen in the nodule. It occurs in the plant cell cytoplasm.[75,76] By facilitated diffusion of oxygen, leghemoglobin provides a balanced flow of oxygen toward the bacteroids[77,78] and probably also toward the host cytoplasm.[79] Generally, leghemoglobin is detectable just before nitrogenase activity can be measured.[80-82]

Leghemoglobin is supposed to be a true "symbiotic protein" in the sense that the heme moiety is a presumed product of the bacteroid,[83] whereas the globin part is plant-genome encoded (see Bisseling et al.[84] for discussion). In contrast to a *R. meliloti* strain carrying a mutation in the gene encoding δ-aminolevulinic acid synthetase that induces white, ineffective nodules,[85] *Bradyrhizobium japonicum* strain MGL1, mutated in the same gene, still induces fully effective nodules that apparently do not suffer any heme deficiency.[86] Moreover, there is no evidence for heme transport across the peribacteroid membranes. Therefore, it is still an open question as to whether the bacterium indeed excretes the heme required for leghemoglobin synthesis.

In all legumes studied to date, more than one leghemoglobin is found in the root nodule, and the leghemoglobins are encoded by more than one gene. Soybean nodules contain four major leghemoglobins in addition to several minor components.[87] The minor components are probably the result of posttranslational modifications of the major leghemoglobins.[88] Slight differences have been observed in the time of synthesis of the different leghemoglobins during nodule development for both soybean[81,89] and pea.[90] The functional significance of the occurrence of different leghemoglobins and the differences in timing of the expression of the leghemoglobin genes is unclear. It has been suggested[90,91] that the increase in nitrogen-fixing activity in nodules is paralleled by an increase in the amount of the leghemoglobin component with the higher oxygen affinity, resulting in a more efficient nitrogen fixation.

The nonlegume *Parasponia* has only one hemoglobin gene.[92] Two hemoglobins are found in *Parasponia* nodules, a major and a minor component, that by consequence are both derived from the unique hemoglobin gene. The oxygen affinity of the major component was found to be sufficiently high to allow the nonlegume hemoglobin to function in a similar way to that of leghemoglobin.[93]

2. Uricase

The second most abundant protein in the cytoplasm of soybean (*G. max*) root nodules is Ngm-35.[32] By protein purification it was shown that Ngm-35 is the 33-kDa subunit of *n*-uricase (or uricase II),[94] a key enzyme in the ureide biosynthetic pathway used in soybean to assimilate ammonia. Uricase activity in root or leaf tissue is due to a diamine-oxidase/peroxidase system,[95] requiring a soluble cofactor.[96] This two-enzyme system has no immunological cross-reactivity with *n*-uricase. Therefore, *n*-uricase is the product of a totally different gene than the root and leaf uricase. However, using an antiserum directed against *n*-uricase, low concentrations of *n*-uricase have been observed in soybean roots[94] and callus tissue.[97] In contrast, no hybridization with RNA from any uninfected soybean tissue could be detected using an *n*-uricase cDNA clone as probe.[98] The reason for the apparent discrepancy between the results obtained with the *n*-uricase antiserum and the *n*-uricase cDNA clone is not clear. Southern blot hybridizations of soybean genomic DNA using an *n*-uricase cDNA clone as probes indicated several fragments homologous with Ngm-35 sequences, suggestive of the existence of a small number of genes.[98]

By immunocytochemistry, soybean *n*-uricase was shown to be localized in the peroxisomes of the uninfected cells.[94,98] The apparent metabolic specialization of uninfected cells in determinate nodules is also indicated by biochemical[99] and ultrastructural[100-102] data. Sequence analysis of the soybean *n*-uricase cDNA clone did not reveal a signal peptide, so information for transport into peroxisomes must reside in the protein itself. Two hydrophobic domains in the amino acid sequence may facilitate translocation across the peroxisomal membrane.[98]

3. Glutamine Synthetase

Various glutamine synthetase (GS) isozymes are found in different organs and cell compartments of plants. These octameric proteins catalyze the first reaction in the assimilation of ammonia into organic nitrogen.[103] The GS isozymes of leguminous plants are immunologically related with the GS isozymes of nonlegume plants,[104,105] but not with bacterial or mammalian GS.[106] In nodules, GS is located in the cytoplasm of the infected cells.[107] Two isozymes of GS are present in the nodules of French bean (*Phaseolus vulgaris*).[108,109] Several different GS subunits have been identified, ranging from 41 to 45 kDa, only one of which is primarily found in nodules although low levels of this subunit are also detectable in other plant organs, with the notable exception of roots.[327] One of the nodule enzymes is composed primarily of a subunit also found in the root, but the other GS is composed mainly of the "nodule-specific" subunit.[110,111] All GS subunits found in pea nodules are also detected in roots and leaves, while conflicting data have been reported for soybean.[104,107,113,114]

Sequence analysis of various GS cDNA clones from leguminous plants revealed 70 to 90% homology in the coding sequence for different GS from the same species and for GS from different species, in agreement with the immunological data. This high homology has hindered the identification of cDNA clones representing the gene for the n-GS subunit. A French bean GS cDNA clone hybridized exclusively to nodule RNA under high stringency conditions, indicating that this clone is derived from a gene for a n-GS subunit.[115] In addition, it was found that there are at least two n-GS genes.[116] In contrast to the coding sequences, the 5' and 3' untranslated regions of GS genes have diverged highly. The divergence in these regions of the GS cDNA clones has been used as a confirmation for the existence of nodule-specific GS genes. An alfalfa n-GS cDNA clone has been isolated that could be distinguished from other GS sequences because of a unique sequence in the 3' untranslated region.[117]

The amino acid homology between root and nodule GS subunits, and heterooligomeric composition of the GS proteins,[111] are both in agreement with the remarkably similar biochemical properties of these GS isozymes.[108] The differential expression of GS genes in different parts of the plant, therefore, results in functionally similar enzymes.

4. Other Metabolic Nodulins

Robertson and Farnden[118] have presented a list of enzymes assayed in the plant and bacteroid fraction from nodule tissue, a list that has been extended steadily. Several of these enzymes have a substantially higher activity in nodules compared to roots. Although the conclusion seems justified that nodulins are involved not only in nitrogen assimilation, but in all aspects of nodule metabolism, only in a few cases has the occurrence of nodule-specific forms of these enzymes been ascertained.

The soybean (G. max) nodulin Ngm-100 has been shown to be the subunit of sucrose synthase.[119] Nodule-specific forms of enzymes that differ in physical, kinetic, or immunochemical properties from the corresponding enzymes in roots have been found for phosphoenolpyruvate carboxylase,[120] choline kinase,[121] xanthine dehydrogenase,[122] purine nucleosidase,[123] and malate dehydrogenase.[124] They may prove to be the result of the expression of nodule-specific genes, but at the moment it is too early to conclude whether these nodule-specific forms are true nodulins or the result of nodule-specific modifications of root enzymes, derived from constitutively expressed genes.

5. Peribacteroid Membrane Nodulins

In soybean, a number of nodulins have been characterized that are associated with the peribacteroid membrane, but have not yet been assigned a clear biochemical function. The peribacteroid membrane is formed during release of rhizobia from the infection thread. After endocytosis, the total amount of peribacteroid membrane increases extensively, indicating a very active membrane synthesizing apparatus. The peribacteroid membrane is initially derived from the plasmalemma, but its chemical composition suggests that the endoplasmic reticulum and the Golgi apparatus also (see Chapter 5) contribute to peribacteroid membrane biogenesis.[17,125-128] The peribacteroid membrane is the physical and metabolic interface between the Rhizobium and its eukaryotic partner.[129] As such, the involvement of peribacteroid membrane-associated nodulins in the symbiosis seems self-evident.

Nodulin Ngm-24 was proven to be part of the peribacteroid membrane through the use of an antiserum directed against a synthetic peptide representing the repeated hydrophobic region of the Ngm-24 protein.[130] Ngm-24 was first identified as the 24-kDa, hybrid-released translation product of a soybean nodulin cDNA clone.[46] The protein contains a signal peptide that can cotranslationally be cleaved off *in vitro* to yield a 20-kDa polypeptide.[131] Protein blot analysis of nodule proteins with the antiserum against Ngm-24 showed that the *in vivo* Ngm-24 protein has a molecular mass of 33 kDa, which suggests that *in vivo* Ngm-24 is subject to major posttranslational modifications.[130] Ngm-24 is only located in the peribacteroid membrane and is not detected in the plasma membrane. Because Ngm-24 is incorporated in only one of both membranes, a specific targeting mechanism is able to discriminate between the plasma membrane and the peribacteroid membrane.[130] Secondary structure analysis of the Ngm-24 protein suggests that it resides as an α-helix in the surface of the peribacteroid membrane, facing the peribacteroid space.[132] Therefore, Ngm-24 is not likely to function as a transmembrane transport molecule.

Another peribacteroid membrane nodulin is Ngm-26.[132] Analysis of the secondary structure of the Ngm-26 protein indicates that it is a transmembrane protein, which is embedded in the peribacteroid membrane and faces both the peribacteroid space and the plant cytoplasm. Ngm-26 may thus function in the exchange of metabolites between *Rhizobium* and the plant.[132]

An antiserum directed against peribacteroid membrane material precipitated the 23-kDa, hybrid-released translation product of a nodulin cDNA clone, thus identifying Ngm-23 as a peribacteroid membrane nodulin.[133] Under low-stringency hybridization conditions, several other nodule-specific cDNA clones cross-hybridized with the Ngm-23 cDNA clone, indicating the presence of common characteristics. Four of these cross-hybridizing clones, Ngm-20, Ngm-26b, Ngm-27, and Ngm-44, have been characterized in some detail.[133,134] Unfortunately, not all these nodulins have been named consistently with the recommendations for nodulin nomenclature (see Section II.A). The nodulins Ngm-23, Ngm-26b, Ngm-27, and Ngm-44 are correctly named on the basis of the molecular mass of the hybrid-released translation products from the corresponding cDNA clones.[133] Ngm-23 has also been published as nodulin C-51, Ngm-44 as nodulin E-27.[39] Sandal et al.[134] base their nomenclature of Ngm-20 and Ngm-22 on the calculated molecular mass of the proteins predicted from DNA sequence data. This explains why Ngm-22 is identical to Ngm-27. With the exception of Ngm-20, for which only a calculated molecular mass is reported, we will use the nomenclature proposed in Section II.A.

Nucleotide sequence analysis of the cDNA clones that represent Ngm-20, Ngm-23, Ngm-26b, Ngm-27, and Ngm-44,[133] and of the genes encoding Ngm-20 and Ngm-27,[134] revealed that the coding sequences of these nodulins contain two regions with 70 to 90% homology, separated by a third region that is unique to each nodulin. The two conserved regions are centered around four cysteine residues, the spatial distribution of which suggests that these nodulins may be metal-binding polypeptides.[134] All these nodulins possess a potential hydrophobic signal sequence which predicts them to be associated with membranes. The cellular location of Ngm-26b and Ngm-44 has not been established conclusively yet, but the hybrid-released translation product of the cDNA clone for Ngm-27 reacted specifically with antiserum raised against soluble nodule-specific proteins and did not react with antiserum directed against peribacteroid membrane proteins.[133] This result suggests that despite the presence of a potential signal peptide, Ngm-27 is located in the cytosol of nodule cells. The result also shows that Ngm-27 does not share antigenic epitopes with the peribacteroid membrane nodulin Ngm-23; otherwise it would have reacted with the antiserum against peribacteroid membrane proteins as well. The unique region for each of the nodulins Ngm-27 and Ngm-23 results in markedly dissimilar polypeptides which apparently differ in location and presumably differ in function. The observed differences in structure between these two nodulins may thus be more significant for their location and function than their common structural characteristics.

Brewin and co-workers[135-137] have isolated monoclonal antibodies directed against peribacteroid membrane components. At the moment, no monoclonal antibodies have been isolated that recognize a nodule-specific, peribacteroid membrane-associated protein, although some antigens are substantially more abundant in nodules than in uninfected roots.

D. NODULIN GENE STRUCTURE

A number of nodulin genes have been sequenced. Their overall structure does not differ from the general structure of eukaryotic genes. They have introns that obey the intron/exon junction rules, a CAT-, and TATAA-box, and a polyadenylation site. So far, the gene for Ngm-75 is the only nodulin gene without an intron.[49]

The best-characterized nodulin genes are the leghemoglobin genes. In soybean, the structure and chromosomal arrangement of four functional genes, two pseudo- and two truncated leghemoglobin genes have been elucidated.[138-145] In addition, single (leg)hemoglobin genes from French bean,[146] *Parasponia*,[92] *Sebania rostrato*,[238] and pea[49] have been sequenced. In all cases, the coding sequences are interrupted by three introns at corresponding positions. Two introns are found at positions similar to those in the mammalian hemoglobin genes. The third central intron is unique for plant hemoglobins. Analysis of the protein structure of animal hemoglobin revealed the presence of four structural domains, predicting three introns in the DNA sequence.[147] The position of the third intron in the plant hemoglobin DNA sequence coincides exactly with the domain boundary predicted on the basis of protein structure. Therefore, plant globin genes may represent primitive globin genes[148,149] from which the central intron has not (yet?) been removed.

The soybean *n*-uricase gene encompasses almost 5000 base pairs (bp). It has seven introns, ranging in size from 154 to 1341 bp.[98] The gene coding for Ngm-24 has four introns.[131] The structure of the Ngm-24 gene is rather unusual. It contains three almost identical exons. Both the 5′ and 3′ intron sequences flanking these exons are conserved, and two of the introns are also almost identical. As a result, the Ngm-24 gene contains three direct repeats arranged in tandem in the center of its DNA sequence. The repeated sequence has the characteristics of an insertion element that may have been used in evolution to generate genes via duplication.[131] The genes for the three peribacteroid membrane nodulins Ngm-23, Ngm-20, and Ngm-27 all have one intron which separates a small exon of 7 to 15 amino acids at the carboxy terminal end from the rest of the coding sequence.[134,150] The homologies found between these nodulin genes extend to the 3′ noncoding region, suggesting that the genes encoding peribacteroid membrane nodulins have only recently been derived from a common ancestor.

E. ORIGIN OF NODULIN GENES

It seems premature to draw meaningful generalizations from the data available on nodulins and nodulin genes. Data are incomplete and obtained from a too-limited number of *Rhizobium*-legume symbioses. Yet, we will evaluate the function of nodulins from an evolutionary point of view in an attempt to place nodulin genes in the broader context of normal plant genes. It appears that all nodulins that have been assigned a clear biochemical function perform quite common tasks. There may be a nodule-specific form of an enzyme, but the function of the enzyme is an activity found at different parts within a plant, performed by related isozymes. Therefore, nodule specificity may indicate that during the evolution of symbiotic nitrogen fixation, there has arisen a need for a different regulation of already existing genetic information. Genes had to be regulated differently to function in the *Rhizobium*-controlled developmental program of the nodule, or genes adapted to the special requirements imposed by the physiological conditions within the nodule. We hypothesize that nodulin genes have been derived from genes already functioning in the plant by duplication events and have evolved to fit the constraints of the symbiosis, but remained recognizably homologous to the gene from which they were derived. As a consequence of this hypothesis, we expect to find nonsymbiotic counterparts of nodulin genes expressed in nonnodule tissue. In the remainder of this section, we will discuss the extent to which the observations on various nodulin genes examined above comply with this hypothesis.

A good example of a nodulin gene derived from an already existing gene is the gene for the "nodule-specific" subunit of GS. Differential expression of GS genes results in structurally and functionally similar enzymes in different parts of plant, including the nodule.[108] It seems likely that during the evolution of the symbiosis between *Rhizobium* and the Leguminosae, it became advantageous to have the synthesis of GS regulated to fit the conditions of the symbiosis. In both alfalfa[117] and French bean[109,151] the gene for the *n*-GS subunit is expressed in ineffective nodules induced by bacterial strains that are defective in nitrogenase activity. This shows that ammonia, the result of the symbiotic nitrogen fixation, is not involved in the regulation of the expression of the *n*-GS genes. On the other hand, in soybean it has been shown that the expression of a nonnodule-specific form of GS is regulated by ammonia.[114] This illustrates that *n*-GS genes are subject to a different regulation mechanism, due to the presence of *Rhizobium*. Because the French bean"*n*"-GS gene is also expressed in stems, but not in roots,[327] *Rhizobium* may use or have adapted the regulatory mechanisms of the GS gene expression in stems.

Less obviously derived from an already-existing plant gene is the *n*-uricase gene. The *n*-uricase is well suited to function in the special physiological conditions in the nodule, i.e., a high pH due to ammonia and a low free-oxygen tension.[94] The main uricase activity in roots is catalyzed by a diamine-oxidase/peroxidase system[95] that seems less advanced because it uses two enzymes, each with a broad substrate specificity. This two-enzyme system requires a low catalase activity, whereas catalase activity is high in peroxisomes,[152] the place where *n*-uricase is operational. Therefore, it seems that two entirely different systems have evolved, each fulfilling its task under different physiological conditions. However, *n*-uricase is also detected in callus,[97] and low activities are observed in uninfected roots,[94] showing that *n*-uricase activity is not unique for the root nodule. It is not known whether the *n*-uricase found in roots and callus is encoded by the same gene as the protein found in the nodule or by a different gene encoding a similar protein. If it is the same gene, the *n*-uricase gene can no longer be considered a true nodulin gene and need not be considered in an evaluation of the origin of nodulin genes. Assuming that the root and callus "*n*-" uricase activities are derived from different genes, a nonsymbiotic form of uricase is obviously present. The *n*-uricase gene, then, is an example of a plant gene evolved to be expressed at markedly higher levels in the root nodule, most likely as an adaptation to the different physiological conditions in the root nodule. Because *n*-uricase is also synthesized in ineffective nodules,[32,97] these physiological conditions are independent of the nitrogen-fixation process.

The early nodulin Ngm-75 is related to the soybean SbPRPs (see Section II.B6) and bears resemblance to the extensins. In several tissues, different SbPRP genes are expressed,[63,64] indicating that related molecules may function in different developmental programs, controlled by distinct regulatory mechanisms. The striking structural, and therefore, presumably, also functional, similarity of the nodulin Ngm-75 to the nonnodule-specific SbPRPs suggests a relationship between the corresponding genes. The nature of this relationship, however, is not obvious. On the basis of the structural similarities between Ngm-75, the SbPRPs, and extensin, a common ancestor seems likely. The genes for these cell wall proteins may have been derived from a common plant gene encoding a small, proline-rich protein that acquired a function as a cell wall constituent. Numerous gene duplications and divergence then resulted in proteins that function more or less similarly in different developmental situations. On the other hand, on an evolutionary time scale, hypocotyl and root differentiation are older than nodule formation, so the Ngm-75 genes may be derived from the SbPRP genes. Irrespective of the exact evolution of the Ngm-75 genes, it complies with the hypothesis that the genes for the early nodulin Ngm-75 also have been derived from an already-functioning plant gene.

So far, most nodulin genes can be assigned a nonsymbiotic counterpart. What, then, about the archetype of the nodulins, leghemoglobin? Leghemoglobin gene expression has never been detected in any part of a leguminous plant other than the nodule, so the absence of a nonsymbiotic counterpart seems fairly well established. There has been a great deal of speculation whether the leghemoglobin genes were acquired by horizontal genetic transmission either from *Rhizobium* itself or from an animal vector.[148] Detailed analysis of the soybean leghemoglobin gene family[153] suggest, however, that globin genes were already present in the common ancestor of present-day plants and animals. The presence of a hemoglobin gene in the nonlegume *Parasponia,* that has a high sequence homology and the same gene structure as the leghemoglobin genes in legumes,[92] has profound consequences for considering the evolution of these genes. The *Parasponia* hemoglobin sequence substantiates the likelihood of a vertical evolution of the globins starting from a hypothetical ancestor before the radiation of animals and plants. In principle, all plants could have the globin sequence. Indeed, the presence of leghemoglobin-like sequences in various nonlegumes has been reported.[154] In fact, a *Parasponia* cDNA clone for hemoglobin hybridizes to hemoglobin genes in the distantly related *Casuarina,* which has a nitrogen-fixing symbiotic association with the actinomycete *Frankia.*[92] It also hybridizes to presumably related sequences in the DNA of *Trema,* a close relative of *Parasponia,* that does not nodulate.[92] In addition, it has been found that in *Parasponia* and *Trema* roots hemoglobin gene expression can be detected at a low level.[155] Because there occurs only one hemoglobin gene in *Parasponia* DNA, this same gene must be expressed in roots as well as in nodules. Thus, in *Parasponia,* the expression of a hemoglobin gene can be detected in nonsymbiotic tissue. Therefore, also in legumes there may be a nonsymbiotic counterpart for leghemoglobin that is transcribed in nonsymbiotic tissue, although expression of such a gene has not been detected yet.

Various laboratories have encountered problems in identifying leghemoglobin sequences in one species using a cDNA clone from another species (pea/soybean, *Parasponia*/soybean, alfalfa/pea) despite the structural homologies among the various leghemoglobins. For instance, *Parasponia* leghemoglobin is 40% homologous with soybean leghemoglobin, yet a cDNA clone for soybean does not cross-react with a *Parasponia* leghemoglobin cDNA clone.[92] Therefore, it appears conceivable that the hypothetical nonsymbiotic form of a leghemoglobin gene has diverged beyond the point of recognizability. On the other hand, this hypothetical nonsymbiotic form may be expressed at very low levels. Soybean contains a leghemoglobin gene that seems to meet all requirements for a functional gene, but is nevertheless thought to be a pseudogene, because the encoded leghemoglobin contains methionine, and a methionine-containing leghemoglobin has never been found in soybean nodules.[140,143] This Lb pseudogene may, in fact, represent a nonsymbiotic leghemoglobin. Sequence analysis of the 5'-flanking region showed some minor deviations from functional leghemoglobin genes. These alterations may withdraw the gene from nodule-specific regulation of expression.

Thus, the leghemoglobin genes also may fit the hypothesis that there is a form that is nonsymbiotic, with or without a function. In this concept, the encoding nodulin genes have evolved from common plant genes in an adaptation to the specific regulatory and/or physiological requirements of root-nodule formation and symbiotic nitrogen fixation.

Nonsymbiotic counterparts have not been identified for peribacteroid membrane nodulins, but it should be realized that the analysis of plant membranes is a relatively new field of research,[156] and the relevant analogous genes may simply not have been identified yet. To illustrate our point, Ngm-26 has recently been found homologous to the major intrinsic protein of the bovine eye lens fiber membrane.[329] The gene for Ngm-24 is characterized by three direct repeats arranged in tandem, and gene duplication has been implicated in the generation of this gene.[131] Because there is very little sequence divergence among the three repeated units of the Ngm-24 gene, the duplication events must have taken place relatively recently in evolution. The peribacteroid membrane nodulins Ngm-20, Ngm-23, Ngm-26b, Ngm-27, and Ngm-44 all have regions in common and regions unique to each nodulin.[133,134] Whereas the regions in common may originate from duplication events, presumably starting from a common ancestor, the regions unique to each nodulin suggest the occurrence of various and different recombination events.[133] The homologies observed in the 3'-noncoding regions indicate that these recombination events also have taken place relatively recently in evolution. The apparent diversity of recombination events during a short evolutionary period has been suggested to represent a "trial-and-error" mechanism still in progress for the generation of nodule-specific functions.[133] Recombination events are more likely to result in totally new genetic information than duplications and subsequent divergence. The peribacteroid membrane nodulin genes may, therefore, be more unique than the nodulin genes discussed above.

F. NODULIN GENE REGULATION
1. Regulating Sequences
From the viewpoint of gene regulation the sequence of special interest is the 5'-promoter region. The importance of 5'-flanking sequences in the regulation of nodulin gene expression is demonstrated by nodule-specific expression of a chimeric leghemoglobin gene consisting of the chloramphenicol acetyltransferase coding sequence flanked by the 5' region of the soybean Lbc$_3$ gene in transgenic birdsfoot trefoil (*Lotus corniculatus*)[157] or white clover (*Trifolium repens*)[158] plants. It has been found that a 2-kb region upstream of the start of transcription of the soybean Lbc$_3$ gene is sufficient for nodule-specific and developmentally correct expression of the chimeric gene.

Sequence homologies have been found between the promoter regions of different soybean leghemoglobin genes[144] and a French bean leghemoglobin gene,[146] including an approximately 30-bp sequence surrounding the cap site that is conserved between soybean, French bean, and the animal globin genes.[153] On the other hand, this "globin box" and the other homologous sequences are found neither in a pea leghemoglobin gene promoter[49] nor in the *Parasponia* hemoglobin gene promoter.[92] The homologous sequences found between the promoter regions of the soybean and French bean leghemoglobin genes may therefore reflect more the taxonomic relatedness of *G. max* and *Phaseolus vulgaris*, than the requirement of these sequences for nodule-specific gene expression. The homologies found in the promoter sequences of the four functional soybean leghemoglobin genes do not explain why these genes are sequentially activated during nodule development,[89] nor do they explain why the respective genes are transcribed to a different extent.[89] In terms of regulation of leghemoglobin gene expression, the differences in the 5'-flanking regions may be as important as the observed homologies.

The genes for the peribacteroid membrane nodulins Ngm-20, Ngm-23, Ngm-27, Ngm-44, and Ngm-24 are activated at approximately the same stage during nodule development, just before the onset of nitrogen fixation and together with the Lb genes.[131,133,134] Therefore, these genes might be subject to one and the same regulatory mechanism and share common regulating sequences. By comparison of the upstream regions from the Ngm-20, Ngm-23, Ngm-27, and Ngm-44 genes over a length of approximately 200 nucleotides, two sequences, of five and six bases, respectively, are identified that occur in front of the initiation ATG in the 5'-flanking region of all sequenced late nodulin genes,[134] including all Lb genes and the Ngm-24 gene, except the *n*-uricase gene. These consensus motives may indicate that rather small regulatory sequences are involved in the regulation of the nodule-specific expression of nodulin genes.

Comparison of the 5'-flanking region of the soybean Lbc_3 gene with the upstream regions of the Ngm-24 and Ngm-23 genes over a length of approximately 750 nucleotides in front of the coding sequence has led to the identification of three other consensus sequences of 23, 9, and 8 nucleotides, respectively, that are putative regulatory sequences involved in induction of the expression of these three late nodulin genes.[150] These sequences are claimed to be missing in the 5'-flanking sequence of the n-uricase gene, probably due to the fact that the n-uricase gene is expressed in another cell type, and as such is subject to another regulatory mechanism. However, only 150 nucleotides in front of the initiation ATG of the n-uricase gene have been considered.[98] Also, the DNA sequences from the promoter regions of all other sequenced late nodulin genes have not been determined over a sufficient length in front of the initiation ATG to allow comparison, so the identified consensus sequences may occur in these genes.

Recently, two sequences of 16 and 24 nucleotides, respectively, have been identified in the promoter region of a soybean leghemoglobin gene that are capable of binding one and the same transacting factor.[159] Interestingly, this transacting factor occurs in relatively high concentrations in nodules and is nodule specific. This factor may, therefore, represent a "regulatory" nodulin. The two sequences involved in binding the transacting factor are located up to 250 nucleotides in front of the start of transcription. Deletion studies show that deletion of both binding sites abolishes promoter activity completely, but deletion of only the most upstream binding site has only a slight effect.[330] There may be a cooperative interaction between both binding sites. The positions of these sequences do not correspond to, nor coincide with, any of the consensus motives found by aligning sequences of the different nodulin genes. Thus, it appears that sequence comparisons are less meaningful in predicting promoter sequences involved in nodule-specific gene expression. For a definitive identification of the sequences responsible for nodule specificity of gene expression, nodulin genes other than the soybean Lbc_3 gene should be used to transform leguminous species amenable to regeneration,[160-164] together with deletion studies of the promoters of the genes.

2. Other Regulating Factors

Although the ultimate control over nodule specificity of gene expression will be exerted at the level of the gene, the physiological conditions within the root nodule are repeatedly claimed to play a role in the regulation of nodulin gene activity. In particular, oxygen, ammonium, heme, and phytohormones have been implied in the regulation of nodulin gene expression.

The late nodulin genes are first expressed just before the onset of nitrogen fixation. It is arguable that the actively dividing rhizobia create within the nodule an environment with a low oxygen concentration, which may act upon late nodulin gene expression, in particular upon leghemoglobin gene expression.[84] Due to nodule structure, bacterial respiration, and leghemoglobin, the free oxygen concentration is kept low, thus allowing the extremely oxygen-sensitive enzyme nitrogenase to function.[165] In an attempt to mimic the conditions for nodulin gene induction, pea roots have been grown under different oxygen concentrations.[79] None of the genes, the expression of which is enhanced in pea roots grown under low oxygen concentrations, belonged to the genes previously identified as nodulin genes. Therefore, a low free-oxygen concentration alone is not sufficient for the induction of nodulin gene expression.

Moreover, it was found that in root nodules alcohol dehydrogenase genes are not expressed at high levels,[79] in agreement with alcohol dehydrogenase activity not being significantly higher in nodules compared to roots.[166] The expression of alcohol dehydrogenase genes is a well-recognized parameter of oxygen stress, because these genes are expressed at markedly higher levels under microaerobic conditions. The low alcohol dehydrogenase activity in nodules shows that, apparently, the nodules do not experience a low free-oxygen concentration. Hence, it is unlikely that a low oxygen concentration within the nodule acts upon late nodulin gene expression.

Oxygen has been postulated to be a factor regulating *n*-uricase gene expression in soybean root nodules. In callus tissue, the specific activity of *n*-uricase was enhanced maximally threefold by lowering the oxygen concentration.[97] However, in these experiments the *n*-uricase activity was already detected in callus cultures kept under normal oxygen concentrations; therefore, it is unclear whether the observed effect of oxygen concerns a true nodulin gene. Furthermore, in nodules the specific activity of *n*-uricase increases at least 150-fold. The physiological significance of the observed threefold enhancement seems limited; hence, the role of oxygen in inducing *n*-uricase gene expression remains doubtful. On the other hand, whereas the oxygen concentration in the nodule does not seem to be of importance for plant gene expression, the *nifA* gene of both *R. meliloti*[167] and *B. japonicum*[168] have been shown to be induced in free-living bacteria when the oxygen concentration is reduced to microaerobic levels.

Heme has been shown to regulate the expression in yeast of a chimeric gene consisting of the 5'-flanking region of the soybean Lbc_3 gene and the coding sequence of the neomycin phosphotransferase gene at the posttranscriptional level.[169] Free heme may decrease the enzyme activity of the nodulin sucrose synthase, because it was found that *in vitro* the enzyme sucrose synthase dissociates rapidly in the presence of free heme.[119] Although heme may thus influence the activity of nodulins, the significance of any heme-mediated mechanism *in vivo* is unknown. There is no evidence that heme is involved in the induction of nodulin gene expression.

Ammonia has been implicated in the regulation of expression of GS genes in soybean,[144] but the GS genes investigated appeared not to be nodulin genes. Three lines of evidence strongly suggest that ammonia is not involved in the induction of the expression of nodulin genes. First, in normal nodule development, all nodulin genes are expressed before nitrogen fixation starts.[38,40] Second, all nodulin genes are expressed in nodules induced by *Rhizobium* and *Bradyrhizobium nif* and *fix* mutants.[38,40] Third, on the roots of cowpea (*Vigna unguiculata*) grown in an argon/oxygen environment containing negligible amounts of nitrogen gas, nodules could be induced by *Rhizobium* in which both nitrogenase activity and leghemoglobin were detectable.[170] Neither the availability of nitrogen nor the result of nitrogenase activity, ammonia, are thus essential for the induction of nodulin gene expression.

Phytohormones affect processes of a wide diversity in plants, and they are considered crucial to the developmental programming of plants.[171] As such, the involvement of phytohormones in nodule development does not seem controversial. Exogenous application of hormones to pea roots and root explants resulted in the induction of cortical cell divisions similar to those found in early nodule development.[172,173] Auxin, auxin-like substances, and kinetin cause hypertrophies on the roots of leguminous and nonleguminous plants that are easily mistaken for nodules.[174-176] Interestingly, in the structures formed on alfalfa roots by the auxin transport inhibitors *N*-(1-naphthyl)phtalamic acid and 2,3,5-triiodobenzoic acid, the early nodulin genes MsENOD2 ans Ms-30 are expressed.[331] Cytological analysis also shows that the structures resemble *Rhizobium*-induced nodules in several aspects.[331] Apparently, these auxin transport inhibitors mimic rhizobial signals in some way.

The root nodule contains up to 100-fold higher concentration of three major groups of plant hormones, auxins,[177,178] cytokinins,[179-181] and gibberillic acid,[182-183] relative to the hormone content of the uninfected root. Because auxins[184,185] and cytokinins[186] are known to be produced by *Rhizobium* in pure culture, the hypothesis is appealing that these phytohormones are produced by *Rhizobium* and trigger plant cell division.[1,172] Analysis of the signaling between plant and *Rhizobium* has indicated that the *Rhizobium* genes required for nodulation may produce signals that interfere with the hormone housekeeping of the root (see Section IV.B). A group of soybean proteins which appeared 3 d after infection with *Bradyrhizobium* also appeared after treatment of roots with auxin.[187] On the other hand,

Rhizobium mutants completely ineffective in phytohormone synthesis have not been identified. Mutants producing only a small amount of auxin are for the greater part not symbiotically defective.[184,188] Therefore, the data available to date do not allow a conclusion with respect to the role of *Rhizobium*-produced phytohormones in the regulation of nodulin gene expression.

3. *Rhizobium* Signals

When nodulin gene expression is not induced via direct alterations in the physiological conditions of the nodule, the invading *Rhizobium* itself will be the factor that does deliver inducing signals for the expression of nodulin genes. In the sections to follow, we will survey the bacterial genome as the origin of causative signals that induce and regulate nodulin gene expression. The nature of those signals is unknown, but the mechanism of nodulin gene expression and, hence, the signals regulating them appear to be conserved in different plant species.

The most convincing evidence for the conservation of the regulation of Lb gene expression is the nodule-specific and developmentally correct expression of a chimeric soybean Lb gene (see Section II.F.1) in transgenic birdsfoot trefoil[157] and clover[158] plants. These results indicate a conserved mechanism for the induction of leghemoglobin gene expression independent from differences in the developmental program of determinate (soybean and birdsfoot trefoil) and indeterminate (clover) nodules. Some naturally occurring, broad host-range rhizobia induce nitrogen-fixing, i.e., with all their nodulin genes properly expressed, root nodules on the roots of a variety of legumes as well as on the roots of the nonlegume *Parasponia*.[26] The very same *Rhizobium nod* genes are involved in the nodulation of all these host plants (see Section IV.B). Therefore, irrespective of their exact nature, the signals from the invading rhizobia that regulate the induction of nodulin gene expression should be very alike in different plants.

III. NODULE DEVELOPMENT AND NODULIN GENE EXPRESSION

Two classes of nodulin genes, early and late nodulin genes, have been revealed by a first analysis of wild-type nodule development. The analysis can be refined by examining in more detail the coupling of histological and molecular biological data through the study of nodules blocked at different stages of development.[189] Correlations between nodule structure and nodulin gene expression may provide clues to the role of nodulins in the developmental program of the nodule. To date, analyses have predominantly been done with nodules that are formed by mutated and engineered bacterial strains. This is because the *Rhizobium* genome can easily be manipulated in comparison to the plant genome.

By classic genetic experiments, several plant genes involved in nodulation and symbiotic nitrogen fixation have been identified in pea,[190,191] soybean,[28,191-194] clover,[195,196] and alfalfa.[197,198] Similar to the control of nodule development by *Rhizobium*, mutations in the plant genome can also result in disturbed nodule development, varying from the absence of nodules to the development of wild-type-like, but ineffective, nodules.[189] Generally, the nature of the plant mutation is not known, and nodulin gene expression has not been analyzed. Therefore, we will not take these plant-conditioned disturbances of nodule development into account.

In the following paragraphs we will discuss the relationship between nodule development and nodulin gene expression, itemized for the different plant species pea, vetch, alfalfa, and soybean. Despite the many reports of clover nodules disturbed in development, the plant species clover is omitted in the following discussions because knowledge about nodulin gene expression in clover is lacking.

A. PEA AND VETCH

The correlation between nodule structure and nodulin gene expression has been extensively studied in the plant species vetch[46,47] and pea,[39,41] both belonging to the same cross-inoculation group. On the roots of these plants, *R. leguminosarum* induces the formation of indeterminate nodules. In nodules induced on vetch (*Vicia sativa* subsp. *nigra*) by wild-type *R. leguminosarum*, two early nodulins, ENOD2 and Nvs-40, and 15 late nodulins have been identified.[47] In nodules induced on pea (*P. sativum*), two early, ENOD2 and Nps-40', and 20 late nodulins have been found.[38]

Figure 3 presents the overall morphology for nodules induced on vetch by the bacterial strains indicated. The pattern of nodulin gene expression in the various nodule types is shown for the early nodulins ENOD2 and Nvs-40 and the late nodulins Nvs-65 and leghemoglobin (Table 1). In case a particular nodule phenotype has been found exclusively for pea and not for vetch, pea nodule morphology and the pattern of pea nodulin gene expression have been included in the following discussions.

Rhizobium strains mutated in one of the *nif* or *fix* genes (see Section IV.A) induce the formation of nodules on pea and vetch that are morphologically similar to nodules induced by wild-type *Rhizobium*. In these nodules, rhizobia differentiate into the characteristic bacteroidal shape, and all nodulin genes are expressed.[38,41,47] Nitrogen fixation per se is apparently not essential for the induction of the expression of nodulin genes.

Strain P8 is a *Rhizobium* wild-isolate that induces ineffective nodules on pea, in which bacteria are released from the infection threads, but do not differentiate into the characteristic Y-shaped bacteroids as do wild-type bacteria. All pea nodulin genes are expressed in the nodules induced by P8.[38] Thus, induction of the expression of nodulin genes does not depend on bacteroid development.

Strain 248ᶜ(pMP104) is *R. leguminosarum* 248,[199] containing essentially a 12-kb *nod* region from a *R. leguminosarum* Sym plasmid.[200] Strain 248ᶜ(pMP104) induces nodules on pea and vetch that have the same histological organization as wild-type nodules, including the development of infected and uninfected cells. In this case, all early and late nodulin genes identified are also expressed.[46] The same is true for strain ANU845(pMP104), which contains the 12-kb *nod* region in a *R. trifolii* chromosomal background. These results show conclusively that the *nod* genes are the only Sym plasmid genes required for the induction of nodulin gene expression (see Section IV.B).

Strain ANU845(pRt032) (*nodE* K11::Tn5) contains essentially a 14-kb *nod* region from a *R. trifolii* Sym plasmid[201] with a Tn5 insertion in the *nodE* gene. The mutation in *nodE* extends the host range of this *R. trifolii* strain to the pea/vetch cross-inoculation group.[202] The nodules formed on vetch by this strain deviate from nodules induced by wild-type strains in the structure of the late symbiotic zone.[49] The late symbiotic zone contains two to four layers of infected and uninfected cells. In the proximal part of the nodule, a large area of senescing tissue is present almost without any bacteria. In these nodules, both early nodulin genes ENOD2 Nvs-40 are expressed. Only the late nodulin gene Nvs-65 is transcribed, but no mRNA from the other late nodulin genes, including the leghemoglobin genes, is detectable. Apparently the class of late nodulin genes must be divided in two subclasses that seem to be regulated in a different manner. The absence of Lb transcripts in the nodules induced by ANU845(pRt032) (*nodE* K11::Tn5) suggests that the Lb genes are first expressed when the nodule meristem cells are fully differentiated into infected and uninfected cells. Immunocytological localization of leghemoglobin in pea wild-type nodules supports this suggestion. In these nodules Lb is not detectable in the early symbiotic zone, nor in the first two cell layers of the late symbiotic zone.[203]

Strain LBA4301(pMP104) is a Ti plasmid-cured *Agrobacterium tumefaciens*[204] containing the same 12-kb *nod* region from a *R. leguminosarum* Sym plasmid as strain 248ᶜ(pMP104).[46]

Strain LBA4301(pMP104) induces nodules in which bacteria are released from the infection threads. The bacteria then become surrounded by a peribacteroid membrane. However, upon release from the infection thread, the bacteria are degraded despite the presence of a peri-bacteroid membrane, and they never develop into bacteroid-shaped structures. At the same time, the organelles in the plant cytoplasm of these "infected" cells also disintegrate. The uninfected cells appear normal, as judged by electron microscopic observations, and the early nodulin genes ENOD2 and Nvs-40 are expressed.[46] On the other hand, none of the late nodulin transcripts, including Nvs-65 and Lb, are detectable. Because the differentiation into uninfected cells in the LBA4301(pMP104)-induced vetch nodules appears normal, all late nodulin genes that are not expressed in these nodules are probably expressed in the infected cells and not in the uninfected cells of nodules induced by wild-type *Rhizobium*. Absence of the late nodulin Nvs-65 in the nodules formed by LBA4301(pMP104) suggests that release of bacteria from the infection thread is not sufficient for the induction of expression of the Nvs-65 gene. Comparison between the nodules formed by ANU845(pRt032) (*nodE* K11::Tn5) and LBA4301(pMP104) indicates that the Nvs-65 gene is probably first expressed in the youngest cells of the late symbiotic zone that are completely filled with bacteria.[46]

To date, no bacterial strains are available that induce on vetch roots the formation of nodules in which release of the bacteria from the infection threads is not observed. Such nodules are formed on pea roots[43] by the *Agrobacterium* transconjugant LBA2712, containing a complete *R. leguminosarum* Sym plasmid instead of the Ti plasmid. The vetch (*V. sativa*) Nvs-40 can be immunoprecipitated with an antiserum raised against the pea (*P. sativum*) early nodulin Nps-40',[49] and the soybean ENOD2 cDNA clone cross-hybridizes with both a pea[41] and a vetch[47] ENOD2-like early nodulin mRNA. This shows that the early nodulins Nps-40'/Nvs-40 and pea-ENOD2/vetch-ENOD2 are closely related. Therefore, the expression of these early nodulin genes could be studied in pea nodules as a substitute for vetch nodules. Only the ENOD2 gene is expressed in LBA2712-induced nodules, while the Nps-40' gene and all pea late nodulin genes are not transcribed in these nodules.[43] If one assumes that the Nvs-40 and Nps-40' genes are regulated similarly, the difference in Nvs-40/Nps-40' gene expression observed between nodules induced on vetch by LBA4301(pMP104) and nodules induced on pea by LBA2712 suggests that Nvs-40 gene expression is related to release of the bacteria from the infection threads and/or to the subsequent differentitation into infected and uninfected cells.[46] In view of the results obtained by analyses of nodulin gene expression in alfalfa nodules induced by *exo* mutants (see Section III.B), it is most likely that the expression of the Nvs-40 gene is related to the differentiation into infected and uninfected cells.

From the detailed studies on pea and vetch, it can be concluded that now at least four classes of nodulin genes can be distinguished on the basis of their expression in develop-mentally disturbed nodules. Both early and late nodulin genes can each be divided into two subclasses, the expression of which correlates with a stage of root nodule development, the formation of a nodule structure (ENOD2), differentiation into infected and uninfected cells (Nvs-40, Nps-40'), packing of the infected cells (Nvs-65), and subsequent processes (Lb and other late nodulins), respectively.

B. ALFALFA

R. meliloti induces on alfalfa (*M. sativa*) roots the formation of indeterminate nodules. In nodules induced on alfalfa by a wild-type *R. meliloti*, about 20 nodulins have been identified.[117,205-207] Two early nodulin genes, Nms-30 and ENOD2, have been found. The

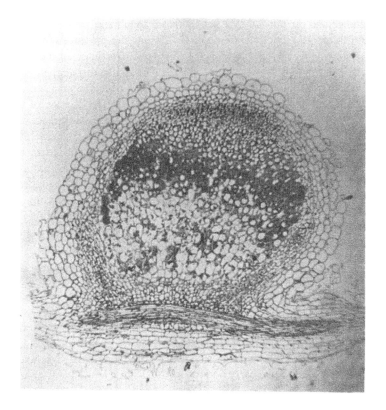

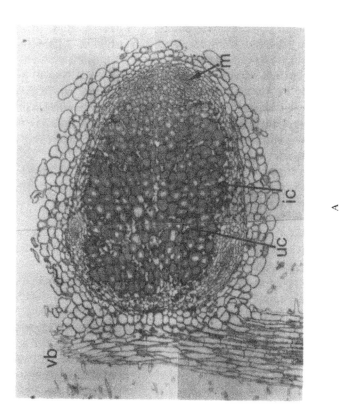

FIGURE 3. Light micrographs of nodules induced on vetch (A, B, and C) and pea (D), and the pattern of expression of two early (ENOD2 and Nvs-40) and two late (Nvs-65 and leghemoglobin) nodulin genes in these nodules. (A) Nodule induced on vetch by strain ANU845 (pMPl04); (B) nodule induced on vetch by strain ANU845 (pRt032) (*nodE* K11 :: Tn5); (C) nodule induced on vetch by strain LBA4301 (pMPl04); (D) nodule induced on pea by strain LBA2712. Because there in no evidence for a pea counterpart for the vetch Nvs-65, a blank is incorporated in the figure on this position. Lb, leghemoglobin; m, meristem; ic, infected cell; uc, uninfected cell; vb, vascular bundle.

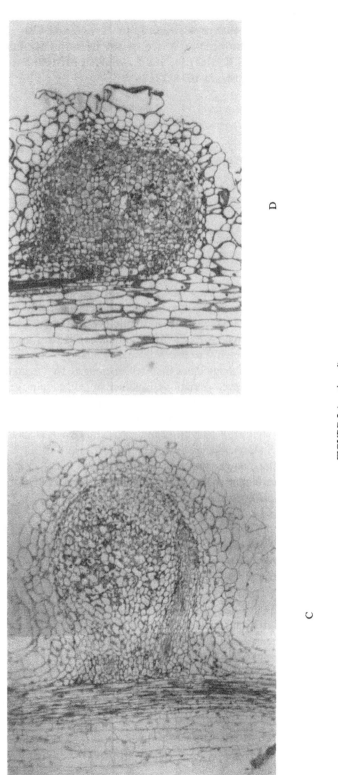

D

C

FIGURE 3 (continued)

TABLE 1

**Pattern of Early (ENOD2 and Nvs-40/Nps-40′) and Late (Nvs-65 and Lb)
Nodulin Gene Expression in Nodules Induced on Vetch by the Bacterial Strains
248ᶜ(pMP104), ANU845(pRt032) *nodE* K11::Tn5), and LBA4301(pMP104) and
on Pea by the Strain LBA 2712**

		Nodulin gene expression			
		Early		Late	
Bacterial strain	Host plant	ENOD2	Nps-40 Nps-40′	Nvs-65	Lb
248ᶜ(pMP104)	Vetch	+	+	+	+
ANU845 (pRt032) (*nodE* Kll::Tn5)	Vetch	+	+	+	−
LBA 4301 (pMP104)	Vetch	+	+	−	−
LBA 2712	Pea	+	−	?	−

alfalfa ENOD2-like early nodulin has approximately 80% amino acid homology with the soybean Ngm-75 early nodulin.[48]

Detailed electron microscopic observations on alfalfa nodules induced by various *R. meliloti nif* and *fix* mutants revealed some minor deviations in structure compared to wild-type nodules.[208,209] These deviations occur after release and maturation of the bacteroids. Nodules remain small due to rapid senescence. A *R. meliloti nifA* mutant induces nodules in which bacteroid maturation appears interrupted. The bacteroids rarely attain the dimensions or appearance of wild-type bacteroids,[209] but the nodules themselves elongate for the greater part to wild-type nodule dimensions. Whatever deviations in structure, all nodulin genes were expressed in nodules induced by the *nif* and *fix* mutants that have been investigated,[117,206] as is the case in pea and vetch nodules induced by *nif* and *fix* mutants.

The *R. meliloti* mutants *exoA* through *exoF* lack an acidic extracellular polysaccharide in their cell wall.[210,211] These strains induce nodules on alfalfa totally devoid of infection threads and intracellular bacteria.[211] This nodule phenotype will be referred to as "empty." The rhizobia are restricted to the intracellular spaces in the nodule outer cortex. Only the early nodulin genes ENOD2 and Nms-30 were expressed in these empty nodules,[48,117] whereas none of the late nodulin gene transcripts were detected.[117,206] The same pattern of nodulin gene expression is found in nodules induced by *Agrobacterium* strains carrying the *R. meliloti* Sym plasmid,[206] or the cloned *nod* genes,[48] which both induce empty nodules resembling the nodules induced by *exo* mutants, and in nodules induced by a *R. meliloti exoH* mutant,[48] which fails to succinylate its acidic extracellular polysaccharide and induces nodules containing aborted infection threads and infrequently released bacteria.[212]

The alfalfa early nodulin Nms-30 (also named Nms-38[207,212]) immunoprecipitates with antiserum directed against pea Nps-40′,[49] just as the vetch nodulin Nvs-40 does.[49] Apparently, a structurally similar early nodulin occurs in all three species. Assuming that pea Nps-40′, vetch Nvs-40, and alfalfa Nms-30 are not only structurally but also functionally related, the pattern of expression of Nvs-40 and Nms-30 in vetch and alfalfa nodules, respectively, allows a potential role for this nodulin to be deduced. From the studies on vetch nodulin gene expression, it is concluded that the expression of this nodulin is related to the release of bacteria from the infection thread and/or to the differentiation into infected and uninfected cells. Infection threads and infected cells are not present in the nodules induced on alfalfa by *exo* mutants, but the Nms-30 gene is expressed. Therefore, the expression of this nodulin gene is related to the differentiation into uninfected cells.

The absence of Nps-40′ in the nodules induced on pea by *Agrobacterium* LBA2712,[43] in which bacteria are not released from the infection threads, seems contradictory to this relationship. The apparent contradiction can, however, be explained in the following way.

The cells in the LBA2712-induced pea nodule may be without bacteria and morphologically resemble the uninfected cells in the empty alfalfa nodule, but differ on the molecular level. It is conceivable that the cells without bacteria in the pea nodule have not differentiated into uninfected cells, whereas the cells without bacteria in the alfalfa nodule are similar to the uninfected cells occurring in mature nodules. This is consistent with the localization of ENOD2 gene expression in the inner cortex of the nodule.[325]

Because the ENOD2 gene is expressed in nodules without infection threads, the early nodulin gene ENOD2 is involved in the establishment of a nodule structure and not in the infection process per se.

C. SOYBEAN

Bradyrhizobium japonicum (and some *Rhizobium fredii* strains) induces on soybean (*Glycine max*) roots the formation of determinate nodules. In nodules induced on soybean by a wild-type *B. japonicum,* more than 20 nodulin mRNAs have been identified.[40,213] At least four nodulins, Ngm-75, Ngm-44b (nomenclature has been changed to distinguish Ngm-44b from the peribacteroid membrane nodulin Ngm-44; see Section II.C.5), Ngm-41, and Ngm-38, are already found at the time of development when a globular meristem has been formed. These early nodulin genes are transiently expressed during nodule development. Except for the early nodulin Ngm-44b, their mRNAs increase in concentration up to the stage in which the complete nodule structure is established, and then decrease in concentration. The concentration of Ngm-44b mRNA remains constant in this period and then decreases as well. Because Ngm-44b gene expression follows meristematic activity, the expression of the Ngm-44b gene may be correlated with meristematic activity in the nodule.[40] Also, several late nodulin cDNA clones have been isolated.[37,39]

Nodules induced on soybean by *B. japonicum* strains mutated in genes for *nif* and *fix* functions develop similar to wild-type nodules,[214] just as in the case of pea, vetch, and alfalfa. All nodulin genes are expressed in these ineffective nodules.[37,39,40] This also applies to nodules formed by a *B. japonicum nifA* mutant,[215] but the nodules induced by this strain are severely disturbed in nodule development.[216] The bacteria are released from the infection thread, but do not multiply extensively. Moreover, the infected cells collapse at an early stage of development. On the other hand, the uninfected cells in these *nifA* mutant-induced nodules appear normal. Although all nodulin genes are expressed, the concentration of leghemoglobin and of leghemoglobin mRNA is reduced drastically.[215] These data will be discussed in Section IV.A.

Strain HS124 is an ill-defined *B. japonicum* mutant obtained by UV irradiation.[217] In the nodules formed by this mutant, bacteria are rarely released from the infection thread, and the few cells that have become infected appear to degenerate immediately.[40,217] The uninfected cells in the HS124-induced nodules appear normal. All early nodulin genes are expressed, but interestingly their expression is no longer transient.[40] Most of the identified late nodulin genes are expressed at approximately the same level of expression as found in wild-type nodules, but the expression of five late nodulin genes is not or is hardly detectable, e.g., leghemoglobin mRNA is hardly detectable. Apparently, late nodulin genes in soybean nodule development can be subdivided into two groups: the group of late nodulin genes that is expressed in these nodules and the group of genes that is not or barely expressed in HS124-induced nodules. Because infected cells develop scarcely or not at all in nodules induced by HS124, the nodulin genes not expressed in HS124-induced nodules are most likely transcribed in infected cells during normal nodule development. The group of nodulin genes fully expressed in HS124-induced nodules is most likely transcribed in uninfected cells, or in both cell types. However, *n*-uricase, a marker for the uninfected cells, could not be detected immunologically.[39] Therefore, the expression of some nodulin genes expressed in uninfected cells is affected as well. The concept of two subclasses of late nodulin genes in soybean has been confirmed by the isolation of late nodulin cDNA clones, some of which represented mRNA absent in HS124-induced nodules.[39]

B. japonicum mutant T8-1 induces normal-sized nodules that contain infection threads, but almost completely lack intracellular bacteria due to a block in bacterial release.[218] Because the marker proteins for the infected and uninfected cells, leghemoglobin and *n*-uricase, respectively, could be detected, the differentiation into these cell types appears to be undisturbed. Whereas the mRNA for most late nodulins was present at reduced levels in these nodules, mRNA for Ngm-26 was present at concentrations also found in wild-type nodules.[132,218] In wild-type nodules, nodulin Ngm-26 is associated with the peribacteroid membrane. However, the expression of the Ngm-26 gene in these developmentally disturbed nodules shows that this gene is also expressed when the peribacteroid membrane is not formed.[132,218] Although these results have been taken to suggest that there are at least two developmental stages in peribacteroid membrane biosynthesis,[218] the apparent contradiction of the expression of the peribacteroid membrane nodulin gene Ngm-26 in nodules induced by *B. japonicum* T8-1 questions the exact location of this nodulin. Ngm-26 gene expression is not detected in nodule-like structures devoid of infection threads,[218] therefore, Ngm-26 may be located in the infection-thread membrane as well as in the peribacteroid membrane.

D. CONCLUSIONS

Various studies on nodulin gene expression and nodule development with specific *Rhizobium* mutants have shown that nodule formation can be arrested at different stages of development as judged by histological criteria, and that these stages of development can be correlated with the expression of different sets of nodulin genes. On the basis of their expression in developmentally disturbed nodules, early and late nodulins can each be divided into at least two subclasses, the expression of which is regulated differently. Each subclass may reflect the occurrence of a different step in the developmental program of the root nodule.

IV. *RHIZOBIUM* GENES INVOLVED IN NODULIN GENE EXPRESSION

Both early and late nodulin genes can be subdivided into subclasses and each subclass correlates with the attainment of a defined stage in root nodule development beyond which further development is blocked by the mutation in the *Rhizobium*. This suggests that the bacterium delivers signals to the plant for the induction of expression of the successive subclasses of nodulin genes. For a characterization of these putative *Rhizobium* signals, the *Rhizobium* genes required for nodulin gene induction have to be identified. In fast-growing rhizobia (genus *Rhizobium*), the majority of the genes essential for nodulation and symbiotic nitrogen fixation are located on a large plasmid,[219] the so-called Sym plasmid, whereas in slow-growing rhizobia (genus *Bradyrhizobium*)[220] the genes involved in the symbiosis are located on the bacterial chromosome.[221,222] The bacterial strains used to correlate nodulin gene expression with nodule development (Section III) will now be discussed with the aim to identify the *Rhizobium* genes required for the induction of the expression of nodulin genes.

A. NITROGEN-FIXATION GENES

The *Rhizobium* genes essential for symbiotic nitrogen fixation are the *nif* and *fix* genes. *Nif* genes have been defined on the basis of structural homology with the *nif* genes in the free-living, nitrogen-fixing species *Klebsiella pneumoniae*.[223,224] *Fix* genes are also required for nitrogen fixation because nodules induced by strains mutated in these genes do not fix nitrogen, but *fix* genes share no homology with *K. pneumoniae* genes. In both *Rhizobium*[225-229] and *Bradyrhizobium*,[222,230] clusters of *nif* and *fix* genes have been identified.

All nodulin genes investigated are expressed in nodules induced on various legume plants by all *nif* and *fix* mutants examined so far. Therefore, these *Rhizobium nif* and *fix* genes are not essential for the induction of nodulin gene expression. The observation that all nodulin genes are expressed in nodules formed on pea and vetch by *Rhizobium* strain 248ᶜ(pMP104), containing only a small region of the Sym plasmid without *nif* and *fix* genes (see Section III.A), is the most conclusive evidence that *nif* and *fix* genes are not required for the induction of nodulin gene expression.[46]

The *nif* and *fix* genes do, however, appear to influence the level of expression of late nodulin genes. In nodules formed on pea by *nif* and *fix* mutants the amount of mRNA of the late nodulin genes is 10 to 40% of the amount found in wild-type nodules.[38] This phenomenon might for the greater part be attributed to impaired nodule growth resulting in a change in the ratio of different cell types. On the other hand, in pea the amount of Lb protein in ineffective nodules is not in proportion to the amount of Lb mRNA.[38] Therefore, a posttranscriptional regulation of leghemoglobin formation is likely.

Whereas *nif* and *fix* mutants of *Rhizobium* do not influence the development of the root nodules they induce, a few notable exceptions have been described for *Bradyrhizobium nif* and *fix* mutants. A *B. japonicum nifA* mutant induces nodules severely disturbed in the later stages of nodule development[216] (see Section III.C). All nodulin genes are expressed, but leghemoglobin is present at an extremely reduced concentration compared to the concentration in nodules induced by other *nif* and *fix* mutants.[215] Whereas in other *B. japonicum nif* and *fix* mutants investigated, the concentration of leghemoglobin protein is about 50% of the wild-type, the leghemoglobin concentration in the *nifA* mutant-induced nodules is less than 1% of the concentration found in nitrogen-fixing nodules. The decrease in leghemoglobin concentration is partly due to a decrease in transcription, since the relative concentration of leghemoglobin mRNA in the *nifA* mutant-induced nodules is about 5 to 10% of that found in wild-type nodules. The relative concentration of two other late nodulin mRNAs in the *nifA* mutant-induced nodules, Ngm-23 and *n*-uricase, is reduced to about 30% of that in wild-type nodules, while the expression of the early nodulin gene Ngm-75 is not decreased at all.[215] The very strong decrease in concentration might thus be unique for leghemoglobin. The dramatic effect of the *B. japonicum nifA* mutation on nodule development and, in particular, on the accumulation of leghemoglobin suggests that the *nifA* gene product not only regulates the expression of the *nif* genes in the *Bradyrhizobium*, but also is in some way involved in the regulation of nodulin gene expression in the plant cells. Because all nodulin genes are expressed in the *nifA* mutant-induced nodules, it can be excluded that the *nifA* product is required for the actual induction of nodulin gene expression. However, the *nifA* product appears to be required for the accumulation of leghemoglobin during nodule development. Besides the *nifA* mutant, a few *B. japonicum fix* mutants have been isolated that are phenotypically impaired in free-living nitrogen fixation, but have no auxotrophic defects.[231] Counterparts for this type of *fix* genes have not been found in fast-growing rhizobia. Just as the *nifA* mutant, these *fix* mutants induce nodules severely disturbed in development, but nodulin gene expression in these nodules has not yet been studied.

The effect of *nifA* and certain *fix* genes on soybean nodule development shows that gene products involved in building up the nitrogen-fixing system have a regulatory role in nodule development and nodulin gene expression in the *B. japonicum*-soybean symbiosis, whereas there is no evidence for such a regulatory role of *nif* and *fix* gene products in the symbiosis of fast-growing rhizobia and legumes.

B. NODULATION GENES

The *Rhizobium* genes required for, or involved in, nodulation of legume hosts, the *nod* genes, have been identified by a variety of genetic means.[232] So far, the genes *nodA* through *nodJ* have been identified in at least one of the various *Rhizobium* and *Bradyrhizobium* species studied[233-242] (see Chapter 6). Of these *nod* genes, the *nodDABC* genes are found in all species. These four genes are functional interchangeable among different species of

Rhizobium[235-238,243] and are therefore called common *nod* genes. The common *nod* genes appear to be absolutely essential for nodulation in all *Rhizobium*-legume symbioses, because mutations in these genes result in a Nod⁻ phenotype.[232] The same four genes are also essential for nodulation of the nonlegume *Parasponia*.[244] The common *nod* genes fall within one region of about 14 kb in *R. leguminosarum*[236] and *R. trifolii*,[201] and within two regions separated by about 12 kb in *R. meliloti*.[233]

The other *nod* genes, called host-specificity *nod* genes, delay or reduce nodulation or alter host range when mutated, but these mutations do not cause a complete inability to form nitrogen-fixing root nodules.[245-250] Therefore, the genes *nodE* through *nodJ* are involved in the fine tuning of the regulation of nodulation, but they are *not* essential for the induction of nodulin gene expression.

Thus, the *nodDABC* genes are the most prominent candidates for generating the signal(s) involved in the induction of the expression of nodulin genes. The *nodABC* genes constitute one operon; the constitutively expressed *nodD* is transcribed separately in the reverse direction.[251-255] In the presence of flavonoids excreted by the root, the *nodD* gene product induces the expression of all other *nod* genes,[255-261] possibly as a positive transcriptional activator.[262] The *nodD* gene products have been shown to differ in responsiveness to different flavonoids in a host-specific way.[263,264] These observations imply that the interchangeability, hence the common *nod* gene status, of *nodD* has to be questioned.

Genetic analyses have indicated already the pivotal role of the *nodDABC* genes in the establishment of the symbiosis. The *nodDABC* genes have been shown to be essential for root hair curling,[234,236,265] formation of the infection thread,[202,249] and the induction of cortical cell division,[266] thus, for the earliest steps in the developmental program of the root nodule in which no nodulin gene expression has been identified yet.

In reaction to the plant-excreted flavonoids, *Rhizobium* produces low molecular weight, soluble factors that cause a thick and short root (Tsr) phenotype on vetch,[267,268] and root hair deformations on vetch[269] as well as on clover.[270] The branching factor produced by *R. trifolii* also causes root hair deformations of vetch.[271] Mutations in the common *nod* genes abolish the ability of the bacterium to produce these factors,[269,272] demonstrating a direct effect of *nodABC* gene products in the production of a return signal from bacterium to plant. In view of the reaction of the plant to this signal, it is likely that the signal is hormone-like in nature. The sequence of the *Rhizobium nod* genes does not resemble sequences of phytohormone synthesis genes, but complementation of a *R. meliloti* Nod⁻ mutant with an *A. tumefaciens* cytokinin gene resulted in a strain able to induce an empty nodule. Also the finding that nodule-like structures can be elicited by application of auxin transport inhibitors,[331] suggests that indeed alterations in the hormone housekeeping of the legume root, in some way brought about by the *nod* gene products, are sufficient for the formation of a nodule structure.[273] Overproduction of the *nodABC* gene products either by increased gene copy number or from strong promoters proved deleterious to nodulation.[274] Thus, the concentration of the *nodABC* gene products is critical for the proper development of the symbiosis.

Upon introduction of a fragment carrying exclusively the *nodDABC* region into a Sym plasmid-cured *Rhizobium* strain, the recipient strain acquired the ability to curl root hairs, but a nodule structure is not formed. Whereas the *nodDABC* region by itself is not sufficient for the formation of a nodule structure, upon introduction of a 12-kb *nod* region from a *R. leguminosarum* Sym plasmid into a Sym plasmid-cured *Rhizobium*, the recipient strain regains the ability of the donor strain to induce nodules on pea and vetch. In these nodules, the expression of all nodulin genes is induced.[46] This result proves conclusively that the 12-kb *nod* region in a *Rhizobium* chromosomal background is sufficient for the induction of the expression of all nodulin genes identified. Apparently, the host-specificity *nod* genes present on this 12-kb region, in addition to *nodDABC*, pave the way for development, without being essential in themselves, because mutations in these additional host-specificity *nod* genes do not result in a Nod⁻ phenotype.

The strain obtained by introducing the same 12-kb *nod* region into a Ti plasmid-cured *Agrobacterium* chromosomal background induces nodules on vetch in which only the early nodulin genes, Nvs-40 and ENOD2, are expressed, but no late nodulin mRNAs are detectable.[46] Thus, the presence of the 12-kb *nod* region in an *Agrobacterium* chromosomal background is sufficient for the induction of nodulin gene expression. The *Agrobacterium* chromosome itself is unlikely to contribute signals specifically involved in the induction of nodulin gene expression, because nodulin gene expression is not detectable in tumors formed on the stem of vetch plants after wounding with *A. tumefaciens*.[47] Therefore, the 12-kb *nod* region must be involved in the induction of the expression of early nodulin genes. In view of the results of the mutation analyses, the *nodABC* genes must be the genes responsible for this induction. Because the expression of the identified early nodulin genes first becomes detectable after the nodule primordia have been formed, the induction of their expression is part of the developmental stage following the induction of cortical cell divisions. Therefore, the *nodDABC* genes are also involved in a stage of development beyond the induction of cortical cell divisions.

The *nodABC* genes seem insufficient for the induction of the expression of the late nodulin genes, because late nodulin gene expression is not detectable in the nodules induced by the strain containing the 12-kb *nod* region in an *Agrobacterium* chromosomal background. Although this result suggests that chromosomal or non-Sym plasmid genes are essential for the induction of late nodulin gene expression, Section V.B will discuss that such a conclusion cannot be drawn.

Indirect evidence indicates that the *nodABC* genes are, indeed, involved in the induction of late nodulin gene expression. Two *Rhizobium* strains that only differ in the origin of the cloned *nod* genes they contain were used to induce nodules on one and the same host plant, vetch. Strain ANU845(pMP104) contains the 12-kb *nod* region discussed above in a *R. trifolii* chromosomal background. Strain ANU845(pRt032) (*nodE* K11::Tn5) contains a 14-kb *nod* region from a *R. trifolii* Sym plasmid in the same *R. trifolii* chromosomal background. Due to a Tn5 mutation in *nodE*, the host range of this *R. trifolii* strain is extended to vetch.[202] Analyses of the pattern of nodulin gene expression in nodules induced on vetch by both strains showed that in nodules induced by ANU845(pMP104), the expression of all nodulin genes is induced, whereas in nodules induced by ANU845(pRt032), (*nodE* K11::Tn5), the late nodulin gene transcripts (except for Nvs-65 mRNA) were not detectable.[46] The marked difference between these two strains with respect to late nodulin gene expression will be related to the only difference between the strains: the apparently distinguishing characteristics of the 12-kb *nod* region in the one strain vs. the 14-kb *nod* region in the other. Irrespective of the exact cause of the observed difference in the induction of late nodulin gene expression, this difference indicates an involvement of *nod* genes in a stage of nodule development associated with the induction of late nodulin gene expression. This conclusion is supported by the observation that the *nodA* and *nodC* genes are expressed in *R. meliloti* bacteroids.[275]

C. SURFACE-DETERMINING GENES

The *Rhizobium* surface has been supposed to be involved in adhesion to the root hair surface, determination of host range, and nodulation.[276-278] A firm relation between the *Rhizobium* surface and the development of a nitrogen-fixing root nodule was established by the isolation of well-defined mutants that fail to produce extracellular polysaccharide and form a developmentally disturbed nodule.[279] Like other Gram-negative bacteria, *Rhizobium* has an outer membrane outside the peptidoglycan cell wall. External to the outer membrane, but tightly associated with it, are the lipopolysaccharides (LPS). More loosely bound are the extracellular polysaccharides which consist of two types, defined by the tightness of adhesion to the bacterial surface: exopolysaccharides (EPS), and the more tightly bound capsular polysaccharides (CPS). The EPS contain a fraction of heteropolysaccharides, the majority of these being acidic, and a fraction of homopolysaccharides which are neutral and mainly glucans.[280] The various surface polysaccharides are a complex mixture of different oligomers and polymers, every one of which may be of importance in the symbiosis.

Several genetic loci for *Rhizobium* surface determinants have been identified, and the genes are being characterized as detailed as the *nif, fix,* and *nod* genes. Eight loci affecting acidic EPS have been identified in *R. meliloti*.[210,211] Strains mutated in the seven loci *exoA* through *exoF* fail to produce a particular acidic EPS.[211] In addition to acidic EPS, the *exoC* mutant also lacks cyclic glucan.[48] These *exo* mutants induce nodules devoid of intracellular bacteria and infection threads. Some of the mutations could be complemented by genes from *A. tumefaciens*.[281] A mutation in the eighth *exo* locus, *exoH*, resulted in a strain that produces a slightly modified acidic EPS in which the succinyl modifications are absent.[212] In the nodules induced by the *exoH* mutant, infection threads are present but bacterial release is rarely observed.[212] Similarly, Pühler et al. have isolated *R. meliloti* mutants that either lack acidic EPS or have an acidic EPS without pyruvate modifications, and that elicit empty or *exoH* mutant-like nodules, respectively.[282] These results show that the acidic EPS and its non-carbohydrate substitutions have a role in the infection process. In all *exo* mutant-induced nodules on alfalfa, the expression of two early nodulin genes, ENOD2 and Nms-30, was induced, but late nodulin gene transcripts were absent.[48,212]

In addition to the *exo* loci, two *ndv* (nodule *development*) loci, *ndvA* and *ndvB,* have been identified in *R. meliloti*.[283] These genes are homologous to and functionally inter-changeable with the chromosomal virulence genes *chvA* and *chvB* of *A. tumefaciens*.[283] *R. meliloti ndv* mutants induce nodules with a morphology similar to that of the *exo* mutant-induced nodules.[40,283] In contrast to *R. meliloti exo* mutants, these mutants are not impaired in the synthesis of acidic EPS, but they are defective in the biosynthesis of cyclic glucan,[48,284] just as the *Agrobacterium chv*-mutants.[285] As is the case in the *exo* mutant-induced nodules, the expression of the early nodulin genes, Nms-30 and ENOD2, is induced, but late nodulin gene expression could not be detected.[48]

In other *Rhizobium*-legume symbioses, a relationship between EPS and the formation of a nodule has been demonstrated.[286-289] Also surface components like LPS and CPS have been shown to affect nodule development.[290-292] In all these cases, nodulin gene expression has not been studied, so the involvement of these surface components in the induction of the expression of nodulin genes is not known.

Besides phenotypical analysis of mutations, complementation studies with mixed in-oculations have demonstrated the importance of surface determinants in nodule development and nodulin gene expression. Coinoculation of a Sym plasmid-cured Exo[+] *nod* mutant of the broad host-range, fast-growing *Rhizobium* strain NGR234 with a Nod[+] *exo* mutant of the same *Rhizobium* that formed severely disturbed nodule-like structures on *Leucaena* plants resulted in the induction of nitrogen-fixing root nodules on *Leucaena* plants.[293] Both original mutants could be isolated from the nodules induced. This observation suggests that the EPS contributed by the Exo[+] *nod* mutant complements the defect of the Nod[+] *exo* mutant. Moreover, the addition of the correct EPS, purified from the parent strain NGR234, to the *exo* mutant has been reported to cause the formation of nitrogen-fixing root nodules.[289] This confirms that the EPS is essential for effective nodulation. Also, on alfalfa, coinoculation of *R. meliloti nod* mutants with *exo* mutants results in the induction of nitrogen-fixing root nodules,[282,294] but suppression of symbiotic deficiency by the addition of EPS purified from the parental strain failed in the case of *R. meliloti exo* mutants.[212]

D. OTHER GENES

It has been shown that the *R. meliloti nod* genes are expressed at normal levels in various *Rhizobium* chromosomal backgrounds and in *Agrobacterium tumefaciens*, but not in other

Gram-negative bacteria.[295] Apparently, the *Agrobacterium* chromosome contains genes that are essential for the induction of the *nod* genes. A mutation in these genes will result in a Nod⁻ phenotype. Introduction of the *Rhizobium nod* region in an *Agrobacterium* chromosomal background results in a strain that is able to induce a nodule structure in which early nodulin genes are expressed (see Section IV.B). It cannot be excluded that the *Agrobacterium* chromosome contains genes essential for the induction of early nodulin gene expression. However, if the *Agrobacterium* chromosome were to contain essential genes for this induction, this implies that *Agrobacterium* has retained genetic information it never uses in its natural situation.

Because the expression of late nodulin genes is not detected in nodules induced by *Agrobacterium* containing the *nod* region, the *Agrobacterium* chromosome is not sufficient for the induction of late nodulin gene expression, which suggests the *Rhizobium* chromosomal genes are involved in the induction of the expression of late nodulin genes (see, however, Section V. B). Various other genes have been shown to be involved in nodule development. A *R. meliloti leu⁺* mutant induces on alfalfa roots small, white nodules in which bacteria are not released from the infection threads.[296] When leucine or one of its precursors is added to the plant growth medium, bacterial release from the infection threads is restored, and nitrogen-fixing root nodules develop. This result suggests an involvement of leucine in nodule development. Several drug-resistant mutants, carbohydrate metabolism mutants, and other auxotrophic mutants have also been reported to induce symbiotically deficient nodules,[297,298] suggesting a role in nodule development for the mutated gene. In most cases, neither nodule morphology nor nodulin gene expression has been studied.

It seems doubtful that all these various genes are responsible for signals toward the plant involved in nodule development. Obviously, certain basic physiological requirements must be met in order for the *Rhizobium* to grow. Active growth of *Rhizobium* seems a prerequisite for proper development of nitrogen-fixing root nodules. Mutations that affect these basic physiological requirements will only, as a secondary consequence, result in a symbiotically deficient strain. The role of the *Rhizobium* chromosome in nodule development and the induction of nodulin gene expression is probably for the greater part the support of the basic physiology of the bacterium.

E. CONCLUSIONS

Several *Rhizobium* genes that affect nodule development and nodulin gene expression have been identified by phenotypic analysis of mutant-induced nodules. The effect of some auxotrophic mutants on nodule development shows that the mere disturbance of nodule development does not necessarily imply that the *Rhizobium* gene mutated is actually responsible for a signal essential for the induction of nodulin gene expression. Of the *Rhizobium* genes identified, the *nod* and surface-determining genes are the most obvious candidates to encode proteins providing signals for induction of the successive phases of nodule development. In Section V we will discuss the role of these genes in relation to the induction of nodulin gene expression.

V. *RHIZOBIUM* AND THE REGULATION OF NODULIN GENE EXPRESSION

The correlation between nodulin gene expression and nodule development (Section III), together with the identification of the *Rhizobium* genes affecting nodule development (Section IV), provides the basis for the discussion on how many *Rhizobium* signals are involved in the induction of nodulin gene expression and which genes generate them. However, in the previous sections the possibility of another plant reaction interfering with nodule formation, a defense response, has been left out of consideration. Yet, such an alternative plant reaction may be an essential factor in identifying the (rhizobial) signals for the induction of nodulin gene expression. Therefore, we will first take the role of a defense response during nodule development into account. We will discuss briefly the role of a plant defense response in nodule development and outline the consequences of that role for *Rhizobium* signals in the induction of nodulin gene expression. Then, we will discuss the role of the *Rhizobium nod* and surface-determining genes in the interplay of plant and bacterium resulting in a root nodule.

A. DEFENSE RESPONSE

Plants are able to defend themselves against plant-pathogens by a variety of means.[299,300] Some of the defenses are general, in that they provide protection from infection by a range of pathogens. These defenses include phytoalexin accumulation,[301] extensin accumulation, and other responses collectively called the hypersensitive response.[299] Other defense responses are highly specific, detected only in response to attack by a particular pathogen. Although root nodule development has repeatedly been considered as a special kind of plant-pathogen interaction,[302-305] the establishment of a nitrogen-fixing root nodule does not appear to provoke any known defense response. In soybean root nodules the concentration of the phytoalexin glyceollin is even lower than in uninfected roots.[306] In preliminary studies in our laboratory no increase of extensin-related RNA in nodules compared to uninoculated roots was observed. Also, stress-related RNA, detected by using a soybean general stress cDNA clone[307] as probe, was not enhanced in concentration.[49] These results suggest that the known defense mechanisms are not operating during normal root nodule development. Furthermore, none of the nodulins identified in vetch and pea nodules are detectable in tumors formed on the stem of vetch and pea plants by *A. tumefaciens*,[47] while *Agrobacterium* is a well-recognized plant-pathogen. This result shows that none of the identified nodulins functions in defense mechanisms.

A perturbation of the normal situation during nodule development appears to elicit a defense response in the host plant. In nodules formed on soybean by a *B. japonicum* mutant that forms unstable peribacteroid membranes, the phytoalexin concentration increases 50-fold.[306] The nodules senesce prematurely, and the necrotic appearance of the degenerated nodule is reminiscent of a hypersensitive response to pathogenic infection. Up to now, this is the only case in which a recognized parameter of defense has been measured. Further evidence for a defense response is circumstantial. A *R. trifolii* mutant that overproduces EPS induces a disturbed infection process in which infection thread growth is aborted in the root hair cell.[308] The reaction of the plant is interpreted as a hypersensitive response, because electron-dense material is deposited around the infection site. Nodules induced on the roots of vetch plants by an *Agrobacterium* transconjugant carrying the complete *R. leguminosarum* Sym plasmid exhibit a clearly dark center in the nodules. This dark center may be similar to the phenomenon of browning, associated with the hypersensitive response.[309] Structural analyses at the light and electron microscopic levels indicated that some nodule cells contain bacteria, but the bacteria degenerate directly after release from the infection thread and plant cells collapse.[203] Thus, although in normal nodule development no indications of a defense response are apparent, such a defense response seems to be present in the development of some ineffective nodules. It can be argued that in these ineffective nodules the symbiosis must be considered as a classic parasitic interaction.[303]

In view of the apparent absence of a plant defense response during normal nodule development, we hypothesize the presence of a system in the plant, to which we will refer as the sensor system, that is probing the performance of the symbiosis. The bacterial surface is constantly, or at defined stages of development, evaluated. The available data on nodule development indicate that the sensor system is active at least at two different stages of development: first, at the initial growth of the infection thread, and second, at the release of bacteria from the infection thread. When this sensor system detects an aberration of the permitted surface, a defense response is elicited. As a result, further nodule development is impaired.

B. *RHIZOBIUM* SIGNALS

The triggering of a defense response in root nodules induced by deviating bacteria puts the communication between *Rhizobium* and the plant in a different perspective, because it implies the existence of two types of signals. On the one hand, there are signals that actively cause the induction of nodule differentiation and nodulin gene expression. We will refer to these signals as inductive signals. On the other hand, there are signals which permit nodule development and achieve avoidance of a defense response. Because the latter type of signal will only be a passive one, we prefer the term avoidance determinant rather than signal. Avoidance determinants turn *Rhizobium* into a "parasite in disguise".

In terms of *Rhizobium* signals involved in the induction of nodulin gene expression, the requirement of correct avoidance determinants runs up a fundamental limitation of what can be concluded from developmentally disturbed nodules induced by engineered rhizobia and agrobacteria. When a mutation changes an avoidance determinant into a component that triggers the defense mechanism, the developmental program of the nodule will be aborted, while the mutant has all genetic information for the inductive signals. In this case, the blockade of development does not indicate a lack of an inductive signal, but is only due to the unmasking of the engineered bacterium as a result of incorrect avoidance determinants. In genetic terms that may represent an epistatic interaction.

We will now discuss the consequences of an interfering defense response for the roles of the *Rhizobium* nodulation and surface-determining genes in the induction of nodulin gene expression. Are these genes merely encoding avoidance determinants, or are their products more directly involved in the induction of the developmental program of the root nodule?

1. Surface-Determining Genes

A relationship between the microbial outer surface and plant defense responses has been well established.[310] For instance, cell surface oligo- and polysaccharides of *Phytophthora* are able to elicit plant defense responses.[301] In the interactions between *Pseudomonas solanacearum* and potato or tobacco, a plant defense response can be generated by living or dead bacteria or their LPS.[311,312] In normal nodule development, a defense mechanism is not elicited. Therefore, the *Rhizobium* surface components are likely to function as avoidance determinants. On the other hand, however, it has been shown that oligosaccharins can regulate some developmental programs in plants directly.[313] Hence, surface determinants may also have inductive capacities. Legume roots excrete enzymes that are able to degrade *Rhizobium* polysaccharides,[314] which appears a plausible way of producing oligosaccharins. We will show that most data available can be explained by assuming that the surface-determining genes code for avoidance determinants.

A mutation in the chromosomal *pss* (polysaccharide synthesis) gene of *R. phaseoli* in combination with the *R. phaseoli* sym plasmid does not affect the capacity to form nitrogen-fixing nodules on French bean,[315] but when the same chromosomal mutation is combined with a *R. leguminosarum* sym plasmid, the ability to nodulate peas is completely blocked. Normally, a strain containing a *R. leguminosarum* Sym plasmid in a *R. phaseoli* chromosomal background nodulates pea. Therefore, the *pss* gene product would be essential for the induction of nodule formation and nodulin gene expression on pea, while on the natural host French bean, this gene product is not important. In view of the apparent conservation of the mechanisms involved in nodulin gene expression (see Section II.F.3), it is unlikely that a gene product absolutely essential for nodulation of one legume (pea) has no importance at all for nodulation of another legume (French bean). Therefore, *pss* most likely codes for an avoidance determinant and is not involved in the generation of an inductive signal. The failure to nodulate pea shows that only the heterologous host pea plant does not tolerate the bacterial surface components exhibited in the absence of the *pss* gene product.

The results obtained with the *pss* gene may show that some plants accept more differences in the bacterial surfaces than others, or react more slowly to an aberration. An *Agrobacterium* transconjugant harboring a *R. leguminosarum* Sym plasmid induces nodules on vetch, in which most likely a defense response is elicited after release of bacteria from the infection thread.[203] The same *Agrobacterium* transconjugant forms nodules on pea plants that are totally devoid of intracellular bacteria, despite the presence of infection threads.[43] Thus, on two different legumes, *Agrobacterium* transconjugant-induced nodules differ in the stage where development is arrested. On alfalfa roots, an *Agrobacterium* transconjugant with the *R. meliloti* sym plasmid, or cloned *nod* genes, forms nodule-like structures that are even more disturbed in development than the nodules formed on pea, because infection threads are only occasionally observed in the root hair cell.[48, 316-318] In clover, infected cells are observed in nodules induced by a transconjugant carrying the *R. trifolii* sym plasmid.[319] A transconjugant harboring the *R. phaseoli* Sym plasmid is capable of inducing nitrogen-fixing root nodules on the roots of French bean plants, unless plants were grown at 21°C instead of 26°C.[320,321] A similar range in nodule morphology has been described for nodules induced by *Agrobacterium* transconjugants on other leguminous plants.[267,322] These results can be interpreted as a different tolerance of the sensor system of different legumes toward the (slightly) deviating surface determinants of *Agrobacterium* compared to the surface of the taxonomically closely related *Rhizobium*.

The best-studied surface-determining genes are the *R. meliloti* genes for the synthesis of acidic EPS. Mutants are available that either lack acidic EPS or have a modified form.[210-212,282] Most *exo* mutants of *R. meliloti* form empty nodules on alfalfa. In coinoculation experiments using a *nod* mutant of *R. meliloti* in combination with an *exo* mutant, wild-type-like nodules were obtained, showing that the defects are mutually restored.[282] However, coinoculation of the *nod* mutant with a strain having EPS that lacks pyruvate residues was unsuccessful.[282] Although the correct EPS is contributed by the *nod* mutant, the presence of modified EPS is apparently sufficient to arrest development. This observation strongly suggests that the blockade in development is not due to the absence of an EPS-derived signal molecule. It is most likely that EPS functions as an avoidance determinant.

The nodules induced on alfalfa by *exo* mutants lacking acidic EPS, by an *Agrobacterium* transconjugant carrying the *R. meliloti* Sym plasmid or cloned *nod* genes, or by *ndv* mutants impaired in cyclic glucan synthesis, are all devoid of intracellular bacteria. Only early nodulin gene expression is induced in these nodules.[48] The different bacterial strains appear to differ only in their outer surface. Because nodule morphology and the pattern of nodulin gene expression in these nodules are similar, all these different bacteria would produce the signals required at the same stage of development. It seems very unlikely that the various surface components are all involved in the generation of signals required at the same time of development. A more simple explanation is that all these surface components are required as avoidance determinants at the same stage of nodule development. In view of the phenotype of the *ndv* mutation and the interpretation here described, it should be noted that the designation *nodule development* for this gene[283] is unfortunate and misleading. The observed nodule morphology is most likely only an indirect effect of the mutation.

In the interpretation of all data available, it cannot be excluded that the bacterial surface components, in addition to functioning as avoidance determinants, are also responsible for the generation of inductive signals. These two possible modes of action of surface determinants are not mutually exclusive. A change in a surface determinant can change an avoidance determinant and at the same time destroy a (saccharide) signal that is essential in the developmental program of the nodule. At the moment, it seems premature to assume that bacterial surface determinants have such an active role. Satisfactory evidence has yet to be provided that surface-determining genes are responsible for signals involved in the induction of nodulin gene expression. The only indication that surface components have an active role is the observation that the oligosaccharide repeat unit of acidic EPS complements the symbiotic inability of an *exo* mutant of the broad-host range *Rhizobium* strain NGR234,[289] when added to the culture medium. Future experiments need to be designed in which the differences between avoidance determinants and inductive signals can be more clearly assessed.

2. Nodulation Genes

Mutations in the common *nod* genes, *nodDABC*, abolish the ability of the bacterium to induce nodules and nodulin gene expression completely. These studies already suggested that the *nodDABC* genes are essential for establishing a nodule and for the induction of nodulin gene expression. The most conclusive evidence that the *nod* genes are involved in nodulin gene expression can be inferred from DNA transfer studies. Transfer of a limited piece of sym plasmid DNA, carrying essentially the *nod* genes to *Agrobacterium*, conferred upon the recipient stain the ability to induce a nodule structure in which early nodulin genes are expressed.[46,48] The *nod* genes are the only *Rhizobium* genes so far for which such a positive correlation between the presence of genetic information and the induction of the expression of (early) nodulin genes has been established. Of the *nod* genes, only *nodDABC* are essential for nodulation (see Section IV.B). The *nodDABC* genes may, therefore, well be the minimum genetic requirement needed for the induction of the developmental program up to and including early nodulin gene expression. In view of the regulatory role of *nodD*, the *nodABC* genes are thus the most likely candidates for the generation of one or more signals toward the host plant that result in the expression of early nodulin genes. It cannot be totally excluded, however, that chromosomal genes or other *nod* genes are also responsible for inductive signals.

The mode of action of the *nodABC* gene products is still largely unknown. The *nodC* protein is a hydrophobic protein that is an integral part of the bacterial outer membrane, and this protein may be involved in transmembrane signaling.[323] Although the *nodA* gene product contains hydrophobic regions,[324] it has been localized in the cytosol.[275] Upon induction, the common *nod* gene products cause the production of low molecular weight, soluble factors, which are excreted by the bacterium. Both the *nodA* and *nodC* gene products are detectable in bacteriods,[275] suggesting that these gene products also function in the mature nodule. The different processing of the *nodC* gene product in free-living bacteria in which the *nod* genes are induced, compared to bacteroids,[275] may indicate that varying *nodC* gene products have different functions at subsequent stages of nodule development. These observations support the notion that the *nod* gene products are actively involved in later stages of nodulin gene expression as well. Because the late nodulin genes are not expressed in the nodules formed by *Agrobacterium* carrying essentially the *nod* region, direct evidence for the involvement of the *nodABC* genes in the induction of late nodulin gene expression is lacking. Just as the surface-determining gene products, the *nodA* and *nodC* gene products in bacteroids may be concerned with the outer surface structure of *Rhizobium*. The role of the *nodABC* genes in late nodulin gene expression may thus be a role in avoiding plant defense reactions.

C. CONCLUSIONS

During development of effective nodules, *Rhizobium* succeeds in bypassing the plant defense mechanisms that normally protect a plant against invading pathogens. However, when plants are infected by mutated or engineered strains, a plant defense response can be induced. *Rhizobium* genes involved in nodule formation and nodulin gene expression can therefore be involved in either the avoidance of the plant defense mechanism (avoidance

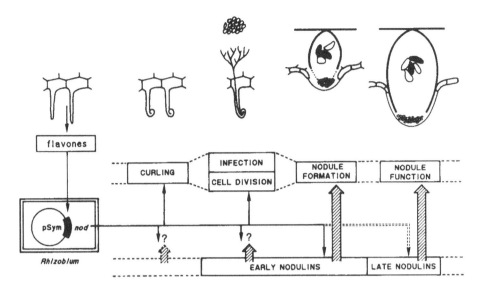

FIGURE 4. Schematic representation of the relationships between nodulins, the *Rhizobium nod* genes, and successive steps in root nodule development. See text and Figure 1.

determinants) or in the generation of a signal responsible for nodulin gene expression (inductive signal), or in both at the same time. A serious consequence of the involvement of a host defense mechanism is that conclusions with respect to the genetic potentials of a bacterial strain are no longer allowed for developmental stages this strain does not induce. A strain may have the genetic information for all inductive signals, but at the same time lack or possess an incorrect avoidance determinant. As a result, expression of the genetic information for the inductive signals is obscured. Put differently, not every blockage in development needs to be due to the absence of a signal of *Rhizobium*.

Evaluation of the data available indicate that the genes responsible for the bacterial outer surface most likely code for avoidance determinants, while the *nodABC* genes are more likely responsible for the induction of the expression of early nodulin genes and possible also late nodulin genes. A model outlining the involvement of the *Rhizobium nod* genes is presented in Figure 4. Plant root secreted flav(an)ones induce, via *nodD*, the expression of the other *nod* genes, upon which the *nod* gene products produce a return signal. The *nodABC* genes are essential for root hair curling, the infection process and the induction of cortical cell divisions. Early nodulins have not been identified in these very early stages of the interaction, and it is not clear whether nodulins are involved in these early stages. The expression of the early nodulin genes that have been identified, is first detectable when the nodule primordia have been formed, so the induction of their expression is part of a stage of development succeeding to the induction of cortical cell division and the formation of a nodule meristem. The *nod* genes are also responsible for the induction of the expression of these early nodulin genes. In Figure 4 this is indicated by a solid arrow for the signal and a hatched arrow for the resulting early nodulin gene products. The involvement of the *nod* genes in the induction of the late nodulin genes is less clear. Therefore, this relationship is indicated by a dashed arrow in the figure. The involvement of late nodulin gene products in the functioning of the nodule is also indicated by a hatched arrow.

VI. CONCLUDING REMARKS

The formation of nitrogen-fixing root nodules is attended by differential expression of nodulin genes. Early nodulins are involved in the organogenesis of the nodule. Late nodulins, on the other hand, most likely function in creating the physiological conditions that allow nitrogen fixation and ammonia assimilation. The expression of the genes encoding early nodulins is first detectable when the nodule meristem is differentiating into a nodule structure. Little is known about specific plant gene expression before that stage. Preliminary data from our laboratory indicate that early nodulins are already present in root hairs as early as 20 h after inoculation of the plant with *Rhizobium* bacteria.[49] These early nodulins may be involved in root hair curling and/or the infection of the root hair cell. Early nodulin gene expression has not been detected in the stages of nodule development, in which cortical cell divisions occur and the formation of the nodule meristem is established. The failure to detect early nodulin gene expression in these stages might be due to technical limitations of the detection methods used, but another possibility is that the nodule meristem does not differ from other plant meristems. In the latter case, nodule specific genes are not yet expressed. If indeed first a normal meristem is generated in the interaction of *Rhizobium* and the leguminous plant, the questions arise of when and how it is determined that this meristem enters the developmental program leading up to a root nodule.

In a first approach to answering these questions, it may be relevant to compare the legume root nodule with the root nodule of the nonlegume *Parasponia*. The morphology of the *Parasponia* root nodule differs substantially from legume root nodule morphology, because the vascular bundle is positioned centrally and not peripherally.[22] Also, nodule growth starts in the pericycle and not in the root cortex.[21] Therefore, the *Parasponia* root nodule is considered to be a modified lateral root. The same *nodDABC* genes of one and the same *Rhizobium* strain are equally essential for legume and for *Parasponia* root nodule induction.[244] Thus, the same bacterial signals might trigger the developmental program for both a legume and the *Parasponia* nodule type. It is unlikely that the same signals trigger two totally different developmental programs. (See next chapter.)

Moreover, *Agrobacterium* and *R. trifolii* transconjugants carrying cloned pieces of the nodulation region of the *R. meliloti* Sym plasmid are capable of inducing the formation of structures on clover roots intermediate between a nodule and a lateral root.[318] Similar hybrid structures have been reported to be formed occasionally on the roots of alfalfa after inoculation with a *R. meliloti* strain that at the same time induces morphologically normal nodules.[266] The structures found on alfalfa roots after treatment with auxin transport inhibitors,[331] or even spontaneously,[332] suggest that the role of *Rhizobium* in establishing a nodule structure may be rather simple and involve alterations of the hormone balance of the legume root. All these observations indicate that the developmental program underlying legume root nodule formation may be more close to the program of lateral root formation than previously thought. It is feasible that the developmental program of a root nodule is the outcome of relatively little changes in the developmental program of a (lateral) root.

The formation of a plant organ is thought to involve huge numbers of tissue-specific genes, which undoubtedly hinders the detailed understanding of the underlying developmental programs. If the differences between (lateral) root and root nodule formation are relatively small, then root nodule formation becomes an attractive system to study plant development. Not because the differentiation into a root nodule will be less complex than other plant differentiation processes, but because the differences between the two developmental programs seem more accessible to understanding than a developmental program as a whole.

A unique feature of root nodule development, as opposed to other plant developmental processes, is the involvement of a prokaryote in the induction and control of development. The regulatory role of *Rhizobium* offers unique possibilities for dissecting this plant differentiation process. Moreover, it offers an entry to the elucidation of the signals that guide root nodule development by allowing the identification of the *Rhizobium* genes responsible for these signals. An amazingly limited number of bacterial genes, the *nod* genes, appear to generate the signal(s) for the induction of early nodulin gene expression. The same genes are also in some way involved in the induction of late nodulin gene expression. Elucidation of the nature and mode of action of the signals involved will contribute to our understanding of root nodule development. Also by virtue of the relative ease of manipulation of the inducing *Rhizobium,* root nodule development is a highly attractive system for the study of plant developmental biology, apart from the intrinsic fascination of symbiotic nitrogen fixation.

ACKNOWLEDGMENTS

We thank all colleagues who have communicated results prior to official publication. We gratefully acknowledge Ab van Kammen, Ann Hirsch, Clemens van de Wiel, Francine Govers, Jeanne Jacobs, and Ton Gloudemans for critical and constructive reading of this manuscript; Gré Heitkönig for excellent secretarial assistance; and Piet Madern for art work. J. P. N. was financially supported by a grant from the Netherlands Organization for the Advancement of Pure Research.

REFERENCES

1. **Libbenga, K. R. and Bogers, R. J.,** Root nodule morphogenesis, in *The Biology of Nitrogen Fixation,* Quispel, A., Ed., North-Holland, Amsterdam, 1974, 430.
2. **Dart, P. J.,** Infection and development of leguminous nodules, in *A Treatise on Dinitrogen Fixation,* Hardy, R. W. F. and Silver, W. S., Eds., John Wiley & Sons, N.Y., 1977, 367.
3. **Goodchild, D. J.,** The ultrastructure of root nodules in relation to nitrogen fixation, in *Studies in Ultra structure,* Bourne, G. H., Danielli, J. F., and Jeon, U. W., Eds., *Int. Rev. Cytol.,* Suppl. 6, Academic Press, New York, 1977, 235.
4. **Bauer, W. D.,** Infection of legumes by rhizobia, *Annu. Rev. Plant Physiol.,* 32, 407, 1981.
5. **Newcomb, W.,** Nodule morphogenesis and differentiation, in *Biology of the Rhizobiaceae,* Giles, K. L. and Atherley, A. G., Eds., *Int. Rev. Cytol.,* Suppl. 13, Academic Press, New York, 1981, 247.
6. **Bhuvaneswari, T. V., Turgeon, B. G., and Bauer, W. D.,** Early events in the infection of soybean (*Glycine max* L. Merr) by *Rhizobium japonicum.* I. Localization of infectible root cells, *Plant Physiol.,* 66, 1027, 1980.
7. **Turgeon, B. G. and Bauer, W. D.,** Early events in the infection of soybean by *Rhizobium japonicum.* Time course and cytology of the initial infection process, *Can. J. Bot.,* 60, 152, 1982.
8. **Turgeon, B. G. and Bauer, W. D.,** Ultrastructure of infection-thread development during the infection of soybean by *Rhizobium japonicum, Planta,* 163, 328, 1985.
9. **Callaham, D. A. and Torrey, J. G.,** The structural basis for infection of root hairs of *Trifolium repens* by *Rhizobium, Can. J. Bot.,* 59, 1647, 1981.
10. **Ridge, R. W. and Rolfe, B. G.,** *Rhizobium* sp. degradation of legume root hair cell wall at the site of infection thread origin, *Appl. Environ. Microbiol.,* 50, 717, 1985.
11. **Ridge, R. W. and Rolfe, B. G.,** Sequence of events during the infection of the tropical legume *Macroptilium atropurpureum* Urb. by the broad-host-range, fast-growing *Rhizobium* ANU240, *J. Plant Physiol.,* 122, 121, 1986.
12. **Newcomb, W. and McIntyre, L.,** Development of root nodules in mung bean (*Vigna radiata*): a rein- vestigation of endocytosis, *Can. J. Bot.,* 59, 2478, 1981.
13. **Libbenga, K. R. and Harkes, P. A. A.,** Initial proliferation of cortical cells in the formation of root nodules in *Pisum sativum* L., *Planta,* 114, 17, 1973.
14. **Newcomb, W., Sippel, D., and Peterson, R. L.,** The early morphogenesis of *Glycine max* and *Pisum sativum* root nodules, *Can. J. Bot.,* 57, 2603, 1979.
15. **Calvert, H. E., Pence, M. K., Pierce, M., Malik, N. S. A., and Bauer, W. D.,** Anatomical analysis of the development and distribution of *Rhizobium* infections in soybean roots, *Can. J. Bot.,* 62, 2375, 1984.
16. **Bassett, B., Goodman, R. N., and Novacky, A.,** Ultrastructure of soybean nodules. Release of rhizobia from the infection thread, *Can. J. Microbiol.,* 23, 573, 1977.
17. **Robertson, J. G., Lyttleton, P., Bullivant, S., and Grayston, G. F.,** Membranes in lupin root nodules. I. The role of Golgi bodies in the biogenesis of infection threads and peribacteroid membranes, *J. Cell Sci.,* 30, 129, 1978.
18. **Robertson, J. G. and Lyttleton, P.,** Division of peribacteroid membranes in root nodules of white clover, *J. Cell Sci.,* 69, 147, 1984.

19. **Chandler, M. R.**, Some observations on infection of *Arachis hypogea* L. by *Rhizobium*, *J. Exp. Bot.*, 29, 749, 1978.
20. **Bhuvaneswari, T. V., Bhagwat, A. A., and Bauer, W. D.**, Transient susceptibility of root cells in four common legumes to nodulation by rhizobia, *Plant Physiol.*, 68, 1144, 1981.
21. **Lancelle, S. A. and Torrey, J. G.**, Early development of *Rhizobium*-induced root nodules of *Parasponia rigida*. I. Infection and early nodule initiation, *Protoplasma*, 123, 26, 1984.
22. **Lancelle, S. A. and Torrey, J. G.**, Early development of *Rhizobium*-induced root nodules of *Parasponia rigida*. II. Nodule morphogenesis and symbiotic development, *Can. J. Bot.*, 63, 25, 1984.
23. **DeFaria, S. M., Sutherland, J. M., and Sprent, J. I.**, A new type of infected cell in root nodules of *Andira* spp. (Leguminosae), *Plant Sci.*, 45, 143, 1986.
24. **Corby, H. D. L., Polhill, R. M., and Sprent, J. I.**, Taxonomy, in *Nitrogen Fixation, Vol. 3: Legumes*, Broughton, W. J., Ed., Clarendon Press, Oxford, 1983, 1.
25. **Price, G. D., Mohapatra, S. S., and Gresshoff, P. M.**, Structure of nodules formed by *Rhizobium* strain ANU289 in the non-legume *Parasponia* and the legume siratro *(Macroptilium atropurpureum)*, *Bot. Gaz.*, 145, 444, 1984.
26. **Trinick, M. J. and Galbraith, J.**, The *Rhizobium* requirements of the non-legume *Parasponia* in relation to the cross-inoculation concept of legumes, *New Phytol.*, 86, 17, 1980.
27. **Sutton, W. D.**, Nodule development and senescence, in *Nitrogen Fixation Vol. 3: Legumes*, Broughton, W. J., Ed., Clarendon Press, Oxford, 1983, 144.
28. **Gresshoff, P. M. and Delves, A. C.**, Plant genetic approaches to symbiotic nodulation and nitrogen fixation in legumes, in *A Genetic Approach to Plant Biochemistry*, Blonstein, A. D. and King, P. J., Eds., Springer-Verlag, Vienna, 1986, 159.
29. **Newcomb, W.**, A correlated light and electron microscopic study of symbiotic growth and differentiation in *Pisum sativum* root nodules, *Can. J. Bot.*, 54, 2163, 1976.
30. **Kamalay, J. C. and Goldberg, R. B.**, Regulation of structural gene expression in tobacco, *Cell*, 19, 935, 1980.
31. **Kamalay, J. C. and Goldberg, R. B.**, Organ-specific nuclear RNAs in tobacco, *Proc. Natl. Acad. Sci. U.S.A.*, 81, 2801, 1984.
32. **Legocki, R. P. and Verma, D. P. S.**, A nodule-specific plant protein (Nodulin-35) from soybean, *Science*, 205, 190, 1979.
33. **Verma, D. P. S. and Brisson, N.**, Eds., *Molecular Genetics of Plant Microbe Interactions*, Martinus Nijhoff, Dordrecht, 1987, various reports.
34. **Van Kammen, A.**, Suggested nomenclature for plant genes involved in nodulation and symbiosis, *Plant Mol. Biol. Rep.*, 2, 43, 1984.
35. **Govers, F., Nap, J. P., Van Kammen, A., and Bisseling, T.**, Nodulins in the developing root nodule, *Plant Physiol. Biochem.*, 25, 309, 1987.
36. **Legocki, R. P. and Verma, D. P. S.**, Identification of "nodule-specific" host proteins (nodulins) involved in the development of *Rhizobium*-legume symbiosis, *Cell*, 20, 153, 1980.
37. **Fuller, F. and Verma, D. P. S.**, Appearance and accumulation of nodulin mRNAs and their relationship to the effectiveness of root nodules, *Plant Mol. Biol.*, 3, 21, 1984.
38. **Govers, F., Gloudemans, T., Moerman, M., Van Kammen, A., and Bisseling, T.**, Expression of plant genes during the development of pea root nodules, *EMBO J.*, 4, 861, 1985.
39. **Sengupta-Gopalan, C., Pitas, J. W., Thompson, D. V., and Hoffman, L. M.**, Expression of host genes during nodule development in soybean, *Mol. Gen. Genet.*, 203, 410, 1986.
40. **Gloudemans, T., De Vries, S. C., Bussink, H.-J., Malik, N. S. A., Franssen, H. J., Louwerse, J., and Bisseling, T.**, Nodulin gene expression during soybean *(Glycine max)* nodule development, *Plant Mol. Biol.*, 8, 395, 1987.
41. **Govers, F., Nap, J. P., Moerman, M., Franssen, H. J., Van Kammen, A., and Bisseling, T.**, cDNA cloning and developmental expression of pea nodulin genes, *Plant Mol. Biol.*, 8, 425, 1987.
42. **Fuller, F., Künstner, P. W., Nguyen, T., And Verma, D. P. S.**, Soybean nodulin genes: analysis of cDNA clones reveals several major tissue-specific sequences in nitrogen-fixing root nodules, *Proc. Natl. Acad. Sci. U.S.A.*, 80, 2594, 1983.
43. **Govers, F., Moerman, M., Downie, J. A., Hooykaas, P., Franssen, H. J., Louwerse, J., Van Kammen, A., and Bisseling, T.**, *Rhizobium nod* genes are involved in inducing an early nodulin gene, *Nature (London)*, 323, 564, 1986.
44. **Nap, J. P., Moerman, M., Van Kammen, A., Govers, F., Gloudemans, T., Franssen, H. J., and Bisseling, T.**, Early nodulins in root nodule development, in *Molecular Genetics of Plant-Microbe Interactions*, Verma, D. P. S. and Brisson, N., Eds., Martinus Nijhoff, Dordrecht, 1987, 96.
45. **Franssen, H. J., Nap, J. P., Gloudemans, T., Stiekema, W., Van Dam, H., Govers, F., Louwerse, J., Van Kammen, A., and Bisseling, T.**, Characterization of cDNA for nodulin-75 of soybean: a gene product involved in early stages of root nodule development, *Proc. Natl. Acad. Sci. U.S.A.*, 84, 4495, 1987.

46. Nap, J. P., Van de Wiel, C., Spaink, H. P., Moerman, M., Van den Heurel, M., Djordjevic, M. A., Van Lammeren, A. A. M., Van Kammen, A., and Bisseling, T., The relationship between nodulin gene expression and the *Rhizobium nod* genes in *Vicia sativa* root nodule development, *Mol. Plant-Microbe Interact.*, 2, 53, 1989.

47. Moerman, M., Nap, J. P., Govers, F., Schilperoort, R., Van Kammen, A., and Bisseling, T., *Rhizobium nod* genes are involved in the induction of two early nodulin genes in *Vicia sativa* root nodules, *Plant Mol. Biol.*, 9, 171, 1987.

48. Dickstein, R., Bisseling, T., Reinhold, V. N., and Ausubel, F. M., Expression of nodule specific genes in alfalfa root nodules blocked at an early stage of development, *Genes Dev.*, 2, 677, 1988.

49. Bisseling, T. and co-workers, unpublished results.

50. Von Heijne, G., Patterns of amino acids near signal-sequence cleavage sites, *Eur. J. Biochem.*, 133, 17, 1983.

51. Cassab, G. I., Arabinogalactan proteins during the development of soybean root nodules, *Planta*, 168, 441, 1986.

52. McNeil, M., Darvill, A. G., Fry, S. C., and Albersheim, P., Structure and function of the primary cell walls of plants, *Annu. Rev. Biochem.*, 53, 625, 1984.

53. Cooper, J. B., Chen, J. A., Van Holst, G.-J., and Varner, J. E., Hydroxyproline-rich glycoproteins of plant cell walls, *Trends Biochem. Sci.*, 12, 24, 1987.

54. Varner, J. E., The hydroxyproline-rich glycoproteins of plants, in *Molecular Biology of Plant Growth Control*, UCLA Symposia on Molecular and Cellular Biology, New Ser., Vol. 44, Fox, J. E. and Jacobs, M., Eds., Alan R. Liss, New York, 1987, 441.

55. Chen, J. and Varner, J. E., An extracellular matrix protein in plants: characterization of a genomic clone for carrot extensin, *EMBO J.*, 9, 2145, 1985.

56. Chen, J. and Varner, J. E., Isolation and characterization of cDNA clones for carrot extensin and a proline-rich 33-kDa protein, *Proc. Natl. Acad. Sci. U.S.A.*, 82, 4399, 1985.

57. Cassab, G. I. and Varner, J. E., Cell wall proteins, *Annu. Rev. Plant Physiol. Plant Mol. Biol.*, 39, 321, 1988.

58. Clarke, A. E., Anderson, R. L., and Stone, B. A., Form and function of arabinogalactans and arabinogalactan-proteins, *Phytochemistry*, 18, 521, 1979.

59. Van Holst, G.-J., Klis, F. M., De Wildt, P. J. M., Hazenberg, C. A. M., Buijs, J., and Stegwee, D., Arabinogalactan protein from a crude cell organelle fraction of *Phaseolus vulgaris* L., *Plant Physiol.*, 68, 910, 1981.

60. Fincher, G. B. F., Stone, B. A., and Clarke, A. E., Arabinogalactan-proteins: structure, biosynthesis, and function, *Annu. Rev. Plant Physiol.*, 34, 47, 1983.

61. Allen, A. K., Desai, N. N., Neuberger, A., and Creeth, J. A., Properties of potato lectin and the nature of its glycoprotein linkages, *Biochem. J.*, 171, 665, 1978.

62. Leach, J. E., Cantrell, M. A., and Sequeira, L., Hydroxyproline-rich bacterial agglutinin from potato; extraction, purification, and characterization, *Plant Physiol.*, 70, 1353, 1983.

63. Hong, J. C., Nagao, R. T., and Key, J. L., Characterization and sequence analysis of a developmentally regulated putative cell wall protein gene isolated from soybean, *J. Biol. Chem.*, 262, 8367, 1987.

64. Hong, J. C. and Key, J. L., personal communication, 1987.

65. Lamport, D. T. A. and Catt, J. W., Glycoproteins and enzymes of the cell wall, in *Plant Carbohydrates. II: Extracellular Carbohydrates*, Encyclopedia of Plant Physiology, Vol. 13b, Tanner, W. A. and Loewus, F. A., Eds., Springer-Verlag, Berlin, 1981, 133.

66. Cooper, J. B. and Varner, J. E., Insolubilization of hydroxyproline-rich cell wall glycoprotein in aerated carrot root slices, *Biochem. Biophys. Res. Commun.*, 112, 161, 1983.

67. Chrispeels, M. J., Sadava, D., and Cho, Y. P., Enhancement of extensin biosynthesis in ageing disks of carrot storage tissue, *J. Exp. Bot.*, 25, 1157, 1974.

68. Esquerré-Tugayé, M. T., Lafitte, C., Mazau, D., Toppan, A., and Touze, A., Cell surfaces in plant-microorganism interactions. II. Evidence for the accumulation of hydroxyproline-rich glycoproteins in the cell wall of diseased plants as a defense mechanism, *Plant Physiol.*, 64, 320, 1979.

69. Hammerschmidt, R., Lamport, D. T. A., and Muldoon, E. P., Cell wall hydroxyproline enhancement and lignin deposition as an early event in the resistance of cucumber to *Cladosporium cucumerinum*, *Physiol. Plant Pathol.*, 24, 43, 1984.

70. Showalter, A. M., Bell, J. N., Cramer, C. L., Bailey, J. A., Varner, J. E., and Lamb, C. J., Accumulation of hydroxyproline-rich glycoprotein mRNAs in response to fungal elicitor and infection, *Proc. Natl. Acad. Sci. U.S.A.*, 82, 6551, 1985.

71. Esquerré-Tugayé, M. T., Mazau, D., Pelissier, B., Roby, D., Rumean, D., and Toppan, A., Induction by elicitors and ethylene of proteins associated to the defense of plants, in *Cellular and Molecular Biology of Plant Stress*, UCLA Symposia on Molecular and Cellular Biology, New Ser., Vol. 22, Key, J. L. and Kosuge, T., Eds., Alan R. Liss, New York, 1985, 459.

72. **Ecker, J. R. and Davis, R. W.**, Plant defense genes are regulated by ethylene, *Proc. Natl. Acad. Sci. U.S.A.*, 84, 5202, 1987.

73. **Cassab, G. I., Nieto-Sotelo, J., Cooper, J. B., Van Holst, G. J., and Varner, J. E.**, A developmentally regulated hydroxyproline-rich glycoprotein from the cell walls of soybean seed coats, *Plant Physiol.*, 77, 532, 1985.

74. **Appleby, C. A.**, Leghemoglobin and *Rhizobium* respiration, *Annu. Rev. Plant Physiol.*, 35, 433, 1984.

75. **Verma, D. P. S. and Bal, A. K.**, Intracellular site of synthesis and localization of leghemoglobin in root nodules, *Proc. Natl. Acad. Sci. U.S.A.*, 73, 3843, 1976.

76. **Robertson, J. G., Wells, B., Bisseling, T., Farnden, K. J. F., and Johnston, A. W. B.**, Immuno-gold localization of leghaemoglobin in cytoplasm in nitrogen-fixing root nodules of pea, *Nature (London)*, 311, 254, 1984.

77. **Wittenberg, J. B., Bergersen, F. J., Appleby, C. A., and Turner, G. L.**, Facilitated oxygen diffusion. The role of leghemoglobin in nitrogen fixation by bacteroids isolated from soybean root nodules, *J. Biol. Chem.*, 249, 4057, 1974.

78. **Sheehy, J. E., Minchin, F. R., and Witty, J. F.**, Control of nitrogen fixation in a legume nodule: an analysis of the role of oxygen diffusion in relation to nodule structure, *Ann. Bot.*, 55, 549, 1985.

79. **Govers, F., Moerman, M., Hooymans, J., Van Kammen, A., and Bisseling, T.**, Microaerobiosis is not involved in the induction of pea nodulin gene expression, *Planta*, 169, 513, 1986.

80. **Robertson, J. G., Farnden, K. J. F., Warburton, M. P., and Banks, J. A. M.**, Induction of glutamine synthetase during nodule development in lupin, *Aust. J. Plant Physiol.*, 2, 265, 1975.

81. **Verma, D. P. S., Ball, S., Guérin, C., and Wanamaker, L.**, Leghemoglobin biosynthesis in soybean root nodules. Characterization of the nascent and released peptides and the relative rate of synthesis of the major leghemoglobins, *Biochemistry*, 18, 476, 1979.

82. **Bisseling, T., Moen, L. L., Van den Bos, R. C., and Van Kammen, A.**, The sequence of appearance of leghaemoglobin and nitrogenase components I and II in root nodules of *Pisum sativum*, *J. Gen. Microbiol.*, 118, 377, 1980.

83. **Nadler, K. D. and Avissar, Y. J.**, Heme synthesis in soybean root nodules. II. On the role of bacteroid δ-aminolevulinic acid synthase and δ-aminolevulinic acid dehydrogenase in the synthesis of the heme of leghemoglobin, *Plant Physiol.*, 60, 433, 1977.

84. **Bisseling, T., Van den Bos, R. C., and Van Kammen, A.**, Host-specific gene expression in legume root nodules, in *Nitrogen Fixation, Vol. 4: Molecular Biology*, Broughton, W. J. and Pühler, A., Eds., Clarendon Press, Oxford, 1986, 280.

85. **Leong, S. A., Ditta, G. S., and Helinski, D. R.**, Heme biosynthesis in *Rhizobium*. Identification of a cloned gene coding for δ-aminolevulinic acid synthetase from *Rhizobium meliloti*, *J. Biol. Chem.*, 257, 8724, 1982.

86. **Guerinot, M. L. and Chelm, B. K.**, Bacterial δ-aminolevulinic acid synthase activity is not essential for leghemoglobin formation in the soybean/*Bradyrhizobium japonicum* symbiosis, *Proc. Natl. Acad. Sci. U.S.A.*, 83, 1837, 1986.

87. **Fuchsman, W. H. and Appleby, C. A.**, Separation and determination of the relative concentrations of the homogeneous components of soybean leghemoglobin by isoelectric focusing, *Biochim. Biophys. Acta*, 579, 314, 1979.

88. **Whittaker, R. G., Lennox, S., and Appleby, C. A.**, Relationship of the minor soybean leghaemoglobins d_1, d_2, and d_3 with the major leghaemoglobins c, c_2 and c_3, *Biochem. Int.*, 3, 117, 1981.

89. **Marcker, A., Lund, M., Jensen, E. Ø., and Marcker, K. A.**, Transcription of the soybean leghemoglobin genes during nodule development, *EMBO J.*, 3, 1691, 1984.

90. **Uheda, E. and Syōno, K.**, Physiological role of leghaemoglobin heterogeneity in pea root nodule development, *Plant Cell Physiol.*, 23, 75, 1982.

91. **Uheda, E. and Syōno, K.**, Effects of leghemoglobin components on nitrogen fixation and oxygen consumption, *Plant Cell Physiol.*, 23, 85, 1982.

92. **Landsmann, J., Dennis, E. S., Higgins, T. J. V., Appleby, C. A., Kortt, A. A., and Peacock, W. J.**, Common evolutionary origin of legume and nonlegume plant haemoglobins, *Nature (London)*, 324, 166, 1986.

93. **Kortt, A. A., Burns, J. E., Trinick, M. J., and Appleby, C. A.**, The amino acid sequence of hemoglobin I from *Parasponia andersonii*, a nonleguminous plant, *FEBS Lett.*, 180, 55, 1985.

94. **Bergmann, H., Preddie, E., and Verma, D. P. S.**, Nodulin-35: a subunit of specific uricase (uricase II) induced and localized in the uninfected cells of soybean nodules, *EMBO J.*, 2, 2333, 1983.

95. **Tajima, S. and Yamamoto, Y.**, Enzymes of purine catabolism in soybean plants, *Plant Cell Physiol.*, 16, 271, 1975.

96. **Tajima, S., Kato, N., and Yamamoto, Y.**, Cadaverine involved in urate degrading activity (uricase activity) in soybean radicles, *Plant Cell Physiol.*, 24, 247, 1983.

97. **Larsen, K. and Jochimsen, B. U.**, Expression of nodule-specific uricase in soybean callus tissue is regulated by oxygen, *EMBO J.*, 5, 15, 1986.

98. **Nguyen, T., Zelechowska, M., Foster, V., Bergmann, H., and Verma, D. P. S.,** Primary structure of the soybean nodulin-35 gene encoding uricase II localized in the peroxisomes of uninfected cells of nodules, *Proc. Natl. Acad. Sci. U.S.A.,* 82, 5040, 1985.

99. **Hanks, J. F., Schubert, K., and Tolbert, N. E.,** Isolation and characterization of infected and uninfected cells from soybean nodules. Role of uninfected cells in ureide synthesis, *Plant Physiol.,* 71, 869, 1983.

100. **Newcomb, E. H. and Tandon, S. R.,** Uninfected cells of soybean root nodules: ultrastructure suggests key role in ureide production, *Science,* 212, 1394, 1981.

101. **Newcomb, E. H., Tandon, S. R., and Kowal, R. R.,** Ultrastructural specialization for ureide production in uninfected cells of soybean root nodules, *Protoplasma,* 125, 1, 1985.

102. **Vaughn, U. C.,** Structural and cytochemical characterization of three specialized peroxisome types in soybean, *Physiol. Plant.,* 64, 1, 1985.

103. **Miflin, B. J. and Lea, P. J.,** Ammonia assimilation, in *The Biochemistry of Plants,* Miflin, B. J., Ed., Academic Press, New York, 1980, 169.

104. **Hirel, B., McNally, S. F., Gadal, P., Sumar, N., and Stewart, G. R.,** Cytosolic glutamine synthetase in higher plants. A comparative immunological study, *Eur. J. Biochem.,* 138, 63, 1984.

105. **McNally, S. F., Hirel, B., Gadal, P., Mann, A. F., and Stewart, G. R.,** Glutamine synthetases of higher plants. Evidence for a specific isoform content related to their possible physiological role and their compartmentation with the leaf, *Plant Physiol.,* 72, 22, 1983.

106. **Cullimore, J. V. and Miflin, B. J.,** Immunological studies on glutamine synthetase using antisera raised to the two plant forms of the enzyme from *Phaseolus* root nodules, *J. Exp. Bot.,* 35, 581, 1984.

107. **Verma, D. P. S., Fortin, M. G., Stanley, J., Mauro, V. P., Purohit, S., and Morrison, N.,** Nodulins and nodulin genes of *Glycine max,* a perspective, *Plant Mol. Biol.,* 7, 51, 1986.

108. **Cullimore, J. V., Lara, M., Lea, P. J., and Miflin, B. J.,** Purification and properties of two forms of glutamine synthetase from the plant fraction of *Phaseolus* root nodules, *Planta,* 157, 245, 1983.

109. **Lara, M., Cullimore, J. V., Lea, P. J., Miflin, B. J., Johnston, A. W. B., and Lamb, J. W.,** Appearance of a novel form of plant glutamine synthetase during nodule development in *Phaseolus vulgaris* L., *Planta,* 157, 254, 1983.

110. **Lara, M., Porta, H., Padilla, J., Folch, J., and Sánchez, F.,** Heterogeneity of glutamine synthetase polypeptides in *Phaseolus vulgaris* L., *Plant Physiol.,* 76, 1019, 1984.

111. **Robert, F. M. and Wong, P. P.,** Isozymes of glutamine synthetase in *Phaseolus vulgaris* L. root nodules, *Plant Physiol.,* 81, 142, 1986.

112. **Tingey, S. V., Walker, E. L., and Coruzzi, G. M.,** Glutamine synthetase genes of pea encode distinct polypeptides which are differentially expressed in leaves, roots and nodules, *EMBO J.,* 6, 1, 1987.

113. **Sengupta-Gopalan, C. and Pitas, J.,** Expression of nodule specific glutamine synthetase genes during nodule development in soybeans, *Plant Mol. Biol.,* 7, 189, 1986.

114. **Hirel, B., Bonet, C., King, B., Layzell, D., Jacobs, F., and Verma, D. P. S.,** Glutamine synthetase genes are regulated by ammonia provided externally or by symbiotic nitrogen fixation, *EMBO J.,* 6, 1167, 1987.

115. **Cullimore, J. V., Gebhardt, C., Saarelainen, R., Miflin, B. J., Idler, K. B., and Barker, R. F.,** Glutamine synthetase of *Phaseolus vulgaris* L.: organ-specific expression of a multigene family, *J. Mol. Appl. Genet.,* 2, 589, 1984.

116. **Gebhardt, C., Oliver, J. E., Forde, B. G., Saarelainen, R., and Miflin, B. J.,** Primary structure and differential expression of glutamine synthetase genes in nodules, roots and leaves of *Phaseolus vulgaris,* *EMBO J.,* 5, 1429, 1986.

117. **Dunn, M. K., Dickstein, R., Feinbaum, R., Burnett, B. K., Peterman, T. K., Tholdis, G., Goodman, H. M., and Ausubel, F. M.,** Developmental regulation of nodule-specific genes in alfalfa root nodules, *Mol. Plant-Microbe Interact.,* 1, 65, 1988.

118. **Robertson, J. G. and Farnden, K. J. F.,** Ultrastructure and metabolism of the developing legume root nodule, in *The Biochemistry of Plants,* Vol. 5, Miflin, B. J., Ed., Academic Press, New York, 1980, 65.

119. **Thummler, F. and Verma, D. P. S.,** Nodulin-100 of soybean is the subunit of sucrose synthase regulated by the availability of free heme in nodules, *J. Biol. Chem.,* 262, 14730, 1987.

120. **Deroche, M.-E., Carrayol, E., and Jolivet, E.,** Phosphoenolpyruvate carboxylase in legume nodules, *Physiol. Veg.,* 21, 1075, 1983.

121. **Mellor, R. B., Christensen, T. M. I. F., and Werner, D.,** Choline kinase II is present only in nodules that synthesize stable peribacteroid membranes, *Proc. Natl. Acad. Sci. U.S.A.,* 83, 659, 1986.

122. **Nguyen, J., Machal, L., Vidal, J., Perrot-Rechenmann, C., and Gadal, P.,** Immunochemical studies on xanthinedehydrogenase of soybean root nodules. Ontogenic changes in the level of enzyme and immunocytochemical localization, *Planta,* 167, 190, 1986.

123. **Larsen, K. and Jochimsen, B.,** Expression of two enzymes involved in ureide formation in soybean regulated by oxygen, in *Molecular Genetics of Plant-Microbe Interactions,* Verma, D. P. S. and Brisson, N., Eds., Martinus Nijhoff, Dordrecht, 1987, 133.

124. **Appels, M. A. and Haaker, H.,** Identification of cytoplasmic nodule-associated forms of malate dehydrogenase involved in the symbiosis between *Rhizobium leguminosarum* and *Pisum sativum, Eur. J. Biochem.,* 171, 515, 1987.

125. **Verma, D. P. S., Kazazian, V., Zogbi, V., and Bal, A. K.,** Isolation and characterization of membrane envelope enclosing the bacteroids in soybean root nodules, *J. Cell Biol.,* 78, 919, 1978.

126. **Mellor, R. B., Christensen, T. M. I. E., Bassarab, S., and Werner, D.,** Phospholipid transfer from ER to the peribacteroid membrane in soybean nodules, *Z. Naturforsch.,* 40c, 73, 1985.

127. **Mellor, R. B. and Werner, D.,** Peribacteroid membrane biogenesis in mature legume root nodules, *Symbiosis,* 3, 75, 1987.

128. **Tu, J. C.,** Structural similarity of the membrane envelopes of rhizobial bacteroids and the host plasma membrane as revealed by freeze-fracturing, *J. Bacteriol.,* 122, 691, 1975.

129. **Robertson, J. G., Lyttleton, P., and Tapper, B. A.,** The role of peribacteroid membrane in legume root nodules, in *Advances in Nitrogen Fixation Research,* Veeger, C. and Newton, W. E., Eds., Martinus Nijhoff/Junk, The Hague, 1984, 475.

130. **Fortin, M. G., Zelechowska, M., and Verma, D. P. S.,** Specific targeting of membrane nodulins to the bacteroid-enclosing compartment in soybean nodules, *EMBO J.,* 4, 3041, 1985.

131. **Katinakis, P. and Verma, D. P. S.,** Nodulin-24 gene of soybean codes for a peptide of the peribacteroid membrane and was generated by tandem duplication of a sequence resembling an insertion element, *Proc. Natl. Acad. Sci. U.S.A.,* 82, 4157, 1985.

132. **Fortin, M. G., Morrison, N. A., and Verma, D. P. S.,** Nodulin-26, a peribacteroid membrane nodulin is expressed independently of the development of the peribacteroid compartment, *Nucleic Acids Res.,* 15, 813, 1987.

133. **Jacobs, F. A., Zhang, M., Fortin, M. G., and Verma, D. P. S.,** Several nodulins of soybean share structural domains but differ in their subcellular locations, *Nucleic Acids Res.,* 15, 1271, 1987.

134. **Sandal, N. N., Bojsen, K., and Marcker, K. A.,** A small family of nodule specific genes from soybean, *Nucleic Acids Res.,* 15, 1507, 1987.

135. **Brewin, N. J., Robertson, J. G., Wood, E. A., Wells, B., Larkins, A. P., Galfre, G., and Butcher, G. W.,** Monoclonal antibodies to antigens in the peribacteroid membrane from *Rhizobium*-induced root nodules of pea crossreact with the plasma membranes and Golgi bodies, *EMBO J.,* 4, 605, 1985.

136. **Bradley, D. J., Butcher, G. W., Galfre, G., Wood, E. A., and Brewin, N. J.,** Physical association between the peribacteroid membrane and lipopolysaccharide from the bacteroid outer membrane in *Rhizobium*-infected pea root nodule cells, *J. Cell Sci.,* 85, 47, 1986.

137. **Brewin, N. J., Bradley, D. J., Wood, E. A., Galfre, G., and Butcher, G. W.,** A study of surface interactions between *Rhizobium* bacteroids and the peribacteroid membrane using monoclonal antibodies, in *Recognition in Microbe-Plant Symbiotic and Pathogenic Interactions,* Lugtenberg, B., Ed., Springer-Verlag, Berlin, 1986, 153.

138. **Sullivan, D., Brisson, N., Goodchild, B., Verma, D. P. S., and Thomas, D. Y.,** Molecular cloning and organization of two leghaemoglobin genomic sequences of soybean, *Nature (London),* 289, 516, 1981.

139. **Jensen, E. Ø., Paludan, K., Hyldig-Nielsen, J. J., Jørgensen, P., and Marcker, K. A.,** The structure of a chromosomal leghaemoglobin gene from soybean, *Nature (London),* 291, 677, 1981.

140. **Brisson, N. and Verma, D. P. S.,** Soybean leghemoglobin gene family: normal, pseudo, and truncated genes, *Proc. Natl. Acad. Sci. U.S.A.,* 79, 4055, 1982.

141. **Brisson, N., Pombo-Gentile, A., and Verma, D. P. S.,** Organization and expression of leghaemoglobin genes, *Can. J. Biochem.,* 60, 272, 1982.

142. **Jørgensen, P. and Marcker, K. A.,** The primary structures of two leghemoglobin genes from soybean, *Nucleic Acids Res.,* 10, 689, 1982.

143. **Wiborg, O., Hyldig-Nielsen, J. J., Jensen, E. Ø., Paludan, K., and Marcker, K. A.,** The structure of an unusual leghemoglobin gene from soybean, *EMBO J.,* 2, 449, 1983.

144. **Lee, J. S., Brown, G. G., and Verma, D. P. S.,** Chromosomal arrangement of leghemoglobin genes in soybean, *Nucleic Acids Res.,* 11, 5541, 1983.

145. **Bojsen, K., Abildsten, D., Jensen, E. Ø., Paludan, K., and Marcker, K. A.,** The chromosomal arrangement of six soybean leghemoglobin genes, *EMBO J.,* 2, 1165, 1983.

146. **Lee, J. S. and Verma, D. P. S.,** Structure and chromosomal arrangement of leghemoglobin genes in kidney bean suggest divergence in soybean leghemoglobin gene loci following tetraploidization, *EMBO J.,* 3, 2745, 1984.

147. **Gö M.,** Correlation of DNA exonic regions with protein structural units in haemoglobin, *Nature (London),* 291, 90, 1981.

148. **Lewin, R.,** Evolutionary history written in globin genes, *Science,* 214, 426, 1981.

149. **Blake, C. C. F.,** Exons and the structure, function and evolution of haemoglobin, *Nature (London),* 291, 616, 1981.

150. **Mauro, V. P., Nguyen, T., Katinakis, P., and Verma, D. P. S.,** Primary structure of the soybean nodulin-23 gene and potential regulatory elements in the 5'-flanking region of nodulin and leghemoglobin genes, *Nucleic Acids Res.,* 13, 239, 1985.

151. **Padilla, J. E., Campos, F., Conde, V., Lara, M., and Sánchez, F.,** Nodule-specific glutamine synthetase is expressed before the onset of nitrogen fixation in *Phaseolus vulgaris* L., *Plant Mol. Biol.,* 9, 65, 1987.
152. **Hanks, J. F., Tolbert, N. E., and Schubert, K. R.,** Localization of enzymes of ureide biosynthesis in peroxisomes and microsomes of nodules, *Plant Physiol.,* 68, 65, 1981.
153. **Brown, G. G., Lee, J. S., Brisson, N., and Verma, D. P. S.,** The evolution of a plant globin gene family, *J. Mol. Evol.,* 21, 19, 1984.
154. **Hattori, J. and Johnson, D. A.,** The detection of leghemoglobin-like sequences in legumes and non-legumes, *Plant Mol. Biol.,* 4, 285, 1985.
155. **Bogusz, D., Appleby, C. A., Landsmann, J., Dennis, E. S., Trinick, M., and Peacock, W. J.,** Functioning haemoglobin in non-nodulating plants, *Nature (London),* 331, 178, 1988.
156. **Norman, P. M., Wingate, V. P. M., Fitter, M. S., and Lamb, C. J.,** Monoclonal antibodies to plant plasma-membrane antigens, *Planta,* 167, 452, 1986.
157. **Stougaard, J., Marcker, K. A., Otten, L., and Schell, J.,** Nodule-specific expression of a chimaeric soybean leghaemoglobin gene in transgenic *Lotus corniculatus, Nature (London),* 321, 669, 1986.
158. **Marcker, K. A. and Sandal, N. N.,** Evolution of the leghemoglobins, in *Plant Molecular Biology,* Von Wettstein, D. and Chua, N.-H., Eds., Plenum Press, New York, 1987, 503.
159. **Jensen, E. Ø., Marcker, K. A., Schell, J., and De Bruijn, F.,** Interaction of a nodule specific, *trans*-acting factor with distinct DNA elements in the soybean legaemoglobin lbc₃ 5' upstream region, *EMBO J.,* 7, 1265, 1988.
160. **Pederson, G. A.,** *In vitro* culture and somatic embryogenesis of four *Trifolium* species, *Plant Sci.,* 45, 101, 1986.
161. **Petit, A., Stougaard, J., Kühle, A., Marcker, K. A., and Tempé, J.,** Transformation and regeneration of the legume *Lotus corniculatus:* a system for molecular studies of symbiotic nitrogen fixation, *Mol. Gen. Genet.,* 207, 245, 1987.
162. **Vlachova, M., Metz, B. A., Schell, J., and De Bruijn, F. J.,** The tropical legume *Sesbania rostrata:* tissue culture, plant regeneration and infection with *Agrobacterium tumefaciens* and *rhizogenes* strains, *Plant Sci.,* 50, 213, 1987.
163. **Gilmour, D. M., Davey, M. R., and Cocking, E. C.,** Plant regeneration from cotyledon protoplasts of wild *Medicago* species, *Plant Sci.,* 48, 107, 1987.
164. **Hammatt, N. and Davey, M. R.,** Somatic embryogenesis and plant regeneration from cultured zygotic embryos of soybean (*Glycine max* L. Merr), *J. Plant Physiol.,* 128, 219, 1987.
165. **Bergersen, F. J. and Turner, G. L.,** Nitrogen fixation by the bacteroid fraction of breis of soybean root nodules, *Biochim. Biophys. Acta,* 141, 507, 1967.
166. **Smith, A. M.,** Capacity for fermentation in roots and *Rhizobium* nodules of *Pisum sativum* L., *Planta,* 166, 264, 1985.
167. **Ditta, G., Virts, E., Palomares, A., and Kim, C. H.,** The *nifA* gene of *Rhizobium meliloti* is oxygen regulated, *J. Bacteriol.,* 169, 3217, 1987.
168. **Fischer, H. M. and Hennecke, H.,** Direct response of *Bradyrhizobium japonicum nifA*-mediated *nif* gene regulation to cellular oxygen status, *Mol. Gen. Genet.,* 209, 621, 1987.
169. **Jensen, E. Ø., Marcker, K. A., and Villadsen, I. S.,** Heme regulates the expression in *Saccharomyces cerevisiae* of chimaeric genes containing 5'-flanking soybean leghemoglobin sequences, *EMBO J.,* 5, 843, 1986.
170. **Atkins, C. A., Shelp, B. J., Kno, J., Peoples, M. B., and Pate, J. S.,** Nitrogen nutrition and the development and senescence of nodules on cowpea seedlings, *Planta,* 162, 316, 1984.
171. **Hall, M. A.,** Hormones and plant development: cellular and molecular aspects, in *Developmental Control in Animals and Plants,* 2nd ed., Graham, C. F. and Wareing, P. F., Eds., Blackwell, Oxford, 1984, 313.
172. **Libbenga, K. R. and Torrey, J. G.,** Hormone-induced endoreduplication prior to mitosis in cultured pea root cortex cells, *Am. J. Bot.,* 60, 293, 1973.
173. **Libbenga, K. R., Van Iren, F., Bogers, R. J., and Schraag-Lamers, M. F.,** The role of hormones and gradients in the initiation of cortex proliferation and nodule formation in *Pisum sativum* L., *Planta,* 114, 29, 1973.
174. **Allen, E. K., Allen, O. N., and Newman, A. S.,** Pseudonodulation of leguminous plants induced by 2-bromo-3,5-dichlorobenzoic acid, *Am. J. Bot.,* 40, 435, 1953.
175. **Arora, N., Skoog, F., and Allen, O. N.,** Kinetin-induced pseudonodules on tobacco roots, *Am. J. Bot.,* 46, 610, 1959.
176. **Rodriguez-Barrueco, C., Miguel, C., and Palhi, L. M. S.,** Cytokinin-induced pseudonodules on *Alnus glutinosa, Physiol. Plant.,* 29, 277, 1973.
177. **Link, J. K. K. and Eggers, V.,** Avena coleoptile assay of ether extracts of nodules and roots of bean, soybean, and pea, *Bot. Gaz.,* 101, 650, 1940.
178. **Badenoch-Jones, J., Rolfe, B. G., and Letham, D. S.,** Phytohormones, *Rhizobium* mutants, and nodulation in legumes. III. Auxin metabolism in effective and ineffective pea root nodules, *Plant Physiol.,* 73, 347, 1983.

179. **Henson, I. E. and Wheeler, C. T.**, Hormones in plants bearing nitrogen-fixing root nodules: the distribution of cytokinins in *Vicia faba* L., *New Phytol.*, 76, 433, 1976.

180. **Syōno, K., Newcomb, W., and Torrey, J. G.**, Cytokinin production in relation to the development of pea root nodules, *Can. J. Bot.*, 54, 2155, 1976.

181. **Badenoch-Jones, J., Rolfe, B. G., and Letham, D. S.**, Phytohormones, *Rhizobium* mutants, and nodulation in legumes. V. Cytokinin metabolism in effective and ineffective pea root nodules, *Plant Physiol.*, 74, 239, 1984.

182. **Radley, M.**, Gibberellin-like substances in plants, *Nature (London)*, 191, 684, 1961.

183. **Dullaart, J. and Duba, L. P.**, Presence of gibberellin-like substances and their possible role in auxin bioproduction in root nodules and roots of *Lupinus luteus* L., *Acta Bot. Neerl.*, 19, 877, 1970.

184. **Wang, T. L., Wood, E. A., and Brewin, N. J.**, Growth regulators, *Rhizobium* and nodulation in peas. Indole-3-acetic acid from the culture medium of nodulating and non-nodulating strains of *R. leguminosarum*, *Planta*, 155, 345, 1982.

185. **Ernstsen, A., Sandberg, G., Crozier, A., and Wheeler, C. T.**, Endogenous indoles and the biosynthesis and metabolism of indole-3-acetic acid in cultures of *Rhizobium phaseoli*, *Planta*, 171, 422, 1987.

186. **Phillips, D. A. and Torrey, J. G.**, Studies on cytokinin production by *Rhizobium*, *Plant Physiol.*, 49, 11, 1972.

187. **Verma, D. P. S., Lee, J., Fuller, F., and Bergmann, H.**, Leghemoglobin and nodulin genes: two major groups of host genes involved in symbiotic nitrogen fixation, in *Advances in Nitrogen Fixation Research*, Veeger, C. and Newton, W. E., Eds., Martinus Nijhoff/Junk, The Hague, 1984, 557.

188. **Hooper, P., Whately, B., and Iyer, V. N.**, A study of *Rhizobium meliloti* JJI mutants in indole acetic acid production, in *Nitrogen Fixation Research Progress*, Evans, H. J., Bottomley, R. J., and Newton, W. E., Eds., Martinus Nijhoff, Dordrecht, 1985, 148.

189. **Vincent, J. M.**, Factors controlling the legume-*Rhizobium* symbiosis, in *Nitrogen Fixation II*, Newton, W. E. and Orme-Johnson, W. H., Eds., University Park Press, Baltimore, 1980, 103.

190. **Holl, F. B.**, Host plant control of the inheritance of dinitrogen fixation in the *Pisum-Rhizobium* symbiosis, *Euphytica*, 24, 767, 1975.

191. **Gresshoff, P. M., Day, D. A., Delves, A. C., Mathews, A. P., Olsson, J. E., Price, G. D., Schuller, K. A., and Carroll, B. J.**, Plant host genetics of nodulation and symbiotic nitrogen fixation in pea and soybean, in *Nitrogen Fixation Research Progress*, Evans, H. J., Bottomley, P. J., and Newton, W. E., Eds., Martinus Nijhoff, Dordrecht, 1985, 19.

192. **Vest, G. and Caldwell, B. E.**, Rj4 — a gene controlling ineffective nodulation in soybean, *Crop Sci.*, 12, 692, 1972.

193. **Caldwell, B. E. and Vest, H. G.**, Genetic aspects of nodulation and dinitrogen fiation by legumes: the macrosymbiont, in *A Treatise on Dinitrogen Fixation. III. Biology*, Hardy, R. W. F. and Silver, W. S., Eds., John Wiley & Sons, New York, 1977, 557.

194. **Carroll, B. J., McNeil, D. L., and Gresshoff, P. M.**, Mutagenesis of soybean (*Glycine max* [L.] Merr.) and the isolation of non-nodulating mutants, *Plant Sci.*, 47, 109, 1986.

195. **Bergersen F. J. and Nutman, P. S.**, Symbiotic effectiveness in nodulated red clover. IV. The influence of the host factors *il* and *ie* upon nodule structure and cytology, *Heredity*, 11, 175, 1957.

196. **Nutman, P. S.**, Hereditary host factors affecting nodulation and nitrogen fixation, in *Current Perspectives in Nitrogen Fixation*, Gibson, A. H. and Newton, W. E., Eds., Elsevier, Amsterdam, 1981, 194.

197. **Vance, C. P. and Johnson, L. E. B.**, Plant-determined ineffective nodules in alfalfa (*Medicago sativa*): structural and biochemical comparisons, *Can. J. Bot.*, 61, 93, 1983.

198. **Vance, C. P., Boylan, K. K. M., Stade, S., and Somers, D. A.**, Nodule-specific proteins in alfalfa (*Medicago sativa* L.), *Symbiosis*, 1, 69, 1985.

199. **Josey, D. P., Beynon, J. L., Johnston, A. W. B., and Beringer, J.**, Strain identification in *Rhizobium* using intrinsic antibiotic resistance, *J. Appl. Microbiol.*, 46, 434, 1979.

200. **Spaink, H. P., Okker, R. J. H., Wijffelman, C. A., Pees, E., and Lugtenberg, B. J. J.**, Promoters in the nodulation region of the *Rhizobium leguminosarum* sym plasmid pRL1JI, *Plant Mol. Biol.* 9, 27, 1987.

201. **Schofield, P. R., Ridge, R. W., Rolfe, B. G., Shine, J., and Watson, J. M.**, Host-specific nodulation is encoded on a 14 kb DNA fragment in *Rhizobium trifolii*, *Plant Mol. Biol.*, 3, 3, 1984.

202. **Djordjevic, M. A., Schofield, P. R., and Rolfe, B. G.**, Tn5 mutagenesis of *Rhizobium trifolii* host-specific nodulation genes result in mutants with altered host-range ability, *Mol. Gen. Genet.*, 200, 463, 1985.

203. **Van de Wiel, C., Nap, J. P., Van Lammeren, A., and Bisseling, T.**, Histological evidence that a defence response of the host plant interferes with nodulin gene expression in *Vicia sativa* root nodules induced by an *Agrobacterium* transconjugant, *J. Plant Physiol.*, 132, 466, 1988.

204. **Hooykaas, P. J. J., Snijdewint, F. M., and Schilperoort, R.**, Identification of the sym plasmid of *Rhizobium leguminosarum* strain 1001 and its transfer to and expression in other rhizobia and *Agrobacterium tumefaciens*, *Plasmid*, 8, 73, 1982.

205. **Lang-Unnasch, N., Dunn, K., and Ausubel, F. M.,** Symbiotic nitrogen fixation: developmental genetics of nodule formation, in *Molecular Biology of Development, Cold Spring Harbor Symp. Quant. Biol.,* 555, 1985.

206. **Lullien, V., Barker, D. G., De Lajudie, P., and Huguet, T.,** Plant gene expression in effective and ineffective root nodules of alfalfa *(Medicago sativa), Plant Mol. Biol.,* 9, 469, 1987.

207. **Morris, J. H., Macol. L. A., and Hirsch, A. M.,** Nodulin gene expression in effective alfalfa nodules and in nodules arrested at three different stages of development, *Plant Physiol.,* 88, 321, 1988.

208. **Hirsch, A. M., Bang, M., and Ausubel, F. M.,** Ultrastructural analysis of ineffective alfalfa nodules formed by *nif*::Tn5 mutants of *Rhizobium meliloti, J. Bacteriol.,* 155, 367, 1985.

209. **Hirsch, A. M. and Smith, C. A.,** Effects of *Rhizobium meliloti nif* and *fix* mutants on alfalfa root nodule development, *J. Bacteriol.,* 169, 1137, 1987.

210. **Leigh, J. A., Signer, E. R., and Walker, G. C.,** Exopolysaccharide-deficient mutants of *Rhizobium meliloti* that form ineffective nodules, *Proc. Natl. Acad. Aci. U.S.A.,* 82, 6231, 1985.

211. **Finan, T. M., Hirsch, A. M., Leigh, J. A., Johansen, E., Kuldau, G. A., Deegan, S., Walker, G. C., and Signer, E. R.,** Symbiotic mutants of *Rhizobium meliloti* that uncouple plant from bacterial differentiation, *Cell,* 40, 869, 1985.

212. **Leigh, J. A., Reed, J. W., Hanks, J. F., Hirsch, A. M., and Walker, G. C.,** *Rhizobium meliloti* mutants that fail to succinylate their Calcofluor-binding exopolysaccharide are defective in nodule invasion, *Cell,* 51, 579, 1987.

213. **Auger, S. and Verma, D. P. S.,** Induction and expression of nodule-specific host genes in effective and ineffective root nodules of soybean, *Biochemistry,* 20, 1300, 1981.

214. **Hahn, M., Meyer, L., Studer, D., Regensburger, B., and Hennecke, H.,** Insertion and deletion mutants within the *nif* region of *Rhizobium japonicum, Plant Mol. Biol.,* 3, 159, 1984.

215. **Studer, D., Gloudemans, T., Franssen, H. J., Fischer, H. M., Bisseling, T., and Hennecke, H.,** In the *Bradyrhizobium japonicum*-soybean symbiosis the bacterial nitrogen fixation regulatory gene *(nifA)* is also involved in control of nodule-specific host plant gene expression, *Eur. J. Cell. Biol.,* 45, 177, 1987.

216. **Fischer, H. M., Alvarez-Morales, A., and Hennecke, H.,** The pleiotropic nature of symbiotic regulatory mutants: *Bradyrhizobium japonicum nifA* gene is involved in control of *nif* gene expression and formation of determinate symbiosis, *EMBO J.,* 5, 1165,1986.

217. **Noel, K. D., Stacey, G., Tandon, S. R., Silver, L. E., and Brill, W. J.,** *Rhizobium japonicum* mutants defective in symbiotic nitrogen fixation, *J. Bacteriol.,* 152, 485, 1982.

218. **Morrison, N. and Verma, D. P. S.,** A block in the endocytosis of *Rhizobium* allows cellular differentiation in nodules but affects the expression of some peribacteroid membrane nodulins, *Plant Mol. Biol.,* 9, 185, 1987.

219. **Beringer, J. E., Brewin, N. J., and Johnston, A. W. B.,** Genetics, in *Nitrogen Fixation, Vol. 2, Rhizobium,* Broughton, W. J., Ed., Clarendon Press, Oxford, 1982, 167.

220. **Jordan, D. C.,** Transfer of *Rhizobium japonicum* Buchanan 1980 into *Bradyrhizobium* gen. nov. a genus of slow-growing, root nodule bacteria from leguminous plants, *Int. J. Syst. Bacteriol.,* 32, 136, 1982.

221. **Stacey, G. Paau, A. S., Noel, K. D., Maier, R. J., Silver, L. E., and Brill, W. J.,** Mutants of *Rhizobium japonicum* defective in nodulation, *Arch. Microbiol.,* 132, 219, 1982.

222. **Hennecke, H., Alvarez-Morales, A., Betancourt-Alvarez, M., Ebeling, S., Filser, M., Fischer, H. M., Gubler, M., Hahn, M., Kaluza, K., Lamb, J. W., Meyer, L., Regensburger, B., Studer, D., and Weber, J.,** Organization and regulation of symbiotic nitrogen fixation genes from *Bradyrhizobium japonicum,* in *Nitrogen Fixation Research Progress,* Evans, H. J., Bottomley, P. J., and Newton, W. E., Eds., Martinus Nijhoff, Dordrecht, 1985, 157.

223. **Ruvkun, G. B. and Ausubel, F. M.,** Interspecies homology of nitrogenase genes, *Proc. Natl. Acad. Sci. U.S.A.,* 77, 191, 1980.

224. **Dixon, R. A.,** The genetic complexity of nitrogen fixation, *J. Gen. Microbiol.,* 130, 2745, 1984.

225. **Corbin, D., Barran, L., and Ditta, G.,** Organization and expression of *Rhizobium meliloti* nitrogen fixation genes, *Proc. Natl. Acad. Sci. U.S.A.,* 80, 3005, 1983.

226. **Schetgens, T. M. P., Bakkeren, G., Van Dun, C., Hontelez, J. G. J., Van den Bos, R. C., and Van Kammen, A.,** Molecular cloning and functional characterization of *Rhizobium leguminosarum* structural *nif* genes by site-directed transposon mutagenesis and expression in *Escherichia coli* minicells, *J. Mol. Appl. Genet.,* 2, 406, 1984.

227. **Schetgens, T. M. P., Hontelez, J. G. J., Van den Bos, R. C., and Van Kammen, A.,** Identification and phenotypical characterization of a cluster of *fix* genes, including a *nif* regulatory gene, from *Rhizobium leguminosarum* PRE, *Mol. Gen. Genet.,* 200, 368, 1985.

228. **Grönger, P., Manian, S. S., Reiländer, H., O'Connell, M., Priefer, U., and Pühler, A.,** Organization and partial sequence of a DNA region of the *Rhizobium leguminosarum* symbiotic plasmid pRL6JI containing the genes *fixABC, nifA, nifB,* and a novel open reading frame, *Nucleic Acids Res.,* 15, 31, 1987.

229. **Earl, C. D., Ronson, C. W., and Ausubel, F. M.,** Genetic and structural analysis of the *Rhizobium meliloti fixA, fixB, fixC,* and *fixK* genes, *J. Bacteriol.,* 169, 1127, 1987.

230. **Ebeling, S., Hahn, M., Fischer, H.-M., and Hennecke, H.,** Identification of *nifE, nifN* and *nifS*-like gene in *Bradyrhizobium japonicum, Mol. Gen. Genet.,* 207, 503, 1987.

231. **Regensburger, B., Meyer, L., Filser, M., Weber, J., Studer, D., Lamb, J. W., Fischer, H. M., Hahn, M., and Hennecke, H.,** *Bradyrhizobium japonicum* mutants defective in root-nodule bacteroid development and nitrogen fixation, *Arch. Microbiol.,* 144, 355, 1986.

232. **Long, S.,** Genetics of *Rhizobium* nodulation, in *Plant-Microbe Interactions,* Kosuge, T. and Nester, E. Eds., Macmillan, New York, 1984, 265.

233. **Long, S. R., Buikema, W., and Ausubel F. M.,** Cloning of *Rhizobium meliloti* genes by direct complementation of Nod⁻ mutants, *Nature (London),* 298, 485, 1982.

234. **Jacobs, T. W., Egelhoff, T. T., and Long, S. R.,** Physical and genetic map of a *Rhizobium meliloti* nodulation gene region and nucleotide sequence of *nodC, J. Bacteriol.,* 162, 469, 1985.

235. **Kondorosi, E., Banfalvi, Z., and Kondorosi, A.,** Physical and genetic analysis of a symbiotic region of *Rhizobium meliloti:* identification of nodulation genes, *Mol. Gen. Genet.,* 193, 445, 1984.

236. **Downie, J. A., Knight, C. D., Johnston, A. W. B., and Rossen, L.,** Identification of genes and gene products involved in the nodulation of peas by *Rhizobium leguminosarum, Mol. Gen. Genet.,* 189, 255, 1985.

237. **Wijffelman, C. A., Pees, E., Van Brussel, A. A. N., Okker, R. J. H., and Lugtenberg, B. J. J.,** Genetic and functional analysis of the nodulation region of the *Rhizobium leguminosarum* sym plasmid pRL1JI, *Arch. Microbiol.,* 143, 255, 1985.

238. **Djordjevic, M. A., Schofield, P. R., Ridge, R. W., Morrison, N. A., Bassam, B. J., Plazinski, J., Watson, J. M., and Rolfe, B. G.,** *Rhizobium* nodulation genes involved in root hair curling (*Hac*) are functionally conserved, *Plant Mol. Biol.,* 4, 147, 1985.

239. **Russell, P., Schell, M. G., Nelson, K. K., Halverson, L. J., Sirotkin, K. M., and Stacey, G.,** Isolation and characterization of the DNA region encoding nodulation functions in *Bradyrhizobium japonicum, J. Bacteriol.,* 164, 1301, 1985.

240. **Lamb, C. J. and Hennecke, H.,** In *Bradyrhizobium japonicum* the common nodulation genes, *nodABC,* are linked to *nifA* and *fixA, Mol. Gen. Genet.,* 202, 512, 1986.

241. **Scott, D. B., Chua, K. Y., Jarvis, B. D. W., and Pankhurst, C. E.,** Molecular cloning of a nodulation gene from fast- and slow-growing strains of *Lotus* rhizobia, *Mol. Gen. Genet.,* 201, 43, 1985.

242. **Scott, K. F.,** Conserved nodulation genes from the non-legume symbiont *Bradyrhizobium* sp. *(Parasponia), Nucleic Acids Res.,* 14, 2905, 1986.

243. **Fisher, R. F., Tu, J. K., and Long, S. R.,** Conserved nodulation genes in *Rhizobium meliloti* and *Rhizobium trifolii, Appl. Environ. Microbiol.,* 49, 1432, 1985.

244. **Marvel, D. J., Kuldau, G., Hirsch, A., Richards, E., Torrey, J. G., and Ausubel, F. M.,** Conservation of nodulation genes between *Rhizobium meliloti* and a slow-growing *Rhizobium* strain that nodulates a nonlegume host, *Proc. Natl. Acad. Sci. U.S.A.,* 82, 5841, 1985.

245. **Djordjevic, M. A., Innes, R. W., Wijffelman, C. A., Schofield, P. R., and Rolfe, B. G.,** Nodulation of specific legumes is controlled by several distinct loci in *Rhizobium trifolii, Plant Mol. Biol.,* 6, 389, 1986.

246. **Downie, J. A., Hombrecher, G., Ma, Q.-S., Knight, C. D., Wells, B., and Johnston, A. W. B.,** Cloned nodulation genes of *Rhizobium leguminosarum* determine host-range specificity, *Mol. Gen. Genet.,* 190, 359, 1983.

247. **Horvath, B., Kondorosi, E., John, M., Schmidt, J., Török, I., Györgypal, Z., Barabas, I., Wieneke, U., Schell, J., and Kondorosi, A.,** Organization, structure and symbiotic function of *Rhizobium meliloti* nodulation genes determining host specificity for alfalfa, *Cell,* 46, 335, 1986.

248. **Debellé, F. and Sharma, S. R.,** Nucleotide sequence of *Rhizobium meliloti* RCR 2011 genes involved in host specificity of nodulation, *Nucleic Acids Res.,* 14, 7453, 1986.

249. **Debellé, F., Sharma, S. B., Rosenberg, C., Vasse, J., Maillet, F., Truchet, G., and Dénarié, J.,** Respective role of common and specific *Rhizobium meliloti nod* genes in the control of lucerne infection, in *Recognition in Microbe-Plant Symbiotic and Pathogenic Interactions,* Lugtenberg, B., Ed., Springer-Verlag, Berlin, 1986, 17.

250. **Debellé, F., Rosenberg, C., Vasse, J., Maillet, F., Martinéz, E., Dénarié, J., and Truchet, G.,** Assignment of symbiotic developmental phenotypes to common and specific nodulation (*nod*) genetic loci of *Rhizobium meliloti, J. Bacteriol.,* 168, 1075, 1986.

251. **Egelhoff, T. T., Fisher, R. F., Jacobs, T. W., Mulligan, J. T., and Long, S. R.,** Nucleotide sequence of *Rhizobium meliloti* 1021 nodulation genes: *nodD* is read divergently from *nodABC, DNA,* 4, 241, 1985.

252. **Rossen, L., Shearman, C. A., Johnston, A. W. B., and Downie, J. A.,** The *nodD* gene of *Rhizobium leguminosarum* is autoregulatory and in the presence of plant exudate induces the *nodA,B,C* genes, *EMBO J.,* 4, 3369, 1985.

253. **Schofield, P. R. and Watson, J. M.,** DNA sequence of *Rhizobium trifolii* nodulation genes reveals a reiterated and potentially regulatory sequence preceding *nodABC* and *nodFE, Nucleic Acids Res.,* 14, 2891, 1986.

254. **Spaink, H. P., Okker, R. J. H., Wijffelman, C. A., Pees, E., and Lugtenberg, B.,** Promoters and operon structure of the nodulation region of the *Rhizobium leguminosarum* symbiosis plasmid pRL1JI, in *Recognition in Microbe-Plant Symbiotic and Pathogenic Interactions,* Lugtenberg, B., Ed., Springer-Verlag, Berlin, 1986, 55.

255. **Djordjevic, M. A., Redmond, J. W., Batley, M., and Rolfe, B. G.,** Clovers secrete specific phenolic compounds which either stimulate or repress *nod* gene expression in *Rhizobium trifolii, EMBO J.,* 6, 1173, 1987.

256. **Mulligan, J. T. and Long, S. R.,** Induction of *Rhizobium meliloti nodC* expression by plant exudate requires *nodD, Proc. Natl. Acad. Sci. U.S.A.,* 28, 6609, 1985.

257. **Innes, R. W., Kuempel, P. L., Plazinski, J., Canter-Cremers, H., Rolfe, B. G., and Djordjevic, M.,** Plant factors induce expression of nodulation and host-range genes in *Rhizobium trifolii, Mol. Gen. Genet.,* 201, 426, 1985.

258. **Peters, N. K., Frost, J. W., and Long, S. R.,** A plant flavone, luteolin, induces expression of *Rhizobium meliloti* nodulation genes, *Science,* 233, 977, 1986.

259. **Redmond, J. W., Batley, M., Djordjevic, M. A., Innes, R. W., Kuempel, P. L., and Rolfe, B. G.,** Flavones induce expression of nodulation genes in *Rhizobium, Nature (London),* 323, 632, 1986.

260. **Firmin, J. L., Wilson, K. E., Rossen, L., and Johnston, A. W. B.,** Flavonoid activation of nodulation genes in *Rhizobium* reversed by other compounds present in plants, *Nature (London),* 324, 90, 1986.

261. **Zaat, S. A. J., Wijffelman, C. A., Spaink, H. P., Van Brussel, A. A. N., Okker, R. J. H., and Lugtenberg, B. J. J.,** Induction of the *nodA* promoter of *Rhizobium leguminosarum* sym plasmid pRL1JI by plant flavanones and flavones, *J. Bacteriol.,* 169, 198, 1987.

262. **Long, S. R., Peters, N. K., Mulligan, J. T., Dudley, M. E., and Fisher, R. F.,** Genetic analysis of *Rhizobium*-plant interactions, in *Recognition in Microbe-Plant Symbiotic and Pathogenic Interactions,* Lugtenberg, B., Ed., Springer-Verlag, Berlin, 1986, 1.

263. **Spaink, H. P., Wijffelman, C. A., Pees, E., Okker, R. J. H., and Lugtenberg, B. J. J.,** *Rhizobium* nodulation gene *nodD* as a determinant of host specificity, *Nature (London),* 328, 337, 1987.

264. **Horvath, B., Bachem, C. W. B., Schell, J., and Kondorosi, A.,** Host-specific regulation of nodulation genes in *Rhizobium* is mediated by a plant-signal, interacting with the *nodD* gene product, *EMBO J.,* 6, 841, 1987.

265. **Downie, J. A., Rossen, L., Knight, C. D., Robertson, J. G., Wells, B., and Johnston, A. W. B.,** *Rhizobium leguminosarum* genes involved in early stages of nodulation, *J. Cell Sci. Suppl.,* 2, 347, 1985.

266. **Dudley, M. E., Jacobs, T. W., and Long, S. R.,** Microscopic studies of cell divisions induced in alfalfa roots by *Rhizobium meliloti, Planta,* 171, 289, 1987.

267. **Van Brussel, A. A. N., Tak, T., Wetselaar, A., Pees, E., and Wijffelman, C. A.,** Small leguminosae as test plants for nodulation of *Rhizobium leguminosarum* and other rhizobia and other agrobacteria harbouring a *leguminosarum* sym plasmid, *Plant Sci. Lett.,* 27, 317, 1982.

268. **Van Brussel, A. A. N., Zaat, S. A. J., Canter Cremers, H. C. J., Wijffelman, C. A., Pees, E., Tak, T., and Lugtenberg, B. J. J.,** Role of plant root exudate and sym plasmid-localized nodulation genes in the synthesis by *Rhizobium leguminosarum* of Tsr factor, which causes thick and short roots on common vetch, *J. Bacteriol.,* 165, 517, 1986.

269. **Zaat, S. A. J., Van Brussel, A. A. N., Tak, T., Pees, E., and Lugtenberg, B. J. J.,** Flavonoids induce *Rhizobium leguminosarum* to produce *nodDABC* gene-related factors that cause thick, short roots and root hair responses on common vetch, *J. Bacteriol.,* 1689, 3388, 1987.

270. **Bhuvaneswari, T. V. and Solheim, B.,** Root hair deformation in the white clover/*Rhizobium trifolii* symbiosis, *Physiol. Plant.,* 63, 25, 1985.

271. **Canter Cremers, H. C. J., Van Brussel, A. A. N., Plazinski, J., and Rolfe, B. G.,** Sym plasmid and chromosomal gene products of *Rhizobium trifolii* elicit developmental responses on various legume roots, *J. Plant Physiol.,* 122, 25, 1986.

272. **Bhuvaneswari, T. V.,** personal communication, 1987.

273. **Long, S. R. and Cooper, J.,** Overview of symbiosis, in Molecular Genetics of Plant-Microbe Interactions 1988, Palacios, R., and Verma, D P. S., Eds, APS Press, St. Paul, Minnesota, 1988, 163.

274. **Knight, C. D., Rossen, L., Robertson, J. G., Wells, B., and Downie, J. A.,** Nodulation inhibition by *Rhizobium leguminosarum* multicopy *nodABC* genes and analysis of early stages of plant infection, *J. Bacteriol.,* 166, 552, 1986.

275. **Schmidt, J., John, M., Wieneke, U., Krüssmann, H.-D., and Schell, J.,** Expression of the nodulation gene *nodA* in *Rhizobium meliloti* and localization of the gene product in the cytosol, *Proc. Natl. Acad. Sci. U.S.A.,* 83, 9581, 1986.

276. **Dazzo, F. B., Napoli, C. A., and Hubbell, D. H.,** Adsorption of bacteria to roots as related to host specificity in the *Rhizobium*-clover symbiosis, *Appl. Environ. Microbiol.,* 32, 166, 1976.

277. **Smit, G., Kijne, J. W., and Lugtenberg, B. J. J.,** Correlation between extracellular fibrils and attachment of *Rhizobium leguminosarum* to pea root hair tips, *J. Bacteriol.,* 168, 821, 1986.

278. **Halverson, L. J. and Stacey, G.,** Signal exchange in plant-microbe interactions, *Microbiol. Rev.,* 50, 193, 1986.

279. **Chakravorty, A. K., Zurkowski, W., Shine, J., and Rolfe, B. G.,** Symbiotic nitrogen fixation: molecular cloning of *Rhizobium* genes involved in exopolysaccharide synthesis and effective nodulation, *J. Mol. Appl. Gen.*, 1, 585, 1982.

280. **Carlson, R. W.,** Surface chemistry, in *Nitrogen Fixation, Vol. 2., Rhizobium,* Broughton, W. J., Ed., Clarendon Press, Oxford, 1982, 199.

281. **Cangelosi, G. A., Hung, L., Puvanesarajah, V., Stacey, G., Ozga, D. A., Leigh, J. A., and Nester, E. W.,** Common loci for *Agrobacterium tumefaciens* and *Rhizobium meliloti* exopolysaccharide synthesis and their roles in plant interactions, *J. Bacteriol.,* 169, 2086, 1987.

282. **Pühler, A., Hynes, M. F., Kapp, D., Müller, P., and Niehaus, K.,** Infection mutants of *Rhizobium meliloti* are altered in acidic exopolysaccharide production, in *Recognition in Microbe-Plant Symbiotic and Pathogenic Interactions,* Lugtenberg, B., Ed., Springer-Verlag, Berlin, 1986, 29.

283. **Dylan, T., Ielpi, L., Stanfield, S., Kashyap, J., Douglas, C., Yanofsky, M., Nester, E., Helinski, D. R., and Ditta, G.,** *Rhizobium meliloti* genes required for nodule development are related to chromosomal virulence genes in *Agrobacterium tumefaciens, Proc. Natl. Acad. Sci. U.S.A.,* 83, 4403, 1986.

284. **Geremia, R. A., Cavaignac, S., Zorreguieta, A., Toro, N., Olivares, J., and Ugalde, R.,** A *Rhizobium meliloti* mutant that forms ineffective pseudonodules in alfalfa produces exopolysaccharide but fails to form β-(1-2) glucan, *J. Bacteriol.,* 169, 880, 1987.

285. **Puvanesarajah, V., Schell, F. M., Stacey, G., Douglas, C. J., and Nester, E. W.,** Role for 2-linked-β-D-glucan in the virulence of *Agrobacterium tumefaciens, J. Bacteriol.,* 164, 102, 1985.

286. **Borthakur, D., Downie, J. A., Johnston, A. W. B., and Lamb, J. W.,** *Psi,* a plasmid-linked *Rhizobium phaseoli* gene that inhibits exopolysaccharide production and which is required for symbiotic nitrogen fixation, *Mol. Gen. Genet.,* 200, 278, 1985.

287. **Chen, H., Batley, M., Redmond, J., and Rolfe, B. G.,** Alteration of the effective nodulation properties of a fast-growing broad-host-range *Rhizobium* due to changes in exopolysaccharide synthesis, *J. Plant Physiol.,* 120, 1985.

288. **Vandenbosch, K. A., Noel, K. D., Kaneko, Y., and Newcomb E. H.,** Nodule initiation elicited by noninfective mutants of *Rhizobium phaseoli, J. Bacteriol.* 162, 950, 1985.

289. **Djordjevic, S. P., Chen, H., Batley, M., Redmond, J. W., and Rolfe, B. G.,** Nitrogen fixation ability of exopolysaccharide synthesis mutants of *Rhizobium* sp. strain NGR234 and *Rhizobium trifolii* is restored by the addition of homologous exopolysaccharides, *J. Bacteriol.,* 169, 53, 1987.

290. **Noel, K. D., Van den Bosch, K. A., and Kulpaca, B.,** Mutations in *Rhizobium phaseoli* that lead to arrested development of infection threads, *J. Bacteriol.,* 168, 1392, 1986.

291. **Puvanesarajah, V., Schell, F. M., Gerhold, D., and Stacey, G.,** Cell surface polysaccharides from *Bradyrhizobium japonicum* and a nonnodulating mutant, *J. Bacteriol.,* 169, 137, 1987.

292. **Gardiol, A. E., Hollingsworth, R. P., and Dazzo, F. B.,** Alteration of surface properties in a Tn5 mutant strain of *Rhizobium trifolii, J. Bacteriol.,* 169, 1161, 1987.

293. **Chen, H. and Rolfe, B. G.,** Cooperativity between *Rhizobium* mutant strains results in the induction of nitrogen-fixing nodules on the tropical legume *Leucaena leucocephala, J. Plant Physiol.,* 127, 307, 1987.

294. **Klein, S., Hirsch, A. M., Smith, C. A., and Signer, E. R.,** Interaction of rhizobial *nod* and *exo* functions in alfalfa nodulation, *Mol. Plant-Microbe Interact.,* 1, 94, 1988.

295. **Yelton, M. M., Mulligan, J. T., and Long, S. R.,** Expression of *Rhizobium meliloti nod* genes in *Rhizobium* and *Agrobacterium* backgrounds, *J. Bacteriol.,* 169, 3094, 1987.

296. **Truchet, G., Michel, M., and Dénarié, J.,** Sequential analysis of the organogenesis of lucerne (*Medicago sativa*) root nodules using symbiotically-defective mutants of *Rhizobium meliloti, Differentiation,* 16, 163, 1980.

297. **Kuykendall, L. D.,** Mutants of *Rhizobium* that are altered in legume interaction and nitrogen fixation, in *Biology of the Rhizobiaceae,* Giles, K. L. and Atherley, A. G., Eds., *Int. Rev. Cytol.,* Suppl. 13, Academic Press, New York, 1981, 299.

298. **Wilson, K. J., Anjaiah, V., Nambiar, P. T. C., and Ausubel, F. M.,** Isolation and characterization of symbiotic mutants of *Bradyrhizobium* sp. (*Arachis*) strain NC92: mutants with host-specific defects in nodulation and nitrogen fixation, *J. Bacteriol.,* 169, 2177, 1987.

299. **Sequeira, L.,** Mechanisms of induced resistance in plants, *Annu. Rev. Microbiol.,* 37, 51, 1983.

300. **Collinge, D. B. and Slusarenko, A. J.,** Plant gene expression in response to pathogens, *Plant Mol. Biol.,* 9, 389, 1987.

301. **Darvill, A. G. and Albersheim, P.,** Phytoalexins and their elicitors—a defense against microbial infection in plants, *Annu. Rev. Plant Physiol.,* 35, 243, 1984.

302. **Jordan, D. C.,** Ineffectiveness in the *Rhizobium*-leguminous plant association, *Proc. Indian Natl. Sci. Acad.,* 40, 713, 1974.

303. **Vance, C. P.,** *Rhizobium* infection and nodulation: a beneficial plant disease?, *Annu. Rev. Microbiol.,* 37, 399, 1983.

304. **Sharifi, E.,** Parasitic origins of nitrogen-fixing *Rhizobium*-legume symbioses. A review of the evidence, *BioSystems,* 16, 269, 1984.

305. **Djordjevic, M. A., Gabriel D. W., and Rolfe, B. G.,** *Rhizobium*: the refined parasite of legumes, *Annu. Rev. Phytopathol.,* 25, 145, 1987.
306. **Werner, D., Mellor, R. B., Hahn, M. G., and Grisebach, H.,** Soybean root response to symbiotic infection. Glyceollin I accumulation in an ineffective type of soybean nodules with an early loss of the peribacteroid membrane, *Z. Naturforsch.,* 40c, 179, 1985.
307. **Czarnecka, E., Edelman, L., Schöffl, F., and Key, J. C.,** Comparative analysis of physical stress responses in soybean seedlings using cloned heat shock cDNAs, *Plant Mol. Biol.,* 3, 45, 1984.
308. **Rolfe, B. G., Redmond, J. W., Batley, M., Chen, H., Djordjevic, S. P., Ridge, R. W., Bassam, B. J., Sargent, C. L., Dazzo, F. B., and Djordjevic, M. A.,** Intercellular communication and recognition in the *Rhizobium*-legume symbiosis, in *Recognition in Microbe-Plant Symbiotic and Pathogenic Interactions,* Lugtenberg, B., Ed., Springer-Verlag, Berlin, 1986, 39.
309. **Anderson, A. J.,** Studies on the structure and elicitor activity of fungal glucans, *Can. J. Bot.,* 58, 2343, 1980.
310. **Sequeira, L.,** Surface components involved in bacterial pathogen-plant host recognition, *J. Cell Sci. Suppl.,* 2, 301, 1985.
311. **Graham, J. S., Sequeira, L., and Huang, T. R.,** Bacterial lipopolysaccharides as inducers of disease resistance in tobacco, *Appl. Environ. Microbiol.,* 34, 424, 1977.
312. **Sequeira, L.,** Recognition systems in plant-pathogen interactions, *Biol. Cell,* 51, 281, 1984.
313. **Tran Tranh Van, K., Toubart, P., Cousson, A., Darvill, A. G., Collin, D. J., Chelf, P., and Albersheim, P.,** Manipulation of the morphogenetic pathways of tobacco explants by oligosaccharins, *Nature (London),* 314, 615, 1985.
314. **Solheim, B. and Fjellheim, K. E.,** Rhizobial polysaccharide-degrading enzymes from roots and legumes, *Physiol. Plant.,* 62, 11, 1984.
315. **Borthakur, D., Barber, C. E., Lamb, J. W., Daniels, M. J., Downie, J., and Johnston, A. W. B.,** A mutation that blocks exopolysaccharide synthesis prevents nodulation of peas by *Rhizobium leguminosarum* but not of beans by *R. phaseoli* and is corrected by cloned genes from *Rhizobium* or the phytopathogen *Xanthomonas, Mol. Gen. Genet.,* 203, 320, 1986.
316. **Wong, C. H., Pankhurst, C. E., Kondorosi, A., and Broughton, W. J.,** Morphology of root nodules and nodule-like structures formed by *Rhizobium* and *Agrobacterium* strains containing a *Rhizobium meliloti* megaplasmid, *J. Cell Biol.,* 97, 787, 1983.
317. **Truchet, G., Rosenberg, C., Vasse, J., Julliot, J. S., Camut, S., and Dénarié, J.,** Transfer of *Rhizobium meliloti* pSym genes into *Agrobacterium tumefaciens*: host-specific nodulation by atypical infection, *J. Bacteriol.,* 157, 134, 1984.
318. **Hirsch, A. M., Drake, D., Jacobs, T. W., and Long, S. R.,** Nodules are induced on alfalfa roots by *Agrobacterium tumefaciens* and *Rhizobium trifolii* containing small segments of the *Rhizobium meliloti* nodulation region, *J. Bacteriol.,* 161, 223, 1985.
319. **Hooykaas, P. J. J., Van Brussel, A. A. N., Den Dulk-Ras, H., Van Slogteren, G. M. S., and Schilperoort, R. A.,** Sym plasmid of *Rhizobium trifolii* expressed in different rhizobial species and *Agrobacterium tumefaciens, Nature (London),* 291, 351, 1981.
320. **Hooykaas, P. J. J., Den Dulk-Ras, H., Regensurg-Twink, A. J. G., Van Brussel, A. A. N., and Schilperoort, R. A.,** Expression of a *Rhizobium phaseoli* sym plasmid in *R. trifolii* and *Agrobacterium tumefaciens*: incompatibility with a *R. trifolii* sym plasmid, *Plasmid,* 14, 47, 1985.
321. **Martínez, E., Palacios, R., and Sánchez, F.,** Nitrogen-fixing nodules induced by *Agrobacterium tumefaciens* harboring *Rhizobium phaseoli* plasmids, *J. Bacteriol.,* 169, 2828, 1987.
322. **Broughton, W. J., Wong, C. H., Lewin, A., Samrey, U., Myint, H., Meyer, H., Dowling, D. N., and Simon, R.,** Identification of *Rhizobium* plasmid sequences involved in recognition of *Psophocarpus, Vigna,* and other legumes, *J. Cell Biol.,* 102, 1173, 1986.
323. **John, M., Schmidt, J., Wieneke, U., Kondorosi, E., Kondorosi, A., and Schell, J.,** Expression of the nodulation gene *nodC* of *Rhizobium meliloti* in *Escherichia coli*: role of the *nodC* gene product in nodulation, *EMBO J.,* 4, 2425, 1985.
324. **Török, I., Kondorosi, E., Stepkowski, T., Pósfai, J., and Kondorosi, A.,** Nucleotide sequence of *Rhizobium meliloti* nodulation genes, *Nucleic Acids Res.,* 12, 9509, 1984.
325. **Franssen, H.J., Scheres, B., Van de Wiel, C., and Bisseling, T.,** Characterization of soybean (hydroxy)proline-rich early nodulins, in *Molecular Genetics of Plant-Microbe Interactions 1988,* Palacios, R. and Verma, D. P. S., Eds., APS Press, St. Paul, Minnesota, 1988, 321.
326. **Averyhart-Fullard, V., Datta, K., and Marcus, A.,** A hydroxyproline-rich protein in the soybean cell wall, *Proc. Natl. Acad. Sci. U.S.A.,* 85, 1082, 1988.
327. **Bennett, M. J., Lightfoot, D. A., and Cullimore, J. V.,** cDNA sequence and differential expression of the gene encoding the glutamine synthetase γ-polypeptide of *Phaseolus vulgaris* L., *Plant Mol. Biol.,* 12, 553, 1989.

328. **Metz, B. A., Welters, P., Hoffman, H. J., Jensen, E. O., Schell, J., and De Bruijn, F. J.,** Primary structure and promoter analysis of leghemoglobin genes of the stem-nodulated tropical legume *Sesbania rostrata:* conserved coding sequences, *cis*-elements and *trans*-acting factors, *Mol. Gen. Genet.*, 214,181, 1988.

329. **Sandal, N. N., and Marcker, K. A.,** Soybean nodulin 26 is homologous to the major intrinsic protein of the bovine lens fiber membrane, *Nucleic Acids Res.*, 16, 937, 1988.

330. **Stougaard, J., Sandal, N. N., Grøn, A., Kühle, A., and Marcker, K. A.,** 5′ Analysis of the soybean leghaemoglobin *lbc₃* gene: regulatory elements required for promoter activity and organ specificity, *EMBO J.*, 6, 1565, 1987.

331. **Hirsch, A. M., Bhuvaneswari, T. V., Torrey, J. G., and Bisseling, T.,** Early nodulin genes are induced in alfalfa root outgrowths elicited by auxin transport inhibitors, *Proc. Natl. Acad. Sci. U.S.A.*, 86, 1244, 1989.

332. **Truchet, G., Vasse, J., Odorico, R., de Billy, F., Camut, S., and Huguet, T.,** Discrimination between nodules and root-derived structures: does alfalfa nodulate spontaneously?, in *Molecular Genetics of Plant-Microbe Interactions 1988*, Palacios, R. and Verma, D. P. S., Eds., APS Press, St. Paul, Minnesota, 1988, 179.

Chapter 9

THE *PARASPONIA-BRADYRHIZOBIUM* SYMBIOSIS

Kieran F. Scott and Gregory L. Bender

TABLE OF CONTENTS

I. INTRODUCTION

A fundamental challenge in the field of symbiotic nitrogen fixation is to determine the genetic and biochemical basis of the recognition of a compatible host by bacteria. This is especially true since the often unstated aim of most research in this area is to be able to manipulate the host range of these organisms to advantage. The association between *Bradyrhizobium* or *Rhizobium* strains and tropical nonlegume woody dicot trees from the genus *Parasponia* has provided a system in which the genetics of symbiotic nitrogen fixation can be easily studied outside the Leguminosae. Although there are at least 15 different genera of nonlegume plants that are capable of symbiotic nitrogen fixation, the microbial symbiont in all of these associations except *Parasponia* (viz., the actinomycete *Frankia;* see chapter 4) has proved difficult to manipulate. In contrast, the ability to manipulate genetically the rhizobia has advanced substantially in the last 10 years.

Since the discovery by Trinick[1] that the microbial symbiont in the root nodules of *Parasponia* trees (initially classified as *Trema*) examined in New Guinea was a *Rhizobium* strain, research has concentrated on two broad areas: (1) the mechanism of invasion of *Parasponia* roots by the bacterium and (2) the identification of both bacterial and plant genes required for nodulation and nitrogen fixation.

The fact that all of the microbial symbionts which nodulate *Parasponia* also, by definition, nodulate certain legume plants provides an opportunity for the comparative analysis of both legume and nonlegume nodulation using the same bacterial strain. Of particular interest in this regard is the ability to compare the genetic requirements in the bacterium for the nodulation and nitrogen fixation of nonlegumes and legumes. Similarly, the system provides the opportunity to examine the plant genes expressed during nodule morphogenesis in a nonlegume species. It is this kind of comparative analysis which will aid in the identification of the essential features of nodule morphogenesis in plants and highlight those genetic determinants which are necessary for the effective nodulation of a nonlegume species. This information will prove invaluable in establishing a conceptual framework for the extension of the nodulation process to other plant species.

This chapter will review progress in our understanding of *Parasponia* nodulation and nitrogen fixation with particular emphasis on a comparative analysis of this knowledge relative to the rapidly advancing legume nodulation field.

II. THE MECHANISM OF *PARASPONIA* INFECTION

A. *PARASPONIA* NODULE STRUCTURE

It has been known for many years, from studies on legume nodule structure, that the morphology of a functional legume nodule is dependent on the nature of the host plant being nodulated rather than the bacterial strain responsible for the infection.[2] For example, strains nodulating temperature legumes such as clover (*Trifolium*) elicit indeterminate cigar-shaped structures with a defined meristematic zone at the tip of the nodule and a peripheral vascular system.[2,3] Strains nodulating tropical legumes such as soybean (*Glycine max*) or *Macroptilium atropurpureum* (cv. Siratro) elicit determinate spherical nodules with a peripheral vascular system and a transient meristematic zone which envelopes the infected cells on the outside of the nodule.[4,5] A section of a typical determinate nodule structure elicited by *Bradyrhizobium parasponia* strain ANU289 on *Macroptilium atropurpureum* is shown in Figure 1A.

The structure of mature *Parasponia* nodules (Figure 1B) appears quite different to that of tropical legume nodules even though both are elicited by the same bacterial strain.[5] In fact, these nodules closely resemble the structure of nodules formed on the other nonlegumes by *Frankia* strains.[6] The most significant similarities are the branched coralloid nodule form and a central vascular bundle with a meristematic tip. As shown in Figure 1B, the infection

233

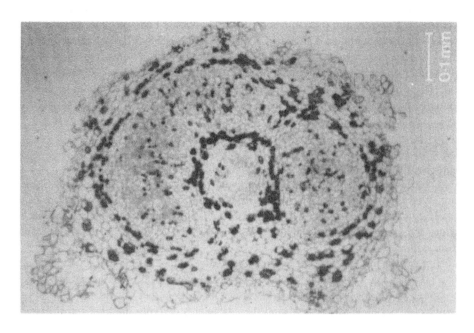

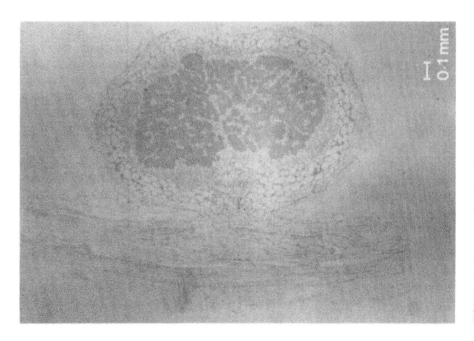

FIGURE 1. Structure of nodules induced by *Bradyrhizobium parasponia* strain ANU289 on (A) the tropical legume species *Macroptilium atropurpureum*; and on (B) the nonlegume species *Parasponia rigida*.

zone of the nodule surrounds the central vascular tissue. This zone contains both infected and uninfected cells as is commonly found in Siratro nodules. *Frankia* is a filamentous actinomycete that spreads throughout nodule tissue and within individual plant cells as septate hyphae (see Chapter 4). In contrast, *Parasponia* infection threads resemble those formed within legumes, being hollow threads with no septae, which are filled with bacteria often in single file.[7,8] The infection thread in most legumes studied buds-off within host cells to form packets of bacteria enclosed in a thin-walled membrane called the peribacteroid membrane.[3] In *Parasponia*, however, infection threads do not bud-off once entering the host cell, but simply differentiate into thinner-walled threads which continue to branch and fill the cell.[7,8,8a] *Parasponia* nodule structure therefore shares similarities with both legume and nonlegume nodules and could be viewed as an intermediate or bridging form between the two.

B. THE *PARASPONIA* INFECTION PROCESS
1. Location of the Infection Zone
Little is known about the early infection events in *Parasponia*. Lancelle and Torrey[9] have suggested that the presence of clumps of multicellular root hairs on the surface of the roots of *P. rigida* are the first visible evidence of infection. They have also shown that the subsequent stages of nodule development are very similar to those found in nonlegume symbioses with *Frankia*.[8a]

In legumes such as soybean, the spot inoculation technique has been used to study early infection events.[10] Spot inoculations involves placing a microscopic anion exchange bead on the surface of the root in the infection zone. The bead is then inoculated with a 1- to 10-nl volume of bacterial suspension, and the agar plates carrying the plants are placed in a plant growth cabinet. The use of this technique means that it is no longer necessary to examine the whole root surface for evidence of infection and that small pieces of root tissue, conveniently marked with a bead, can be sampled by scanning and transmission electron microscopy.

The utility of this approach to examine *Parasponia* infection is dependent on the identification of regions of the nonlegume root which are susceptible to infection by bacteria. Studies using flood-inoculated *Parasponia* plants[11] have shown that, as in the legume infection process,[12] there is a zone of elongating cells directly behind the root tip which is highly susceptible to infection (Figure 2). These cells are also only transiently susceptible to infection during the period of elongation as is observed in legume infection. Delay in inoculation by 3 d (Figures 2B and C) resulted in a decrease in nodule number on the main taproot; however, total nodule number remained constant, since nodulation of emerging lateral roots accounted for the loss of nodule number on the taproot. This observation also parallels the legume nodulation process.[12]

2. Early Infection Events
Although a similar region of the *Parasponia* root is susceptible to nodule initiation, the means by which this zone is infected differs from that of legumes[13] and other nonlegumes.[14] Root hair curling or distortion is not observed in *Parasponia*. *Rhizobium* infect legumes from within the "pockets" formed by curled root hairs, by eroding the surrounding mucilage and primary cell walls followed by the formation of infection threads which enter hair cells.[13] In contrast to legume infection, where root hairs are usually curled or distorted in some way, no changes are observed in the appearance of *Parasponia* root hairs. The apparant immunity of *Parasponia* root hairs means that bacteria must gain entry by an alternative means.

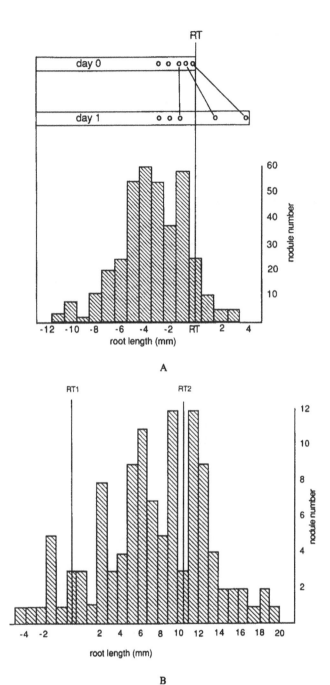

FIGURE 2.   Infection zone of *Parasponia* on flood inoculation with *B. parasponia* strain CP279. (A) Inoculation at day 0. RT, position of root tip at the time
of inoculation (day 0). Small circles represent anion exchange beads placed on
the root at day 0. After 1 d of growth, only the first two beads had moved due
to root elongation with no further movement of the second bead as growth continued past day 1. Negative numbers represent the distance from the RT toward
the shoot. Data were collected from 96 plants. (B) Inoculation at day 3. RT1,
position of the root tip when plants were placed under assay conditions. RT2,
average position of the root tip at time of inoculation (after 3 d growth). Data
were collected from 100 plants. (C) Relative percentage of tap vs. lateral root
nodulation with an increasing delay in inoculation. Lateral roots begin to emerge
after 1 d growth. Percentages were calculated from 48 plants per treatment.

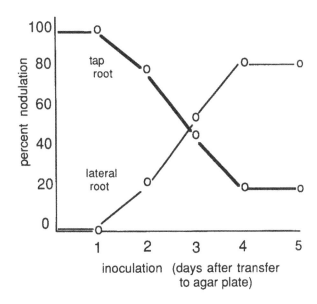

FIGURE 2C.

Time course studies using the spot inoculation technique described above have shown that the infection process can be divided into a series of sequential steps. These steps are

1.   Erosion of the root surface within 24 h
2.   Stimulation of cell division beneath the point of erosion which is visible in 4 d
3.   The emergence of dividing cells through the epidermis after 6 d (which is closely followed by)
4.   The intercellular colonization of the cortex and initiation of infection threads

Bacteria begin to degrade the root surface surrounding the marker bead as early as 24 h after inoculation, with erosion around bacterial colonies being clearly visible by 72 h (Figure 3). Closer examination of erosion sites clearly shows an association between bacterial contact and degradation of the mucilage layers in the primary cell walls of epidermal cells. Erosion continues until only the granular, fibrous remains of mucilage are seen within bacterial colonies. Localized root swelling associated with erosion of the root surface is observed beneath beads 4 d after inoculation. The swelling of the root produces a break in the epidermis near the site of bacterial colonization. Dividing cells then emerge through the break in the epidermis, forming an exposed site for intercellular colonization by bacteria (Figure 4).

Transmission electron microscopy of thin sections of plant tissue taken immediately below inoculation beads does not reveal any intercellular colonization of the outer cortex before the emergence of dividing cells at 6 d. This suggests that emerging exposed cells may be the only avenue through which bacteria can gain entry into the root cortex. However, as shown in Figure 5, sections through a fully developed prenodule structure show the presence of both intercellular and intracellular threads derived from infection threads within the prenodule.

This work and the work of Lancelle and Torrey[9] suggest that it is through the ability of bacteria to induce cell division beneath the epidermis of the root that they gain entry into *Parasponia*. In legumes, bacteria induce cortical cell division, root hair curling, and the formation of infection threads within curled root hairs. Only cell division is observed in *Parasponia*, without root hair curling and the formation of infection threads in curled root

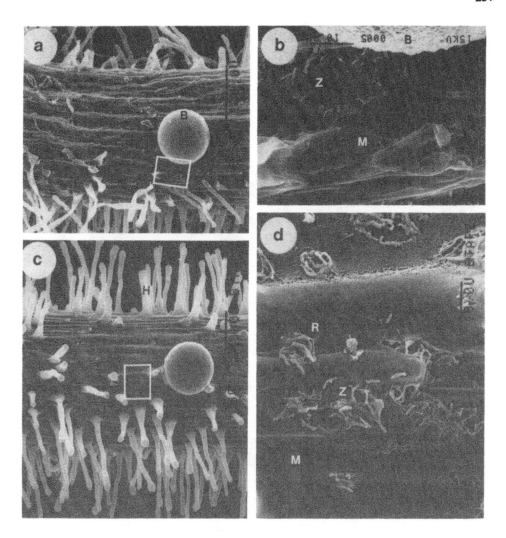

FIGURE 3. Colonization of the root surface. (a) Scanning electron micrograph of the root surface 24 h after spot inoculation using an anion exchange bead (B). (Magnification × 390.) (b) Higher magnification of inset (a) showing bacterial colonization and the beginnings of a zone of erosion (Z) into the mucilage (M) of the root surface. (Magnification × 3900.) (c) Colonization and erosion of the root surface 72 h after inoculation. Note the absence or root hair (H) curling or distortion. (Magnification × 390.) (d) Higher magnification of inset (c) showing erosion of the surface mucilage (M) of epidermal cells by small colonies (R) and the formation of erosion zones (Z) beneath larger accumulations of bacteria. (Magnification × 3900.)

hairs. The initiation of cell division in the cortex of legume roots, without the formation of curled root hairs or infection threads, has been reported for alfalfa[15-18] and soybean.[19] However, in these cases the nodules formed contained no bacteria except in the intercellular spaces of the surface layers. This suggests that the formation of both legume and *Parasponia* nodules, occupied by bacteroids within host cells, requires first the initiation of cell division, then intracellular entry via the formation of infection threads. If a microenvironment equivalent to the pocket formed by curled legume root hairs is needed for infection thread initiation,[13] then, in *Parasponia*, rhizobia could only find these conditions in intercellular spaces within the root. This is the first stage at which bacteria are confined within a "pocket" equivalent to the confinement of bacteria found in the curl of a root hair. Perhaps this is an explanation for the appearance of infection threads at this relatively late stage of infection.

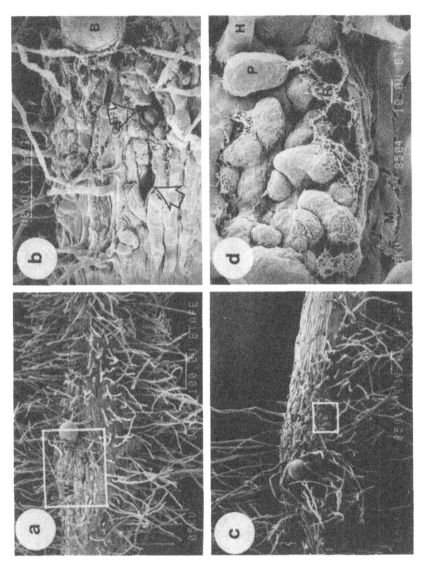

FIGURE 4.   Cell division and prenodule formation. (a) Swelling (W) of the root beneath a spot-inoculated site after 4 d (Magnification × 306.) (b) Higher magnification of inset (a) showing lesions (arrows) within the epidermal layer adjacent to a spot inoculation bead (B). (Magnification × 655.) (c) The emergence of dividing cells through a break in the epidermis after 6 d. (Magnification × 195.) (d) Higher magnification of inset (c) showing dividing plant cells (P) emerging past root hairs which appear to be unaffected by the presence of bacteria (R). The surface mucilage (M) is pushed aside by the emerging cells with the remnants of the overlying erosion zone still evident on the surface. (Magnification × 2040.)

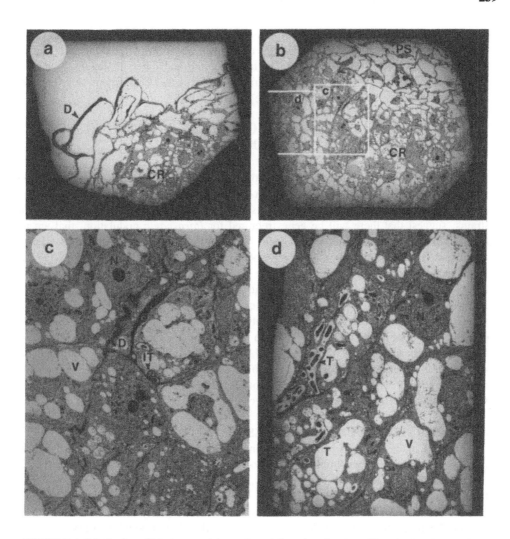

FIGURE 5. Colonization within the prenodule. (a) Prenodule surface showing epidermal cells with deposition of electron-dense material (D) in the cell wall and the presence of cytoplasmic-rich (CR) dividing cells within the prenodule. (Magnification × 120.) (b) Section beneath the prenodule surface (PS) showing the predominance of cytoplasmic-rich (CR) cells. (Magnification × 120.) (c) Inset showing the deposition of electron-dense material (D) in cell walls and the initiation of intercellular infection threads (IT). V, vacuole; N, nucleus. (Magnification × 1950.) (d) Inset showing the presence of intracellular threads (T) derived from infection threads which have entered the cytoplasmic-rich prenodule cells. V, vacuole; N, nucleus. (Magnification × 1950.)

The infection of *Parasponia* by this means in some ways resembles the infection of the legumes *Arachis hypogaea*,[20] *Stylosanthes* spp.,[21] and *Sesbania rostrata*.[22] Root hairs are not involved in the infection of these legumes with bacteria invading by intercellular colonization of cells exposed by the disruption caused by emerging lateral roots. Infection via spot-inoculated emerging lateral roots is not observed in *Parasponia* under the conditions described here. This observation is consistent with the idea that induction of plant cell division by bacteria is needed to break the epidermis of *Parasponia* and that cells exposed by emerging lateral roots are not susceptible to invasion. It is interesting to note that the infectability of peanut epidermal cells, like *Parasponia*, is developmentally restricted and transitory.[23] The penetration mechanism in peanut also involves enclosure of rhizobia in a pocket between two cell walls and the subsequent degradation of one of these walls.[20]

Alternate modes of infection may also be relatively common among woody legumes.[24] Taken together, all these data suggest that the *Parasponia-Bradyrhizobium* symbiosis appears to be less highly evolved than that found in most legume and other nonlegume symbioses where the symbionts gain entry by the relatively sophisticated mechanism of intracellular infection thread formation after inducing the curling of root hairs.

III. BACTERIAL GENES INVOLVED IN THE NODULATION PROCESS

It is clear from the previous section that the infection process in *Parasponia* has features which resemble legume nodulation and other features which are unique to the nodulation of nonlegumes by *Frankia*. These observations, together with genetic studies on *Rhizobium* strains, suggest that there will be genes required for *Parasponia* nodulation, that are also required for legume nodulation and genes that are involved specifically in *Parasponia* nodulation. Therefore, two approaches have been used to identify the bacterial genes involved in the nodulation of *Parasponia*. The first has been to identify structural or functional homologues of previously identified nodulation genes from legume-nodulating organisms and to ask, by creating mutants in the genes, whether they or genes linked to them are required for *Parasponia* nodulation. Mutants created in this fashion can then be used to examine the genetic requirements for each step in the nodulation process. The second approach has been to identify regions of DNA from *Parasponia*-nodulating strain which on transfer to a bacterium that normally fails to nodulate *Parasponia* confers on it the ability to do so. In this way, genes which may be unique to *Parasponia*-nodulating strains or involved in the determination of host range can be isolated and characterized.

A. CONSERVED NODULATION GENES

Analysis of the genetics of nodulation in *Rhizobium* strains has resulted in the identification of some 14 genes implicated in nodule induction.[25-28] Some of these genes are functionally interchangeable between species, since mutations in them can be complemented by an analogous gene from another species. DNA sequence analysis has also shown that there is substantial structural conservation within the coding regions. This functional and structural conservation has provided a means of identification of nodulation genes in strains capable of *Parasponia* nodulation. In this way the *nodA, nodB, nodC* and *nodD* genes have been localized in *Bradyrhizobium parasponia* strains, ANU289[29] and RP501[30].

Marvel et al.[30] isolated a region of DNA from RP501 which, on transfer to a mutant strain of *Rhizobium meliloti* containing the transposon Tn5 in the *nodC* gene, could complement the mutation and allow the strain to nodulate its normal host, alfalfa. This region was subsequently shown by hybridization analysis to contain structural homologues of *nodA, nodB*, and *nodD* in addition to the *nodC* gene.[31]

DNA sequence analysis of the analogous region in *B. parasponia* strain ANU289[29] showed that the *nodA, B,* and *C* genes are linked on a single operon and that *nodD* is located 5′ to and transcribed divergently from *nodABC* as shown in Figure 6a. Interestingly, however, a novel open reading frame (ORF), designated *nodK*, is located between *nodD* and *nodABC*. An ORF of similar size but with little structural homology to *nodK* has also been identified in *Bradyrhizobium japonicum*.[32] Structural homologues to the *nodI* and *nodJ* genes, first identified in *R. leguminosarum*,[33] have also been identified by DNA sequence analysis and are located some 2.6 kb to the 3′ side of *nodC*[34] (Figure 6a). The organization of these genes is also similar to that found in *B. japonicum*.[32,35] In addition to this *nod* gene cluster, there is at least one other region of the *B. parasponia* genome carrying structural homologues of *Rhizobium nod* genes.[36] This region carries sequences homologous to the recently identified *nodL, nodM,* and *nodN* genes from *R. leguminosarum*[37] and *R. trifolii*[38] (Figure 6b). The

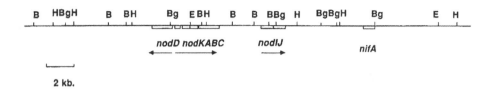

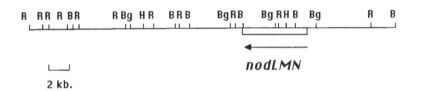

FIGURE 6.   Nodulation genes from *Bradyrhizobium parasponia* strain ANU289. (a) *nodKABC-nodD* cluster. (b) *nodLMN* cluster. Open boxes indicate location of genes. Horizontal arrows indicate direction of transcription. R, *Eco*RI; B, *Bam*HI; H, *Hind*III: Bg, *Bgl*II.

TABLE 1
Nodulation Phenotype of Mutants[40]

	Nodulation phenotype	
Strain	*Parasponia rigida*	*Macroptilium atropurpureum*
ANU289	nod⁺	nod⁺
ANU1271	nod⁺	nod⁺
ANU1272	nod⁺	nod⁺
ANU1273	nod⁻	nod⁻
ANU1274	nod⁻	nod⁻
ANU1292	nod⁻	nod⁻

observation that nodulation genes are dispersed in the *B. parasponia* genome parallels the similar observation made in *B. japonicum*[39] and may be a feature of *Bradyrhizobium* strains. No significant homology to the *nodFE* genes from *R. trifolii* has yet been found in the ANU289 genome by hybridization.[34]

The potential role of these conserved genes in *Parasponia* nodulation has been addressed by creating site-directed mutations in each gene using the transposon Tn5, or fragments thereof containing the selectable kanamycin-resistance gene *nptII*, cloned on suicide vectors[31,40] (see Chapter 2). The nodulation profiles of these mutants on their various host plants can then be determined. Table 1 shows a compilation of the phenotypes of such mutants constructed in *B. parasponia* strain ANU289 on both legume and nonlegume hosts. A detailed map of their location is shown in Figure 7. Mutations in the *nodA, nodB,*[31] and *nodC*[31] ORFs totally abolish nodule formation on both legume and nonlegume hosts. Detailed analysis of the nodulation profile of *nodA* mutants using the spot-inoculation assay has shown that, at a low frequency, these mutants are able to initiate subepidermal cell division in *Parasponia* roots.[41] However, this cell division fails to continue to the point where a prenodule structure is formed. The data obtained from the analysis of mutant phenotypes show

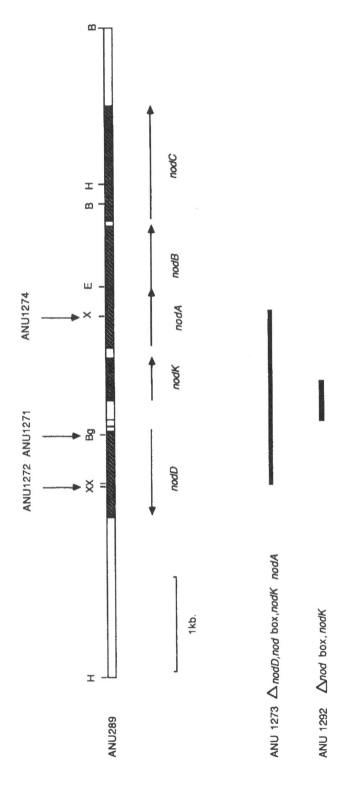

FIGURE 7. Construction of mutants in *B. parasponia* strain ANU289 nodulation genes. Hatched boxes indicate location of coding regions. Horizontal arrows indicate direction of transcription. Solid boxes indicate extent of DNA sequence replaced on insertion of the *nptII* cassette in deletion mutants. Vertical arrows indicate the point of insertion of the *nptII* cassette in insertion mutants. H, *Hind*III; X, *Xho*I; Bg, *Bgl*II; E, *Eco*RI; B, *Bam*HI.

that, as in legume nodulation, expression of the *nodABC* operon is obligatory for the persistence of cell division in the early stages of nodule morphogenesis.

Mutations in the *nodD* gene of both ANU289 and RP501,[31] however, still result in nodulation of both legume and nonlegume plants. This apparent lack of phenotype can be explained by the observation that in *B. parasponia* strain ANU289, at least three copies of the *nodD* gene can be identified by Southern hybridization to genomic DNA.[34] It is likely, therefore, that these extra copies of *nodD* are able to "internally complement" the *nodD* mutation. Multiple copies of the *nodD* gene have also been reported in *R. meliloti*[42,43] and *R. japonicum* strain USDA 191.[44] In addition, *Rhizobium* strain NGR234, a fast-growing strain capable of ineffective nodulation of *Parasponia* and a broad range of tropical legumes, has two copies of the *nodD* gene. One of these (*nodD1*) is essential for the nodulation of both legumes and *Parasponia*.[45]

B. HOST-SPECIFIC NODULATION GENES

While the approach of identifying structural and functional homologues of nodulation genes first identified in legume-nodulating organisms is useful to determine which of these genes are required for nonlegume nodulation, this strategy alone will not lead to the identification of genes which are required exclusively for *Parasponia* nodulation. However, once isolated, regions containing conserved genes can be further characterized to determine if such host-specific genes are linked to them. For example, Marvel et al.[31] have shown that, in addition to the genes identified above, mutagenesis of the *nodD-nodABC* region of *B. parasponia* strain RP501 has identified an as yet uncharacterized sequence following the *nodABC* operon which is required for nodulation of *Parasponia* but is not essential for the nodulation of the legume *Macroptilium atropurpureum*. Alternatively, as shown in Figure 8, cosmid-mediated transfer of libraries of *B. parasponia* genomic DNA to *Rhizobium* strains such as *R. trifolii* which do not normally form nodules on *Parasponia*, followed by assay for nodulation ability on *Parasponia*, has led to the observation that genes linked to the *B. parasponia* strain ANU289 *nodABC-nodD* region are capable of conferring *Parasponia* nodulation ability on *R. trifolii*.[40] Further analysis of this region has shown that the *B. parasponia nodD-nodKAB* genes alone are able to mediate the extension of host range to *Parasponia*, although with only about half the efficiency of the entire region in terms of nodule number.[34] These strains, however, are unable to nodulate the legume host of ANU289, *Macroptilium atropurpureum*.

Bender et al.[41] have shown that there are regions of the *Rhizobium* strain NGR234 symbiotic plasmid which carry host specificity determinants for *Parasponia* nodulation. Also, R-prime plasmids carrying regions determining both *Parasponia* and legume nodulation from NGR234 show relatively poor ability to induce nodulation on *Parasponia* when in different genetic backgrounds such as *Agrobacterium tumefaciens* compared to the nodulation ability of a parental heat-cured strain carrying these same plasmids. Therefore, in addition to the genes resident on the Sym plasmid in this strain, it is clear that chromosomal background genes also influence the efficiency of *Parasponia* nodulation.

Studies on host-specific nodulation genes in the *Rhizobium*-legume symbiosis have shown that there are at least two classes of *Rhizobium* genes that mediate the host-specific response.[25-28] The first class was identified from mutants which failed to nodulate their specific host plant. These mutants were unable to be complemented by analogous genes from other bacterial species even though in some cases, such as the *nodFE* genes, the genes shared substantial structural homology. In some cases, mutations in these genes also led to an extension of the range of plant species nodulated by the strain.[46] The second class of host-specific nodulation gene is the *nodD* gene.[47] This gene is required, together with plant-derived flavone or isoflavone compounds, for the induction of other *nod* operons such as *nodABCIJ*.[48-51] Construction of a chimeric *nodD* gene containing the amino-terminal region

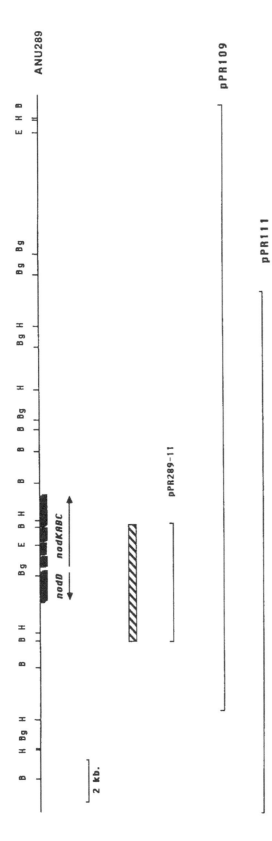

FIGURE 8. Transfer of nodulation ability to *R. trifolii* strain ANU843. Location and identity of cloned DNA sequences capable of conferring *Parasponia* nodulation ability on *R. trifolii* are shown by horizontal lines below map. Shaded boxes indicate location of genes. Hatched box indicates minimum region required for transfer of *Parasponia* nodulation phenotype. Horizontal arrows indicate direction of transcription. R, *EcoRI*; B, *BamHI*; H, *HindIII*; Bg, *BglII*.

of *nodD* from *R. meliloti* and the carboxy terminal region of the *nodD* gene isolated from strain MPIK3030, a strain which nodulates siratro, enabled *R. meliloti* isolates carrying this construct to nodulate siratro.[52] These data suggest that in the *Rhizobium*-legume symbiosis, host-specific responses are mediated at least in part by the control of *nod* gene expression via a specific interaction with *nodD*. It has been proposed that the specificity of this interaction occurs at the level of recognition of different flavone or isoflavone compounds produced by each plant host.

The *nodD* gene product has been shown to bind specifically to a conserved sequence called the "nod box" found preceding all inducible *nod* operons.[53] Since *nodD* also has some homology to other known regulatory proteins, such as *lysR*,[25] which activate transcription via binding to DNA in the promoter region of an operon, *nodD* is suggested to activate transcription of *nod* operons in a similar fashion by binding to the "nod box". This conserved sequence is also found preceding the *nodKABC* operon in *B. parasponia* strain ANU289.[29] Also, a mutant construct which deletes out half the nod box and a portion of *nodK* fails to nodulate *Parasponia* (Table 1, Figure 7). Taken together, these data suggest that the activation of *nod* gene transcription during *Parasponia* nodulation occurs in an analogous fashion to activation during legume nodulation.

While strains that nodulate *Parasponia* have not been studied in such detail as those in the *Rhizobium*-legume symbiosis, it is clear already that there are "host-specific" nodulation genes involved only in *Parasponia* nodulation, and it is likely that host-specific interactions are also mediated through the *nodD* gene as is the case in the nodulation of legume plants. These data predict, then, that the *Parasponia* nodulation process uses substantially the same genetic mechanism in the bacterium as is seen in legume nodulation and that the specificity for this particular nonlegume plant will reside in the precise structure of the host-specific nodulation genes and in the recognition of specific plant-derived *nod* gene inducers.

IV. BACTERIAL GENES INVOLVED IN NITROGEN FIXATION

While a great deal of effort has been put into understanding the molecular basis for infection of *Parasponia*, much less is known about the requirements for nitrogen fixation in nonleguminous plants. Structural studies have shown that nitrogen fixation appears to occur while the bacteria remain enclosed in infection threads once they have penetrated the cortical cell wall (Figure 5).[7,8] This is in contrast to fixation in legume hosts which occurs within bacteroids, resulting from bacteria being budded off from the infection thread surrounded by a plant-derived membrane.[25] Studies on the genetics of nitrogen fixation in *B. parasponia* strains have been directed almost entirely toward characterizing genes which have previously been identified in other nitrogen-fixing organisms and are detectable in *Bradyrhizobium* strains which nodulate *Parasponia*. These studies have, in the early stages, relied heavily on the use of cloned genes from the paradigm for nitrogen-fixing systems, namely, the enteric bacterium *Klebsiella pneumoniae*, since these genes have strong sequence homology with analogous genes from other species.[54]

It is clear from a substantial amount of genetic analysis of the nitrogen-fixation phenotype in the facultative anaerobe *K. pneumoniae*, that synthesis of the enzyme directly responsible for the reduction of nitrogen to ammonia (viz., nitrogenase) requires the expression of 20 genes. These genes are arranged in eight transcription units clustered in a 25-kb region of the *K. pneumoniae* chromosome.[55] Three of the 20 genes, *nifH*, *nifD*, and *nifK*, encode the subunit proteins of nitrogenase while several others are required for the synthesis of the iron-molybdenum cofactor (FeMoco) which is part of the active nitrogenase enzyme complex. Other gene products act as *nif*-specific electron transport proteins donating electrons to nitrogenase. The coordinate regulation of these genes is mediated through the induction of a *nif*-specific activating factor, the product of the *nifA* gene. *nifA* transcription is dependent

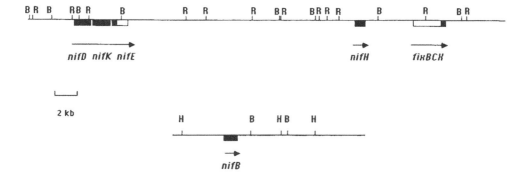

FIGURE 9.   Restriction map of *nif* and *fix* genes in *B. parasponia*. Shaded boxes indicate location of genes which have been characterized at the DNA sequence level. Open boxes indicate location of genes determined by hybridization analysis. Horizontal arrows indicate direction of transcription. R, *Eco*RI; B, *Bam*HI; H, *Hind*III; Bg, *Bgl*II.

on the nitrogen status of the cell, being induced under nitrogen-limiting conditions. This control is achieved through the action of a second positive regulatory protein *ntrC* which modulates the expression of a number of nitrogen metabolism pathways in addition to the *nif* genes.

DNA sequences carrying *K. pneumoniae nif* genes have been used as hybridization probes to determine whether *B. parasponia* has structural homologues of them and to examine the structure and regulation of the genes and their products. As shown in Figure 9, such studies have shown that *B. parasponia* contains homologues of the *nif* structural genes *nifH*, *nifD*, and *nifK*. The *nifH* gene is transcribed on a separate operon to *nifDK* and is located some 20 kb to the 3' side of it.[56,57] In addition, a *nifE* homologue has been located immediately 3' to *nifH* and the *B. parasponia nifB* gene identified at a locus unlinked to *nifH*.[58] As expected, DNA sequence analysis reveals a substantial amount of amino acid sequence homology between these genes and those from other organisms.[58] A structural homologue of the regulatory gene *nifA* has been located near the conserved nodulation gene locus (Figure 6a).[59] Nixon et al.[60] have also cloned and characterized the *ntrB* and *ntrC* genes from *B. parasponia*. These genes also share strong sequence homology with their *K. pneumoniae* counterparts and, interestingly, have regions of conserved sequence with other two-component regulatory gene pairs from Gram-negative bacteria. Whether the expression of *ntrC* is required for the induction of nitrogen fixation in *Parasponia* nodules remains to be determined.

Mutagenesis studies on nitrogen fixation in the *Rhizobium*-legume symbiosis have identified genes which are essential for nitrogen fixation within the nodule but which are not present in *K. pneumoniae*.[61] These genes have been called *fix* genes and are presumed to be involved in processes required during nodule morphogenesis prior to the induction of the *nif* genes or in the induction of symbiosis-specific pathways such as electron transport or metabolite transport. In *B. parasponia* strain ANU289, three such *fix* genes, *fixA*, *fixB*, and *fixX*, have been identified, linked on a single operon to the 3' side of *nifH* (Figure 9).[58]

The location, relative organization, and DNA sequence of these genes is in most cases similar to that found in *Bradyrhizobium japonicum*[35] and in *Bradyrhizobium* (sp. *Vigna*).[62] The similarity suggests that *Bradyrhizobium* strains are evolutionarily closely related regardless of their host range. All of these genes have been shown to be essential for nitrogen fixation in legume hosts;[61] however, whether they are essential for nitrogen fixation in *Parasponia* remains to be determined.

Also, the question as to whether there are *fix* genes which are specifically required for nitrogen fixation in *Parasponia* has not been addressed. This point is of particular interest in view of the recent report of mutants of *Rhizobium* strain NC92 which are able to fix nitrogen on certain legume hosts but not on other hosts that are effectively nodulated by the parental wild-type strain.[63] Since the studies on *Parasponia* nodule structure suggest that nitrogen fixation occurs within intracellular threads rather than in differentiated bacteroids, the possibility of *Parasponia*-specific fixation genes cannot be overlooked.

V. PARASPONIA HEMOGLOBIN GENES

Nodule induction and nitrogen fixation are the result of the coordinated expression of genes derived from both the plant and the bacterium. In addition to the work on the bacterial genes involved in the process, a large amount of data is accumulating concerning the nature and role of a number of plant gene products and metabolic pathways in the controlled development of a functional nodule structure. As with the bacterial genes, these studies have been focused almost exclusively on legume hosts such as soybean (*Glycine max*) and peas (*Pisum sativum*). Several recent reviews[64-66] describe progress in this area.

Work on the plant genes involved in nodulation of *Parasponia* stems from studies on the symbiotically induced oxygen carrier protein, leghemoglobin, which is present in large amounts in legume nodules.[67] This protein is the most well characterized of all the nodule-induced proteins, and the genes encoding it have been cloned from many legume species. However, until recently, it was not clear whether *Parasponia* nodules contained a similar oxygen carrier or whether the regulation of oxygen tension, necessary for the induction of nitrogen fixation, was controlled by the bacterium within the intracellular infection threads. Appleby et al.[68] succeeded in isolating a dimeric hemoglobin from *Parasponia* nodules and later showed that the protein contained about 40% sequence homology with leghemoglobins from soybean and lupin.[69] Isolation of the *Parasponia* hemoglobin gene[70] showed that it has three introns in identical positions to the introns found in legume hemoglobin genes, suggesting that the genes come from a common evolutionary origin.

Interestingly, the *Parasponia* sequences were also shown to cross-hybridize with sequences in a closely related but nonnodulating species *Trema*.[70] The homologous region of the *Trema* genome has been isolated and shown to be a functional hemoglobin gene transcribed in a tissue-specific fashion in roots.[71] Transcription of *Parasponia* hemoglobin has also been demonstrated at low levels in root tissue with a substantial induction occurring within nodules. These data indicate that hemoglobin genes may play a role in normal root metabolism, perhaps in the control of oxygen tension, in the absence of nodulation. The presence of this gene in plant species which are not nodulated by bacteria is the first indication that the extension of host range to such species may not require the transfer of legume hemoglobin genes to the nonlegume species of interest. See also the discussion of nodulins in Chapter 8.

VI. SUMMARY

The structural and genetic studies presented here clearly show that *Parasponia* nodulation occurs through mechanisms which, although apparently quite unique to nonlegumes, are in fact substantially similar to the processes used in legume nodulation. For example, infection does not occur through root hair curling in *Parasponia* as it does in legume infection; however, the early stages of infection require the expression of similar genes in both cases. This observation is not dependent on the bacterial strain used, since on addition of the appropriate genes to legume-nodulating stains such as *Rhizobium trifolii*, infection of *Parasponia* still proceeds through prenodule formation, and root hair curling is not observed. Of greater interest is the fact that these first steps in the infection process, which appear radically different between legume and nonlegume hosts, are mediated at least in part through the expression of the same bacterial genes *nodABC*. There is mounting evidence to suggest that *nodABC* are implicated in the induction of cell division in legumes as well as the induction of the curling response. *Parasponia* root hairs are apparently immune to a bacterial signal(s), the synthesis of which is mediated through the expression of the *nodABC* genes, and which results in root hair curling and cell division in legumes, but only prenodule formation (the induction of cell division) without root hair curling in nonlegumes.

While the infection processes in legume and nonlegume nodulation use substantially the same bacterial genetic mechanism to proceed, there is obviously strain specificity in these interactions. The data accumulated for legume systems suggest that this specificity lies in the recognition of very small differences in the structure of plant signal molecules (flavones or flavanones) which are required for the induction of *nod* gene expression. The second level of specificity is mediated through a set of bacterial genes which are required in a host-specific fashion for nodule formation. The most attractive hypothesis to explain the action of these genes is that they are involved in the synthesis of a bacterial signal which is specific for one plant host only. Specificity in *Parasponia* nodulation appears to be mediated through these same basic mechanisms since the *nodD-nodKAB* region of *B. parasponia* is capable of conferring *Parasponia* nodulation on *R. trifolii*, and mutations, which affect only *Parasponia* nodulation and not legume nodulation, have been found in *B. parasponia*.

Given that the bacterial genetics of nodulation in *Parasponia* is so similar to legume nodulation and that specificity of the interaction is essentially a biochemical phenomenon, the focus of attention with regard to the possible extension of host range to nonlegume plants not normally nodulated must now turn to the biochemical nature of the signal recognition process. It is clear that most plants make flavonoid compounds, but not all plants induce *nod* gene expression. Data are accumulating indicating that the bacterial signal recognition system, which appears to involve the product of the *nodD* gene, is quite specific for particular flavone molecules and that the specificity of the interaction is reflected in the structure of the *nodD* protein from different strains. An understanding of the structural basis of this recognition will allow the engineering of appropriate molecules to respond to specific flavonoid compounds in plants of interest, thus resulting in *nod* gene induction in conditions under which it would not normally occur.

The nature of host-specific bacterial signal molecules which act on the plant remains unknown and will have to be determined before any meaningful progress can be made in extension of host range. It is possible that not all plants carry suitable mechanisms to identify such signals, in which case genes would have to be engineered into a given host plant to allow for the appropriate host response to *nod* gene induction. Alternatively, plants which are not nodulated in nature may carry determinants which make them immune to infection by *Rhizobium* or *Bradyrhizobium* strains.

The data presented here suggest that there is a class of genetic determinants in plants which regulate the infection process. These determinants could positively control the process and, therefore, would be present in plants such as legumes and *Parasponia* which are nodulated, but absent from other nonlegumes and nonnodulating legume species. Alternatively, they may act to negatively control the process by actively excluding infection in those plant species which are not found to be nodulated in nature. In this case, the genes would be absent or inactive in those plants that are nodulated in nature. In any event, extension of host range to other plant species will require a much broader knowledge of the plant component of the process than we now possess.[65a] The finding of hemoglobin genes in nonnodulating species achieved using DNA sequences cloned from *Parasponia* is a tantalizing first step toward this goal.

ACKNOWLEDGMENTS

The authors wish to thank Susan Howitt, Joanne Stanton, Jeremy Weinman, and Michael Djordjevic for communication of data prior to publication, Peter M. Gresshoff for Figure 1, and Caroline Salom for proofreading the manuscript.

REFERENCES

1. **Trinick, M. J.**, Symbiosis between *Rhizobium* and the nonlegume, *Trema aspera, Nature (London)*, 244, 459, 1973.
2. **Vincent, J. M.**, Factors controlling the legume-*Rhizobium* symbiosis, in *Nitrogen Fixation, Vol. II.* Newton, W. E. and Orme-Johnson, W. H., Eds., University Park Press, Baltimore, MD 1980, 103.
3. **Dart, P. J.**, Infection and development of leguminous nodules, in *A Treatise on Dinitrogen Fixation*, Hardy, R. W. F. and Silver, W. S., Eds., John Wiley & Sons, New York, 1977, Vol. 3, 367.
4. **Bauer, W. D.**, Infection of legumes by rhizobia, *Annu. Rev. Plant Physiol.*, 32, 407, 1981.
5. **Ridge, R. W. and Rolfe, B. G.**, Sequence of events during the infection of the tropical legume *Macroptilium atropurpureum* by the broad-host-range, fast growing *Rhizobium* ANU240, *J. Plant Physiol.*, 122, 121, 1985.
6. **Angulo Carmona, A. F., Van Dijk, C., and Quispel, A.**, Symbiotic interactions in non-leguminous root nodules, in *Symbiotic Nitrogen Fixation in Plants*, Nutman, P.S., Ed., Cambridge University Press, Cambridge, 1976, 474.
7. **Trinick, M. J.**, Structure of nitrogen-fixing nodules formed by *Rhizobium* on roots of *Parasponia andersonii* Planch., *Can. J. Microbiol.*, 25, 565, 1979.
8. **Price, G. D., Mohapatra, S. S., and Gresshoff, P. M.**, Structure of nodules formed by *Rhizobium* strain ANU289 in the nonlegume *Parasponia* and the legume Siratro *(Macroptilium atropurpureum), Bot. Gaz.*, 145, 444, 1984.
8a. **Lancelle, S. A. and Torrey, J. G.**, Early development of *Rhizobium*-induced root nodules of *Parasponia rigida*. II. Nodule morphogenesis and symbiotic development, *Can. J. Bot.*, 63, 25, 1985.
9. **Lancelle, S. A., and Torrey, J. G.**, Early development of *Rhizobium*-induced root nodules of *Parasponia rigida*. I. Infection and early nodule initiation, *Protoplasma*, 123, 26, 1984.
10. **Turgeon, B. G., and Bauer, W. D.**, Early events in the infection of soybean by *Rhizobium japonicum*. Time course and cytology of the initial infection process, *Can. J. Bot.*, 60, 152, 1982.
11. **Bender, G. L., Nayudu, M., Goydych, W., and Rolfe, B. G.**, Early infection events in the nodulation of the nonlegume *Parasponia andersonii* by *Bradyrhizobium, Plant Sci.*, 51, 285, 1987.
12. **Bhuvaneswari, T. V., Turgeon, G., and Bauer, W. D.**, Early stages in the infection of soybean (*Glycine max* L. Merr.) by *Rhizobium japonicum*. I. Localization of infectible root cells, *Plant Physiol.*, 66, 1027, 1980.
13. **Turgeon, B. G., and Bauer, W. D.**, Ultrastructure of infection-thread development during the infection of soybean by *Rhizobium japonicum, Planta*, 163, 328, 1985.
14. **Berry, A. M. and Torrey, J. G.**, Root hair deformation in the infection process of *Alnus rubra* Bong, *Can. J. Bot.*, 61, 1983.
15. **Finnan, T. M., Hirsch, A. M., Leigh, J. A., Johansen, E., Kuldau, G. A., Deegan, S., Walker, G. C., and Signer, E. R.**, Symbiotic mutants of *Rhizobium meliloti* that uncouple plant from bacterial differentiation, *Cell*, 40, 869, 1985.
16. **Jacobs, T. W., Egelhoff, T. T., Long, S. R.**, Physical and genetic map of a *Rhizobium meliloti* nodulation gene region and nucleotide sequence of nodC, *J. Bacteriol.*, 162, 469, 1985.
17. **Debelle, F., Rosenberg, C., Vasse, J., Maillet, F., Martinez, E., Denarie, J., and Truchet, G.**, Assignment of symbiotic developmental phenotypes to common and specific nodulation (*nod*) genetic loci of *Rhizobium meliloti, J. Bacteriol.*, 168, 1075, 1986.
18. **Dudley, M. E., Jacobs, T. W., Long, S. R.**, Microscopic studies of cell divisions induced in alfalfa roots by *Rhizobium meliloti, Planta*, 171, 289, 1987.
19. **Calvert, H. E., Pence, M. K., Pierce, M., Malik, N. S. A., and Bauer, W. D.**, Anatomical analysis of the development and distribution of *Rhizobium* infections in soybean roots, *Can. J. Bot.*, 30, 2375, 1984.
20. **Chandler, M. R.**, Some observations on infection of *Arachis hypogaea* L. by *Rhizobium, J. Exp. Bot.*, 29, 749, 1978.

21. **Chandler, M. R., Date, R. A., and Roughley, R. J.,** Infection and root nodule development in *Stylosanthes* species by *Rhizobium, J. Exp. Bot.,* 33, 47, 1982.
22. **Tsien, H. C., Dreyfus, B. L., and Schmidt, E. L.,** Initial stages in the morphogenesis of nitrogen fixing stem nodules of *Sesbania rostrata, J. Bacteriol.,* 156, 888, 1983.
23. **Bauer, W. D., Bhuvaneswari, T. V., and Bhagwat, A. A.,** Transient susceptibility of root cells in five common legumes to infection by rhizobia, *Plant Physiol. Abstr.,* 748, 136, 1980.
24. **Sprent, J. I.,** personal communication.
25. **Appelbaum, E. R.,** The *Rhizobium/Bradyrhizobium-* legume symbiosis, Gresshoff, P., Ed., CRC Press, Boca Raton, FL, 1989.
26. **Djordjevic, M. A., Gabriel, D. W., and Rolfe, B. G.,** *Rhizobium*-the refined parasite of legumes, *Annu. Rev. Phytopathol.,* 25, 145, 1987.
27. **Downie, J. A. and Johnston, A. W. B.,** Nodulation of legumes by *Rhizobium, Plant Cell Environ.,* in press.
28. **Long, S. R.,** Genetics of *Rhizobium* nodulation, in *Plant-Microbe Interactions,* Kosuge, T. and Nester, E. W., Eds., Macmillan, New York, 1, 265, 1984.
29. **Scott, K. F.,** Conserved nodulation genes from the nonlegume symbiont *Bradyrhizobium* sp. *(Parasponia), Nucleic Acids Res.,* 14, 2905, 1986.
30. **Marvel, D. J., Kuldau, G., Hirsch, A., Richards, E., Torrey, J. G., and Ausubel, F. M.,** Conservation of nodulation genes between *Rhizobium meliloti* and a slow-growing *Rhizobium* strain that nodulates a nonlegume host, *Proc. Natl. Acad. Sci. U.S.A.,* 82, 5841, 1985.
31. **Marvel, D. J., Torrey, J. G., and Ausubel, F. M.,** *Rhizobium* symbiotic genes required for nodulation of legume and nonlegume hosts, *Proc. Natl. Acad. Sci. U.S.A.,* 84, 1319, 1987.
32. **Nieuwkoop, A. J., Banfalvi, Z., Deshmane, N., Gerhold, D., Schell, M. G., Sirotkin, M., and Stacey, G.,** A locus encoding host range is linked to the common nodulation genes of *Bradyrhizobium japonicum, J. Bacteriol.,* 169, 2631, 1987.
33. **Evans, I. J. and Downie, J. A.,** The *nodI* gene product of *Rhizobium leguminosarum* is closely related to ATP-binding bacterial transport proteins: nucleotide sequence of the *nodI* and *nodJ* genes, *Gene,* 43, 95, 1986.
34. **Scott, K. F.,** unpublished.
35. **Hennecke, H., Fischer, H.-M., Ebeling, S., Gubler, M., Thony, B., Gottfert, M., Lamb, J., Hahn, M., Ramseier, T., Regensberger, B., Alvarez-Morales, A., and Studer, D.,** *nif, fix* and *nod* gene clusters in *Bradyrhizobium japonicum* and *nifA*-mediated control of symbiotic nitrogen fixation, in *Molecular Genetics of Plant-Microbe Interactions,* Verma, D. P. S. and Brisson, N., Eds., Martinus Nijhoff, Dordrecht, 1987, 191.
36. **Stanton, J. and Scott, K. F.,** unpublished.
37. **Surin, B. P. and Downie, J. A.,** Characterisation of the *Rhizobium leguminosarum* genes *nodLMN* involved in efficient host-specific nodulation, *Mol. Microbiol.,* in press.
38. **Weinman, J. J. and Djordevic, M. A.,** personal communication.
39. **Hahn, M. and Hennecke, H.,** Cloning and mapping of a novel nodulation region from *Bradyrhizobium japonicum* by genetic complementation of a deletion mutant, *Appl. Environ. Microbiol.,* 54, 55, 1988.
40. **Scott, K. F., Saad, M., Dean Price, G., Gresshoff, P. M., Kane, H. and Chua, K. Y.,** Conserved nodulation genes are obligatory for nonlegume nodulation, in *Molecular Genetics of Plant-Microbe Interactions,* Verma, D. P. S. and Brisson, N., Eds., Martinus Nijhoff, Boston, 1987, 238.
41. **Bender, G. L., Goydych, W., Rolfe, B. G., and Nayudu, M.,** The role of *Rhizobium* conserved and host specific nodulation genes in the infection of the nonlegume *Parasponia andersonii, Mol. Gen. Genet.,* 210, 299, 1987.
42. **Gottfert, M., Horvath, B., Kondorosi, E., Putnoky, P., Rodriguez-Quinones, F., and Kondorosi, A.,** At least two *nodD* genes are necessary for efficient nodulation of alfalfa by *Rhizobium meliloti, J. Mol. Biol.,* 191, 411, 1986.
43. **Honma, M. A. and Ausubel, F. M.,** *Rhizobium meliloti* has three functional copies of the *nodD* symbiotic regulatory gene, *Proc. Natl. Acad. Sci. U.S.A.,* 84, 1987.
44. **Appelbaum, E. R., Thomson, D. V., Idler, K., and Chartrain, N.,** *Rhizobium japonicum* USDA 191 has two *nodD* genes that differ in primary structure and function, *J. Bacteriol.,* 170, 12, 1988.
45. **Bender, G. L.,** unpublished.
46. **Djordjevic, M. A., Schofield, P. R., and Rolfe, B. G.,** Tn5 mutagenesis of *Rhizobium trifolii* host-specific nodulation genes result in mutants with altered host-range ability, *Mol. Gen. Genet.,* 200, 463, 1985.
47. **Spaink, H. P., Wijffelman, C. A., Pees, E., Okker, R. J. H., and Lugtenberg, B. J. J.,** *Rhizobium* nodulation gene *nodD* as a determinant of host specificity, *Nature (London),* 328, 337, 1987.
48. **Redmond, J. W., Batley, M., Djordjevic, M. A., Innes, R. W., Kuempel, P. L., and Rolfe, B. G.,** Flavones induce expression of the nodulation genes in *Rhizobium, Nature (London),* 323, 632, 1986.

49. **Peters, N. K., Frost, J. W., and Long, S. R.,** A plant flavone, luteolin, induces expression of *Rhizobium meliloti* nodulation genes, *Science,* 233, 977, 1986.

50. **Firmin, J. L., Wilson, K. E., Rossen, L., and Johnston, A. W. B.,** Flavanoid activation of nodulation genes in *Rhizobium* is reversed by other compounds present in plants, *Nature (London),* 324, 90, 1986.

51. **Kosslak, P. M., Bookland, R., Barkei, J., Paaren, H. E., and Appelbaum, E. R.,** Induction of *Bradyrhizobium japonicum* common *nod* genes by isoflavones isolated from *Glycine max, Proc. Natl. Acad. Sci. U.S.A.,* 84, 7428, 1987.

52. **Horvath, B., Bachem, C. W. B., Schell, J., and Kondorosi, A.,** Host-specific regulation of nodulation genes in *Rhizobium* is mediated by a plant-signal, interacting with the *nodD* gene product, *EMBO J.* 6, 841, 1987.

53. **Hong, G. F., Burn, J. E. and Johnston, A. W. B.,** Evidence that DNA involved in the expression of nodulation (*nod*) genes in *Rhizobium* binds to the product of the regulatory gene *nodD, Nucleic Acids Res.,* 15, 9677, 1987.

54. **Ruvkun, G. B. and Ausubel, F. M.,** Interspecies homology of nitrogenase genes, *Proc. Natl. Acad. Sci. U.S.A.,* 77, 191, 1980.

55. **Arnold, W., Rump, A., Klipp, W., Priefer, U. B., and Pühler, A.,** The nucleotide sequence of a 24 kilobase-pair DNA fragment carrying the entire nitrogen fixation gene cluster of *Klebsiella pneumoniae,* in *Proc. 7th Int. Congr. Nitrogen Fixation,* Cologne, in press.

56. **Scott, K. F., Rolfe, B. G., and Shine, J.,** Nitrogenase structural genes are unlinked in the nonlegume symbiont *Parasponia Rhizobium, DNA,* 2, 141, 1983.

57. **Weinman, J. J., Fellows, F. F., Gresshoff, P. M., Shine, J. and Scott, K. F.,** Structural analysis of the genes encoding the molybdenum-iron protein of nitrogenase in *Parasponia Rhizobium* strain ANU289, *Nucleic Acids Res.,* 12, 8329, 1984.

58. **Weinman, J. J.,** Ph.D. thesis, Australian National University, Canberra, 1986.

59. **Howitt, S. and Scott, K. F.,** unpublished.

60. **Nixon, B. T., Ronson, C. W., and Ausubel, F. M.,** Two-component regulatory systems responsive to to environmental stimuli share strongly conserved domains with the nitrogen assimilation regulatory genes *ntrB* and *ntrC, Proc. Natl. Acad. Sci. U.S.A.* 83, 7850, 1986.

61. **Gussin, G. N., Ronson, C. W., and Ausubel, F. M.,** Regulation of nitrogen fixation genes, *Annu. Rev. Genet.,* 20, 567, 1986.

62. **Yun, A. C. and Szalay, A. A.,** Structural genes of dinitrogenase and dinitrogenase reductase are transcribed from two separate promoters in the broad host range cowpea *Rhizobium* IRc78, *J. Bacteriol.,* 81, 7358, 1984.

63. **Wilson, K. J., Anjaiah, V., Nambiar, P. T. C., and Ausubel, F. M.,** Isolation and characterisation of symbiotic mutants of *Bradyrhizobium* sp. (*Arachis*) strain NC92: mutants with host-specific defects in nodulation and nitrogen fixation, *J. Bacteriol.,* 169, 2177, 1987.

64. **Verma, D. P. S., Fortin, M. G., Stanley, J., Mauro, V. P., Purohit, S., and Morrison, N.,** Nodulins and nodulin genes of *Glycine max, Plant Mol. Biol.,* 7, 51, 1986.

65. **Govers, F., Nap, J.-P., Van Kammen, A., and Bisseling, T.,** Nodulins in the developing root nodules, *Plant Physiol. Biochem.,* 25, 309, 1987.

65a. **Rolfe, B. G. and Gresshoff, P. M.,** Genetic analysis of legume nodule initiation, *Annu. Rev. Plant Physiol. Plant Mol. Biol.,* 39, 297, 1988.

66. **Gresshoff, P. M. and Delves, A. C.,** Plant genetic approaches to symbiotic nodulation and nitrogen fixation in legumes, in *Plant Gene Research,* Vol. 3, Blonstein, A. D. and King, P. J., Eds., Springer-Verlag, 1986, 159.

67. **Appleby, C. A.,** Leghaemoglobin and *Rhizobium* respiration, *Annu. Rev. Plant Physiol.,* 35, 443, 1984.

68. **Appleby, C. A., Tjepkema, J. D., and Trinick, M. J.,** Hemoglobin in a nonleguminous plant, *Parasponia:* possible genetic origin and function in nitrogen fixation, *Science,* 220, 951, 1983.

69. **Kortt, A. A., Burns, J. E., Trinick, M. J., and Appleby, C. A.,** The amino acid sequence of hemoglobin I from *Parasponia andersonii,* a nonleguminous plant, *FEBS Lett.,* 180, 55, 1985.

70. **Landsmann, J., Dennis, E. S., Higgins, T. J. V., Appleby, C. A., Korrt, A. A., and Peacock, W. J.,** Common evolutionary origin of legume and nonlegume plant haemoglobins, *Nature (London),* 324, 166, 1986.

71. **Bogusz, D., Appleby, C. A., Landsmann, J., Dennis, E. S., Trinick, M. J., and Peacock, W. J.,** Functioning haemoglobin genes in nonnodulating plants, *Nature (London),* 331, 178, 1988.

GLOSSARY OF TERMS

Abscissin — Phytohormone eliciting an inhibitory effect in plants. Involved in leaf abscission and bud dormancy. May be a competitive inhibitor of auxin action (for example, abscisic acid).

Agrobacterium — Soil bacterium capable of T-DNA transfer to plants usually causing crown gall disease or hair root disease (most commonly investigated species are *A. tumefaciens* and *A. rhizogenes*).

ATP — Adenosine triphosphate, a nucleotide coenzyme that provides a common source of energy for a range of cellular activities by transferring a phosphate grouping to other substances by enzyme action and simultaneously transferring bond energies.

ATPase — Enzyme involved in the synthesis or breakdown of ATP.

Autoregulation — (1) Self-regulation; e.g., product inhibition in a biochemical reaction; (2) mechanism by which leguminous plants limit the number of cell divisions leading to the formation of nodules.

Auxin — Phytohormone promoting cell division and enlargement (plant growth regulation). Initiates cell division in association with cytokinins (for example, indole acetic acid [IAA], 2-4 dichlorophenoxy-acetic acid[2,4-D], and others).

Bacteriophage — Virus that specifically attacks bacteria (e.g., lambda).

Bacteroidin — Bacterial coded proteins contained within nodules.

Blue-green algae — Bacteria capable of photosynthesis, for example, *Anabaena* or *Anacystis*.

Bradyrhizobium — Like *Rhizobium*, but usually slow-growing and lacking symbiotic plasmids.

Chloroplast — Site of photosynthesis in higher plants and algae. Contains circular DNA molecules.

Cistron — A segment of DNA specifying one polypeptide chain in protein.

Complementation group — Linearly adjacent segments of DNA which supplement each other in phenotypic effect.

Conjugation — Process of sexual reproduction in unicellular organisms (e.g., *E. coli*) in which two organisms of opposite mating type temporarily pair and transfer genetic material.

Cortex — Parenchymal tissue encasing the vascular cylinder (pericycle) in stems and roots (and nodules).

Cytokinin — Phytohormone (purine based) that elicits a stimulatory effect on plant cell division. Also influences bud and leaf growth mobilization of nutrients, and some light responses (for example, kinetin, zeatin, benzylamino-purine).

DNA — Nucleic acid containing sequences of monodeoxyribonucleotide units that function as the carrier of genetic information in most organisms.

DNA polymerase — Enzyme that synthesizes DNA using DNA as a template.

Endodermis — Innermost layer of cells of plants surrounding core of vascular tissue consisting of xylem, phloem, and pericycle

Enhancer — Sequences necessary for enhanced expression of a gene. These sequences are often engineered around the gene (or sequence) of interest, e.g., promoter sequences (such as the 35S promoter).

Epidermis — Outermost layer of cells in plants. Source tissue for root hairs.

Epistasis — Interference of one allele with the detection of the phenotypic expression of a nonallele.

Escherichia coli (E. coli) — Gram-negative bacterium found in mammalian intestines. Has become a major organism in genetics, biochemistry, and molecular biology.

Euchromatin — Coiled or clumped interphase chromatin, transcriptionally active (i.e., capable of directing RNA synthesis).

Eukaryote — Unicellular and multicellular organisms in which the nucleus is separated from the cytoplasm by a nuclear membrane and the genetic material contained in a number of chromosomes.

Exon — The sequences within a eukaryotic gene that encode an amino acid sequence to become part of the final gene product.

Ferridoxin — Small sulfur-rich protein involved in electron transport.

Fix gene — Gene coding for a nitrogen-fixing function.

Flavone — Aromatic compound derived from the amino acid phenylalanine involved in secondary product metabolism (i.e., flavor or pigmentation) and plant microbe interactions.

Frankia — Bacterial genus capable of nodulation of some nonlegumes, e.g., *Alnus, Casuarina, Myrica*. Belongs to actinomycetes.

Gene — The DNA (or RNA in some viral instances) making up a unit of the material of inheritance.

Gene bank — Collection of bacteria or viruses, which carry recombinant DNA molecules of another organism.

Genome — Complete set of genetic material or chromosome genes, inherited as a unit from one parent.

Gibberellin — Phytohormone promoting stem elongation of certain plants. Also involved in plant growth and development (responds to light, temperature, dormancy, flowering, and fruiting).

Glutamine synthetase — Enzyme directly involved in the assimilation of ammonia.

Glycine max — Soybean, a seed and oil legume; nodulated by *Bradyrhizobium japonicum* and *Rhizobium fredii*.

Heterochromatin — Coiled or clumped interphase chromatin, inactive in transcription.

Homogenotization — Marker exchange technique used in bacterial genetics.

HPLC — High performance liquid chromatography — means of separating substances on a column using a liquid carrier, then characterizing the substances on the basis of mobility and absorbance.

Hydrogenase — Bacterial enzyme system capable of assimilating hydrogen gas to yield ATP and water.

Independent assortment — Nonhomologous chromosomes or chromosome segments segregating independently (one of Mendel's discoveries).

Inducer — Chemical or physical agent causing induction of genes.

Infection thread — Tube-like structure associated with the entry of bacteria into root tissue in the early stages of nodule formation.

Intron — The DNA sequence within eukaryotic genes that does not form part of the final makeup of the gene product. These intervening sequences between exons are excised from the original nuclear transcript.

In vitro translation — The use of mRNA in a test-tube translation system to produce protein in isolation from the whole organism (literally "in glass").

In vivo translation —The conversion of mRNA to protein within the living organism, using the organism's own processes. Usually carried out using a radioactively labeled isotope.

Isoelectric focusing (IEF) — Separation of molecule's across a pH gradient on the basis of the molecules isoelectric point.

Isoflavone — Similar phenolic compound as flavone.

Klebsiella pneumoniae — Free-living nitrogen-fixing organism similar to *E. coli*. Used as a reference system for nitrogen-fixation research.

Lac-fusion — Fused DNA insertion containing the β-galactosidase gene.

Lactose operon — Set of three bacterial genes in *E. coli* involved with lactose (milk sugar) metabolism. These genes are coordinately regulated.

Lectin — Sugar-binding proteins that do not have enzymatic activity.

Leonard jar — Means of growing plants within a controlled (sterile) root environment. Used extensively in nodulation tests with specific bacterial strains, free of contamination from other strains.

Linkage — The association of two or more alleles on the same chromosome.

Macroptilium atropurpureum — A tropical forage legume originally bred in Australia (common name: siratro).

Maternal inheritance — The extranuclear inheritance of a trait through cytoplasmic factors of organelles such as chloroplasts and mitochondria contributed by the female gamete.

Maxim-Gilbert sequencing —Method of sequencing DNA.

Medicago sativum — Alfalfa or lucerne; a forage and hay legume (nodulated by *Rhizobium meliloti*).

Mesophyll — Internal cells of a leaf blade differentiated into regularly (palisade) and irregularly (spongy) arranged mesophyll cells.

Mitochondria — Site of eukaryotic respiration. Contains circular DNA molecule.

mRNA — Messenger RNA. Transcription product of a gene and capable of translation to protein.

Nick-translation — Method of radioactively labeling DNA fragments.

Nif gene — Bacterial gene coding for the nitrogenase enzyme directly (such as *nif*HDK) or genes involved in the regulation of nitrogenase (usually has a homologue in *Klebsiella*).

Nitrate reductase — Enzyme that catalyzes the conversation of nitrate to nitrite.

Nitrogen assimilation — Absorption and buildup of nitrogen into nitrogenous compounds within an organism (see glutamine synthetase).

Nitrogenase — Enzyme that converts atmospheric di-nitrogen gas to ammonia. Encoded by *nif*HDK genes, limited only to prokaryotes.

Nod gene — Genes contained within bacteria coding for a nodulation function.

Nodulin — Plant-coded proteins contained preferentially within nodules (e.g., leghemoglobin).

Northern hybridization — Detection of RNA by means of blotting an agarose gel separation of RNA onto a membrane and subsequently probing with a single-stranded DNA probe using autoradiography.

Open reading frame (ORF) — Sequence of DNA that can be translated into a polypeptide.

Operator — Negatively controlled gene that regulates the transcription of structural genes in an operon.

Orthostiches — Interorgan vascular connections in plants (see phloem and xylem).

Oxidative phosphorylation — Coupling of phosphate groupings with ADP to make ATP using energy released during respiration.

Parasponia rigida — One of the *Parasponia* species capable of nodulation by *Bradyrhizobium*; tropical nonlegume and member of the Ulmaceae.

Peribacteroid membrane — Plant-derived membrane encapsulating bacteroids in plant cells within nodules.

Pericycle — Tissue of the vascular cylinder contained within the endodermis consisting of parenchyma, xylem, phloem, and sometimes fibers (elements of sclerenchyma).

Phloem — Vascular tissue that conducts organic compounds (sucrose) and some mineral salts throughout the plant. Characterized by the presence of sieve tubes.

Photosynthesis — Synthesis of organic compounds from water and carbon dioxide using energy absorbed by chlorophyll from sunlight (localized in chloroplasts).

Phytoalexin — Phenolic compound involved in plant-microbe defenses (related to isoflavones).

Phytohormone — Plant hormone, a small molecule that elicits a response in plant tissues, e.g., cell elongation by gibberellic acids.

Pisum sativum — Pea, a seed and vegetable legume. Has excellent genetics (nodulated by *Rhizobium leguminosarum*).

Plasmid — Extrachromosomal, circular DNA molecule capable of independent replication.

Polycistronic message — Messenger RNA which encodes several genes (cistrons) (see lactose operon).

Polypeptide — Chain of amino acids linked by amide bonds (direct product of most genes).

Prokaryote — Cells or organisms having the genetic material contained in simple filaments of DNA and not separated from the cytoplasm by a nuclear membrane (e.g., bacteria and blue-green algae).

Promoter — DNA sequence that binds transcriptase before transcription and translation, respectively.

Quiescent center — Cell cluster in root tips involved in the regulation of cytokinin biosynthesis.

Repressor — Chemical or physical agent causing repression (lack of transcription).

Restriction nuclease — Site-specific endonuclease capable of cleaving sequences of DNA. Used extensively to analyze DNA.

Rhizobium — Soil bacteria capable of eliciting nodules on legumes.

Rhizosphere — Zone immediately surrounding the roots which is modified by the activity of the root.

Ribosome — Subcellular particles of protein and RNA present in all cells. Associated in eukaryotic cells with the endoplasmic reticulum, the site of protein synthesis.

Ribosome binding site — Sequences of RNA in front of coding sequences used by ribosome to initiate translation.

RNA — Nucleic acid containing monoribonucleotide units that function as an intermediate carrier of the genetic information between DNA and proteins in most organisms. May be mRNA (messenger), tRNA (transfer), or rRNA (ribosomal).

RNA polymerase — Enzyme that synthesizes RNA using DNA as a template. Often called transcriptase.

Sanger sequencing — Method of DNA sequencing.

Sigma factor — Polypeptide needed for transcriptase to initiate transcription.

Southern hybridization — Detection of DNA homology in agarose-gel-separated DNA fragments by means of blotting the gel onto a membrane and subsequent probing with a single-stranded DNA probe. Named after E. Southern, the developer of the technique (1975).

Suicide plasmid — Plasmid containing DNA sequence of interest to be inserted into the recipient genome upon suicide of the plasmid.

Transacting element — Protein of regulatory function capable of DNA binding in promoter region.

Transconjugant — Bacterium containing new plasmid after conjugation.

Transcriptional fusion — Gene fusion at the mRNA level. Fused gene maintains its own translational start.

Transformant — Organism containing and expressing foreign DNA.

Translational fusion — Gene fusion giving hybrid protein, i.e., fused gene does not maintain its own translational start.

Transposon — Transposable (moving) genetic element often associated in antibiotic resistance.

Trifolium repens — White clover, a pasture legume (nodulated by *Rhizobium trifolii*).

Two-dimensional electrophoresis — Separation of macromolecules by two consecutive, but different electrophoretic techniques, IEF followed by slab gel polyacrylamide electrophoresis.

Western hybridization — Detection of proteins in a gel separation by means of blotting onto a membrane and probing with antibodies.

Xylem — Vascular tissue that conducts water and mineral salts throughout the plant. Characterized by the presence of tracheids and vessels, fibers, and parenchyma. Also provides the plant with mechanical support.

INDEX

A

Abscisic acid, 170
Accessions, 163
Acetophenone, 144
Acetoranillone, 144
Acetosyringone, 145
Acetyl-CoA, 112—115
Acetylene reduction, nitrate inhibition of, 171—172
N-Acetylgalactosaminyltransferase, 124
Achromopeptidase, 80
Actinorhizal symbiosis, 77—109, see also *Frankia*
Acyl carrier proteins, 141
Adenosine triphosphate, see ATP
Aglycones, 144
Agrobacterium radiobacter, 70
Agrobacterium tumefaciens, 7, 18, 137, 146, 208—209, 243
  chromosome, 209
  transconjugants, 210, 212
  12-kb *nod* region from *R. leguminosarum* Sym plasmid as strain 248^c(pMP104), 199
Akinetes, 60
Alfalfa, 133, 139, 140, 144, 182, 196, see also *nod* genes; Nodulation; Nodules
  nodule development, 199, 202—203, 212, 215, 237
  nodulin gene expression, 199, 202—203
  root structure, 182, 196
Algal packet, 60—61
Allantoinase, 118, 120
Allosteric enzymes, 4
Alnus-compatible strains, 97—99
Alpha-mannosidase, 123
Alpha-mannosidase isoenzyme II, 122
Aluminum, 173
Amide-transporting legumes, 117—120
Amido phosphoribosyl transferase, 141
Amino acids, 115—117, 173
  homology of, 189
  sequences, 246
Aminoglycoside 3′ phosphotransferase II, 17
Ammonia, 197
Ammonium, 195
Amyloplasts, 124—125
Anabaena azollae, 52, 59—60
  cell properties, 59—60
  genetic studies, 66—71
  isolates, 59
  molecular studies, 66—71
  nif gene, 67, see also *nif* genes
  phylogeny of symbiotic strains, 68
  taxonomy, 59
Andira species, 182
Animals, globin genes, 193
Antibiotics
  metalloglycopeptide, 17
  resistance to, 89—90, 133

resistance markers, 19
Anticodon, 2
Apical meristem, 183
Apigenin, 144
Apigenin 7-*O*-glucoside, 144
Aquatic ferns, 52, see also *Azolla-Anabaena* symbiosis
Arabinogalactan proteins, 186
Arginine synthesis, 1
Asparagine synthetase, 117, 119
Aspartate aminotransferase, 124
Aspergillus nidulans, 173
Assays, of host plants, convenience of, 133—134
Assimilation, of fixed nitrogen, 117—118
ATP-binding proteins, 141
ATP formation, 121
ATPase, 122
Attachment, 146, 160
Autoregulation of nodulation, 166, 168—171, 174
  genes, 142
  inhibitor, 169—170
  model for, 168—169
Auxins, 196—197
Auxotrophic mutants, 132, 209
Avoidance determinants, 211—214
Azolla, 52, see also *Azolla-Anabaena* symbiosis
  bacteria in leaf cavities, 70—71
  ecology, 63—64
  growth rate, 63
  leaf structure, 56, 58
Azolla
  morphology, 55—62
  physiology, 63—65
  sexual reproduction, 52—53
  taxonomy, 54—55
  use of, 52
  vegetative reproduction, 53—54
Azolla-Anabaena combinations and recombinations, 65—66
Azolla-Anabaena symbiosis, 51—75, see also *Anabaena azolla; Azolla*
  bacteria in leaf cavities, studies of, 70—71
  DNA polymorphism, 67
  endophyte, 59—60
  genetic studies, 66—71
  hydrogen evolution, 65
  isolates, studies of, 66—70
  life cycle of *Azolla*, 53—54
  molecular studies, 66—71
  morphological aspects, 55—62
  morphology of the symbiosis, 60—62
  nitrogen fixation, 64—65
  photosynthesis, 64—65
  physiological aspects, 63—65
  plant host, 55—59
  root development, 56
Azorhizobium sesbaniae, 132

Printed and bound by CPI Group (UK) Ltd, Croydon, CR0 4YY

22/10/2024

01777600-0009